D1726490

MIND OUT OF MATTER

STUDIES IN COGNITIVE SYSTEMS

VOLUME 20

The titles published in this series are listed at the end of this volume.

MIND OUT OF MATTER

Topics in the Physical Foundations of Consciousness and Cognition

by

GREGORY R. MULHAUSER

Applied Research and Technology,
British Telecom Laboratories,
Martlesham Heath, England,
#http://www.labs.bt.com/people/mulhaug
and
Department of Philosophy,
University of Glasgow, Scotland

KLUWER ACADEMIC PUBLISHERS

DORDRECHT / BOSTON / LONDON

A C.I.P. Catalogue record for this book is available from the Library of Congress.

ISBN 0-7923-5103-7

Published by Kluwer Academic Publishers,
P.O. Box 17, 3300 AA Dordrecht, The Netherlands.

Sold and distributed in North, Central and South America
by Kluwer Academic Publishers,
101 Philip Drive, Norwell, MA 02061, U.S.A.

In all other countries, sold and distributed
by Kluwer Academic Publishers,
P.O. Box 322, 3300 AH Dordrecht, The Netherlands.

Printed on acid-free paper

Printed in the Netherlands

Having written it on the peaceful European soil which he fought to keep free more than half a century ago, I dedicate this book to the memory of my grandfather, Raymond B. Mulhauser.

Table of Contents

Series Preface

This series will include monographs and collections of studies devoted to the investigation and exploration of knowledge, information, and data-processing systems of all kinds, no matter whether human, (other) animal, or machine. Its scope is intended to span the full range of interests from classical problems in the philosophy of mind and philosophical psychology through issues in cognitive psychology and sociobiology (regarding the mental abilities of other species) to ideas related to artificial intelligence and computer science. While emphasis will be placed upon theoretical, conceptual, and epistemological aspects of these problems and domains, empirical, experimental, and methodological studies will also appear from time to time.

In this volume, Greg Mulhauser undertakes the ambitious task of introducing a novel approach to the mind/body problem that not only integrates scientific and philosophical concepts and methods but promises to combine computational and dynamic approaches to understanding the mind. On this approach, a theory that unifies conscious experience with the material world emerges from adopting a perspective according to which the conscious subject (the 'I' or 'self' of personal experience) is a materially instantiated structure of dynamically changing information. The span of this work is broad (from unconscious zombies to ineffable feels and from quantum mechanics to analogue computers). Every reader should find it both challenging and illuminating.

J.H.F.

Detailed Table of Contents

Table of Figures

Foreword

By the time I finished an undergraduate degree in philosophy and mathematics at Willamette University in Salem, Oregon, I had acquired at least my share of zany ideas... Chaos theory, no doubt, could explain free will. Quantum nonlocality, I was certain, would put psychic phenomena on firm scientific footing. I thought it was a good idea to take a job doing classified research for the US government.

All soon gave way to better judgement, and I moved to the University of Edinburgh under a British Marshall Scholarship to pursue a passion for philosophy of mind—an affliction contracted largely through exposure to Willamette philosophy professors Lou Goble and Tom Talbott. For helping shape the foundations which would ultimately make this work possible, as well as for encouraging me away from zaniness all along, I'm grateful to those two as well as to Mark Janeba and Frank Zizza in mathematics and Mike Dunlap in computer science.

After a little less than three years in Edinburgh, with the valuable guidance and support of my first supervisor Stig Rasmussen and his successor Alexander Bird, I submitted the initial draft of *Mind Out of Matter* as my doctoral dissertation. I gratefully acknowledge the support of the Marshall Aid Commemoration Commission and that of the following other groups and individuals whose financial assistance through various gifts and bursaries contributed substantially to my research, travel, or participation in conferences during those three years:

- Analysis Trust
- British Society for the Philosophy of Science
- European Commission Human Capital and Mobility Programme
- Florence Mulhauser and the late Ray Mulhauser
- Oxford Centre for the Environment, Ethics, and Society
- Society for the Study of Artificial Intelligence and Simulation of Behaviour
- University of Edinburgh, Cognitive Science Centre
- University of Edinburgh, Department of Philosophy
- University of Edinburgh, Human Communication Research Centre
- University of St. Petersburg, Russia, Centre for Modern Communications

The year immediately following I spent as the Gifford Research Fellow forty miles west at the University of Glasgow, pursuing loose ends which remained in the first version of the book and rejecting outright extensive portions of it (better judgement again). For the freedom of that fruitful period of research, and for personal assistance frequently well beyond the call of duty, I am deeply grateful to the Gifford Lectureship Committee of the University of Glasgow. During this time, the original manuscript entered a 14-month evaluation process at Kluwer Academic, eventually bringing insightful comments from reviewers in both philosophy of science and philosophy of mind. My thanks go to the reviewers, to the series editor Jim Fetzer, and to the publisher—particularly for their intellectual and financial daring in accepting the first book writing effort of someone then barely 25 years old.

With the end of the Fellowship I began a stint in what is euphemistically called 'independent research', having been rejected for research or teaching posts by virtually every major university in the United Kingdom. This was in many ways a blessing, freeing time for reorganising and rewriting the manuscript almost entirely, incorporating work undertaken during the Gifford Fellowship and writing large portions from scratch. (In a sense, this remains the same book only in name.) For augmenting my dwindling savings while I played starving author, I am indebted to my grandmother Florence Mulhauser, to my parents, and especially to my fiancée's parents, Lin and Kerry Stares. For replacing my 'independence' with more generous resources than I ever could have expected in the academic realm, I am grateful to my new employers at British Telecom Laboratories in Martlesham Heath, England. My colleagues there support a remarkably vibrant, creative, and competent atmosphere for advancing my research in artificial life, machine consciousness, and related areas. Under the auspices of BT Labs, many suggestions about consciousness and cognition first offered here on theoretical grounds may finally find their way into scientific testing and implementation in artificial organisms. Thanks to managers Alan Steventon and Chris Winter for their flexibility in allowing me to 'telework' from Scotland, via the Internet, for my first half year with BT; this made it possible to finalise most of the book in advance of the distractions and delays of an actual geographical move.

Being so new, this manuscript hasn't benefited as directly as earlier versions from the painstaking comments and corrections of professional correspondents. Nonetheless, for especially influential conversations or for direct commentary on the newer material, I am grateful to Selmer Bringsjord, Greg Chaitin, Andy Clark, Dan Dennett, David Dowe, Terry Horgan, Brian Josephson, Adam Morton, Tim van Gelder, and Jim Wright. Thanks to Basil Hiley for a long talk in the fells of Lapland which helped inspire me to shore up my grasp of decoherence (appearing in this book in a guise he will no doubt detest). Earlier drafts were shaped by similar contacts with—in addition to some of those above—my two PhD supervisors, plus Sue Blackmore, Tony Browne, Jim Garson, Jesse Hobbs, Thomas Metzinger, David Rumelhart, and especially Peter Smith, who wrote more comments on parts of the older material than any other single person. Pat Churchland and Christof Koch offered brief but at the time very encouraging feedback on the hideous early drafts which eventually mutated into Chapter 9. Thanks also to the staff of the International Philosophical Preprint Exchange and to the several readers across the world who sent their thoughts after suffering through early versions of this manuscript or related papers downloaded from the service. Finally, I am grateful to many anonymous referees of papers whose themes found their way into this book as well as to conference and seminar audiences in Scotland, England, Australia, Belgium, Finland, Russia, and Spain. After all the thoughtful and patient feedback, errors which persist in the book of course reflect purely my own efforts!

Finally, most dear to my heart are my parents, Gerald and Dawn Mulhauser, and the friends without whose support I should not have lasted through to the end of this project. Those who gave their encouragement, love, and inspiration are many, and I hope none will be offended at a mistaken omission. Among them I especially thank Violet Anderson, Eva Bayerlein, Jeff Carr, Andrew Horne, Olga Illicholva, Mags MacLean, Hélène Marie, Kristina Mathesen, James McCormick, the late Julia Shaffner, John Tosney, and Rachel Weaver. My last debt of gratitude is to my fiancée Kathyrn Stares, without whom neither life nor book would be complete.

GREGORY R. MULHAUSER
SUFFOLK, ENGLAND

Please Note

From time to time, references to neurophysiological research on animals appear in this book. Often involving primates or other mammals, many studies include in vivo *lesioning, ablation, or outright mutilation of the animals concerned. While data from this research, when available, may usefully enhance the development of philosophical and cognitive scientific theories, I do not sanction abusive* in vivo *procedures, and I do not intend my work in any way to promote or legitimise research which depends essentially on them. Broadly speaking, I deplore such exploitation, even if the relevant knowledge could not have been had any other way.*

CHAPTER ONE

Frontal Assault

Nobody has the slightest idea how anything material could be conscious. Nobody even knows what it would be like to have the slightest idea... So much for the philosophy of consciousness. (Fodor 1992)

One of my clearest memories of early childhood finds me sitting alone in my bedroom at twilight when I was about five, pondering a curious family of questions. Why does the universe exist? What if it didn't? What would be left over if it stopped existing? Wouldn't *something* still exist? What colour would it be? Even now, the questions elicit the same peculiar twisting sensation from my stomach. And now, as then, I find the basic mystery of why anything exists the most unfathomable of all.

Probably it was fortunate for the sake of a happy childhood that the gravity of the second most difficult question didn't impress itself on me until years later: how is it that it feel like it does to be me (or anyone else) if that feeling is done with nothing more than a stringy ball of nerve fibres, glia, and other organic structures made mostly of water? How can it *possibly* feel like anything?

It is unusual that I agree significantly with Fodor, especially in print. But I think his pessimistic prognosis that an unassailable conceptual roadblock separates consciousness and the material world is *almost* right. After all, conscious experience seems entirely different from mundane features of material objects like mass or colour or fuzziness. Such attributes generate comparatively few deep metaphysical quandaries. While the road from fundamental physics and chemistry to their macroscopic appearance is, in practice, a very long and complex one which has yet to be explored in every detail, in principle nothing stands in the way of a complete low level account of, say, the particular bulk, reddish-yellow hue, and deceptively appealing fuzziness of a ripe peach. No one puzzles over the question 'how can a material object like a peach possibly be *fuzzy*?'. But on the face of it, at least, the characteristic features of phenomenal experience stand in nothing at all like the same sort of relationship to low level physics and chemistry that renders peach fuzz so metaphysically unintriguing.

While a fuzzy peach looks to be a straightforward example of material substances arranged in the right way, it isn't easy to see, even in principle, how Nature could have built something like a *conscious mind* out of matter. Taking as a working hypothesis the notion that fundamental particles of matter lack phenomenal experience entirely, it isn't easy to see how any set of such particles—however they might be organised, energised, shaken, or stirred—could somehow come to be conscious. Nevertheless, each day we find ourselves surrounded by creatures who seemingly enjoy conscious experience without incorporating within themselves any extra 'secret ingredients' beyond myriad ordinary material particles. (At least, contemporary science has yet to uncover any such extra ingredients.)

After years of flirting with extravagant but, ultimately, explanatorily bankrupt nonphysical candidates for a solution to this apparent mystery, some significant conclusions have finally begun to emerge which lead me to think the instantiation of conscious minds by purely material structures really may be fathomable after all. This book explores some of those conclusions and the lines of thought behind them, constructing a story within which conscious experience and the material world may plausibly be unified.

Taking the position that understanding the puzzles of conscious experience cannot proceed without some grasp of who or what the *subject* of conscious experience might be, I ultimately defend the view that the conscious subject—the *I*—is a materially instantiated structure of dynamically changing information. On this account, phenomenal experience is 'what it is like to be' such a changing structure. While it may be true that it is not identical to any material thing, I suggest that a conscious mind may nonetheless be *implemented* by matter. For now, I will say little more about this central unifying idea, apart from noting that information is understood here in a wholly naturalistic, precise, and objective sense and quantified within a formal framework provided by the field of *algorithmic information theory*. Several chapters of preliminaries—concerning information and representation, problems of perspective, functionalism, supervenience, and mental states—must be completed before much sense can be read into this view of the conscious subject, which doesn't receive full treatment until the latter half of Chapter 6. Drawing on resources from traditional philosophy, computer science, mathematics, physics, and neuroscience, the job of developing and exploring this account amounts to a full frontal assault on the clutch of confusions which philosophers collectively refer to as 'the mind/body problem'.

Many themes underlie this campaign (some of which receive more attention below in section 2), but of paramount concern is my aim of embracing the rich 'ineffable feel' of the phenomenal world while still operating within the constraints set by the laws of physics. The method I advocate here takes the laws of physics as given and then examines what sort of picture of consciousness and of cognition might be built up within some framework consistent with those laws. In other words, we *assume* for the sake of enquiry that the creatures of interest, such as humans, are instantiated in a purely physical world[1]—and then attempt to construct a picture of how (or whether) those creatures could possibly be conscious in such a world. The main task then is to evaluate whether such a picture accounts fully for the way our conscious experience really does seem to us in the actual world or whether some significant feature has been left behind. Only if one or more significant features cannot be accommodated should we then consider rejecting both the framework itself and perhaps even the original assumption that consciousness exists in a purely physical world.

This 'world first', or 'physics first', approach, which begins with the physical and attempts to work toward a phenomenal goal, contrasts starkly with methods traditionally dear to the hearts of philosophers. More often, the preferred starting place includes some set of features of phenomenal experience thought to be 'manifestly clear' to a subject's introspection, and the goal is to reconcile those features with what we observe in the physical world.[2] But frequently these innocuous clear beginnings lead on in very short order to conclusions clearly conflicting with what a physicalist picture can deliver. It shouldn't be too surprising, however, if starting only with what *seems* manifestly clear from heady introspection sometimes leads us to places which are in fact impossible to accommodate with actual mere physics.

For instance, it is not unusual to include in an initial flotilla of assumptions the notion that human minds can generate arbitrarily many well-formed English sentences, or that they can perform addition over the full domain of natural numbers, or that their conception of a valid proof can be infinitely extended. Likewise, it is not uncommon to suppose that human minds are capable, in principle, of perfect rationality—in the sense of being able to find all deductively valid consequences of a set of beliefs—or even that they are capable of determining whether any particular proposition (say, in first order logic) follows from a given set of assumptions. Despite the fact that not one of the above assumptions finds support from a single scrap of empirical evidence, even many naturalistically minded philosophers happily invite them on board. A popular justification for doing so, it seems, boils down to a belief that these 'in principle' ideals are what *really* need explaining, and that actual world deviations from the ideal are little more significant than the noisy boundary conditions which always blur experimental measurements away from the ideal predictions of a good theory. As it happens, however, convincing reasons suggest all the above assumptions are false for any cognizer physically instantiated in the real world—even though the failure of each is consistent with the *appearance*, from the vantage point of the cognizer concerned, that it is true.

Many of the mind/body confusions which this book attempts to clear away find their roots in just such lavish assumptions about what *seems* clear from introspection. For progress to be made on the puzzle Fodor thinks no one should even hope to understand, it is helpful for philosophers to adopt a little more modesty about their own capabilities, acknowledging that—in the absence of some *argument* to the contrary, and mere appearances aside—the default assumption should be that we humans are as subject to the constraints of the laws of physics as any helium atom in the core of a star or any fuzzy peach poised to fall from a tree. This book, I hope, marks significant moves in that direction.

1. STRUCTURE AND OVERVIEW

In view of its interdisciplinary nature, the text has been organised in a way which I hope will accommodate readers with a wide variety of backgrounds and personal interests in the material. Below, I outline my attempts to package far-flung topics in an accessible way, describe the book's anticipated audiences, and briefly preview each chapter.

1.1 Pathways Through the Book

Because perhaps only a minority of readers whose interests happen to coincide with my own will want to pursue the arguments set out here linearly from start to finish, I have structured the book with a view to making other reading strategies as painless as possible. Especially technical remarks are typically flagged in advance, and in each such case I suggest an alternative route through the text which will allow readers with less interest in details to skip ahead without missing central points. Around two hundred cross-references to particular chapters, sections, or page numbers, together with an extensive index, should help those taking a nonlinear path through the book to locate other points supporting key lines of thought.[3]

With a few exceptions, the discussions require little in the way of specialist background knowledge, but by no means do I intend the book as a comprehensive introduction to any of the subdisciplines which feature in it, nor to the mind/body problem itself. Although in places the text still reveals roots in a doctoral dissertation, the overhead dedicated to one dissertation favourite, reviews of existing literature, is substantially reduced—particularly in areas (such as representation) where fresh perspectives are in the offing, perspectives which may fit only awkwardly or not at all into the categories provided by that literature. Likewise, I aim mainly to present views positively, largely dispensing with cautionary lists detailing what they *are not* and contrasting them with the many similar cousins for which they might be mistaken. And while I hope the broad collection of about five hundred references provides a helpful start for readers following up particular threads, it remains far from complete. I have tried to avoid the 'my bibliography is bigger than your bibliography' syndrome—the urge to cite the kitchen sink—which seems to be spreading rampantly through populations of undergraduates and seasoned researchers alike (perhaps under the influence of easy to use electronic abstract databases).

As for exceptions to general accessibility, I do assume familiarity with basic logical connectives, plus a few mathematical and set theoretic symbols, and I assume experience with the philosophical notions of *a priori*, *a posteriori*, intension, extension, possible worlds, and the like. Prior acquaintance with Turing Machines is helpful. The 'worst' exception occurs in Chapter 7; the summary notes on the quantum formalism beginning on page 145, as well as some subsequent sections, will be most useful for those with at least a passing familiarity with linear algebra. However, as always, those portions may safely be skipped by readers preferring to bypass the particulars, and pointers on where to pick up again are of course included.

Overall, I hope the book will be enjoyable for most people interested in materialist cognitive science and the challenges of understanding consciousness, from advanced undergraduates to senior researchers and those whom publishers' marketing departments sometimes dub 'the motivated lay reader'. It may find a place in graduate seminars or as a supplementary text in philosophy, cognitive science, artificial intelligence, or artificial life.

1.2 Chapters Summary

The second chapter is a warm-up exercise. Taking up Dennett's recent challenge to defend the philosophical relevance of zombies without begging important questions about their capabilities (or lack thereof), the chapter includes an architecturally explicit zombie construction and applies it to motivate later explorations not of how conscious subjects *behave* but of what internal processes bring about that behaviour. Main outcome: nonconscious zombies with external behaviour indistinguishable from that of normal subjects are logically possible, so conscious experience requires more than the right external behaviour.

Those later explorations of internal processes depend largely on the precise and objective notions of information and of representation which take centre stage in Chapter 3. Introducing a purely physical view of information based on Gregory Chaitin's version of algorithmic information theory, the chapter describes representation in the general case with a formal measure of *mutual information content* between two physical objects. (Note that this appeal is to a concept of information which differs from Shannon's, used by Dretske 1981.) Along the way, I outline the modern and easily understood information theoretic version of Gödel's incompleteness results, questioning the foundations of claims occasionally made about incompleteness and its bearing on philosophy of mind. Main outcome: representation is objectively linked to the physical world.

The next chapter shows why, far from underwriting a convincing argument against physicalism, the curiosities of Jackson's example of Mary, the colour-deprived neuroscientist, arise naturally within a wholly physicalist setting. Tricky puzzles of perspective, or points of view, first arise here—to return in

Chapter 6—and the relationship between logic and the physically instantiated cognizers using it merits brief remarks. Main outcome: the same information may be physically instantiated in many distinct ways and with disparate ramifications for conscious experience; Mary's ignorance before seeing red for herself is not a matter of information, but of state.

Chapter 5 outlines a new formal framework for understanding functional systems. Problems of mathematical triviality threaten traditional approaches to functionalism based on correlation or correspondence, while teleofunctionalism requires links to facts historically removed from the system in question, rendering it more suitable for questions like 'why is this component here?' than those like 'how does this system work?'. This chapter's objective method of functional decomposition, using a formal measure of process complexity called *functional logical depth* (inspired by work of Chaitin and Charles Bennett), circumvents both difficulties. Main outcome: the revised functionalism is now objective and better disposed for explanatory work.

Within a context featuring arguments about supervenience, perspectives, and mental states, Chapter 6 outlines the first components of a theory of consciousness based on the *self model*, a materially instantiated dynamic data structure which emerges as a promising candidate for the seat of phenomenal experience. On this view, one motivated by the need for a conceptual link between consciousness and the material world yet grounded empirically and thus falsifiable, conscious experience is 'what it is like to be' a particular kind of changing data structure. The self is not identical to any of those physical components underlying it, yet it is implemented by them; *I am a self model.* Main outcome: taking the concept with an appropriate intension, consciousness supervenes logically on the physical world, with the self model as link.

Taking a brief side trip to debunk a competing class of theories, the so-called 'quantum theories of consciousness', the next chapter describes the work of Roland Omnès and the mechanisms of *interactive decoherence* which, automatically and extremely efficiently, eliminate the need for a conscious observer in quantum mechanics and guarantee that special quantum effects are virtually nonexistent at the levels of description where they are sometimes imagined to feature in the human brain. As a bonus, interactive decoherence accounts for the automatic emergence of apparently deterministic quasi-classical reality from a quantum substrate. Main outcome: quantum physics doesn't need consciousness, and consciousness doesn't need quantum physics.

Where Chapter 6 focused mainly on the conceptual territory between the self model and supervening consciousness, Chapter 8 sets out in the opposite direction, from the self model down toward the sorts of lower level materials—neural tissue, in the case of humans and other terrestrial life forms—which may implement such data structures. After tidying the information theoretic description of self models given in Chapter 6 and introducing basic tools from neuroscience, the chapter shows how one particular research programme, Stephen Grossberg's adaptive resonance theory, bears especially on the task of implementing self models in real neural systems. Main outcome: the right neural systems can do the representational work which self models require, and real organisms may plausibly have evolved so as to implement them.

Turning from cognitive models to the mathematical presuppositions underlying them, Chapter 9 examines the relationship between models and properties of the physical systems being modelled. In particular, the chapter examines recent work by Hava Siegelmann and Eduardo Sontag suggesting that analogue models (based on the real numbers) may display computational capabilities which exceed those of digital models (based on the rational numbers). To the extent that the two sorts of models differ in their capabilities, an interesting question then arises as to whether one or the other makes for a better match with reality. Main outcome: contrary to popular dogma, the choice of number systems does make a difference for models of cognitive systems, but the significance of that difference for the real world has yet to be established.

Finally, Chapter 10 recaps some of the text's central themes and reflects on the broader significance of the mind/body problem for understanding who we are, as individuals and as a civilisation. It finishes with some remaining open questions and possible directions for future research.

Readers who find my attempts in each of these chapters to situate the discussion within an overall picture a little too obscure might also want to skim through that last chapter's section 2, starting on page 243, which, unlike the above summary, surveys the principal developments of each discussion on the assumption that readers will already have acquired some familiarity with them.

2. UNDERLYING THEMES AND METHODS

Although the book treats many distinct topics, all are united both in the sense that they relate to some aspect of the mind/body problem and in terms of underlying themes. Below I begin with a theme I specifically attempt to sidestep and move on to two more positive ones. Readers less interested in comparatively dull methodological issues should skip ahead to section 3 below.

2.1 Dynamics and Computation

One fracas I hope to avoid is the war raging between advocates of dynamical and computational methods in cognitive science. At a recent workshop marking the opening of Sussex University's Centre for Computational Neuroscience and Robotics, for example, I was astonished at the number of participants who appeared quite abruptly to have rediscovered dynamical systems and were vigorously promoting a dynamical approach to artificial life and related fields as the greatest new advance—perhaps 'revelation' would be more fitting—in decades. Worse than the hype[4] was the impression that one should be exclusively either dynamicist or computationalist and that for the sake of scientific purity the two must not be mixed—a notion asserted with characteristic vehemence by Tim Smithers, the well known 'non-representationalist' roboticist from the University of the Basque Country. I hardly dared raise a voice for a hybrid view for fear of witnessing my own public stoning!

Typically, allegiances to a particular approach run deep, often it seems to such an extent that what appears at a glance a straightforward observation supporting one or the other position becomes utterly invisible to those favouring the opposing viewpoint; all too frequently, arguments are simply ignored or contradicted rather than rebutted. I hope in this book to challenge both sides while incensing neither. For my part, while generally inclined more in the direction of a dynamical approach, I believe tools of *both* types play a valuable rôle in cognitive science: both dynamics and a computationally defined variety of representation feature crucially in my own view. In Chapter 3, I outline a method of quantifying information content (and representation) which, while itself wholeheartedly computational, applies equally well to all physical systems, whether interpreted as computational or dynamical entities. Later, in Chapter 6, I argue against the typical computationalism-friendly notion of an instantaneous mental state in favour of an inherently temporal replacement, advancing a view of consciousness itself which, while described in 'computational' information theoretic terms, subsequently receives a dynamical account of neural instantiation in Chapter 8. (For readers harbouring strong feelings on the debate and an irrepressible urge to categorise this book in the language of it, the approach I adopt here might be verbosely labelled, with tongue in cheek, as 'computationally described, dynamical semi-representationalism'.)

Contra Smithers, I emphatically *do not* believe that one should decide in advance to adopt one framework or the other, on pain of tainted science. Whether a good explanation of observed phenomena should be dynamical, computational, or hybrid in nature is a question properly evaluated in light of empirical evidence and the resulting matches between candidate theories and reality; it is not, to my mind, a matter of evaluating empirical evidence in light of a pre-existing conviction that explanation is to be found in one and only one particular form. Likewise, *contra* van Gelder (see note 4 above), I find very little *ontological* mileage in the methodological distinction between dynamical and computational approaches to cognitive science. As far as I can see, whether we speak in computational language, dynamical language, or little green men language, we still talk about the *same actual world*. ('World first', not 'words first'!) In the terminology of Chapter 6, the 'things' talked about supervene logically on the physical world. Depending on the needs and aims of any given situation, we might find it useful sometimes to describe cognitive processes computationally, while at other times dynamical or hybrid descriptions will prove most helpful. There may be a *de facto* rough cultural division between cognitive scientists who generally prefer computational tools and those who generally prefer dynamical tools, just as there is a *de facto* rough division between those who prefer reverse Polish notation calculators and those who prefer algebraic calculators, but such a division needn't necessarily reflect ontologically significant facts about the real world of cognizers which they are studying.

2.2 Priorities and the Naturalistic Urge

Closely related to the business of choosing appropriate languages of explanation are questions about the goals of research itself and the choices of tools which those goals motivate. One opinion popular in the United Kingdom, at least, counts the task of furthering the cause of a particular discipline or even department as, if not quite an end in itself, at least a highly significant component of academic research. On this view, a central aim of the philosophy researcher should be to advance the understanding of philosophy for philosophy's sake; the general field of philosophy is primary, while the particular issues to be explored take a back seat. A diametrically opposed tack shapes my own research in general and this book in particular.

My main interest lies not with furthering a particular discipline (or, even less, a department), but with examining fascinating questions, whatever disciplines may count as their own those questions or their solutions. I make no apologies for importing methods or conclusions from outwith traditional philosophy, particularly from scientific fields, in the course of exploring the mind/body problem.[5] Although many philosophical colleagues in the United Kingdom are quick to dismiss my own work as "science, not philosophy!"—as though the two were mutually exclusive—I am only too happy to recruit for 'philosophical' purposes the tools of other fields. (Were I to encounter tomorrow some good reasons for believing the art of pottery held the key to understanding problems of consciousness, I would learn to make pots!) In the spirit of thinkers like Kitcher (1992) and those featured in Kornblith (1994), I take philosophy and the natural sciences as forming a continuum. The eloquent words of E.W. Hobson, Sadleirian Professor of Pure Mathematics and Fellow of Christ's College in Cambridge, speaking at the conclusion of his 1921-22 Gifford Lectures at the University of Aberdeen, express the idea succinctly:

> If we were in possession of, and able to grasp, a unified view of the Universe, in which all the elements of existence and valuation were completely synthesized...we should not require to mark out frontiers between Science and Philosophy or Theology... The untrammelled freedom which must be allowed to workers in all depart-

> ments of the great cultural work of humanity...should not...involve the erection of rigid impassable barriers which shall mark off domains which hold no communication with one another. On the contrary, workers in one department will often receive the most valuable enlightenment, and most important suggestions, from quarters outside their own special line. (Hobson 1923, p. 501)

Borrowing the contemporary words of Sussex researcher Inman Harvey (personal communication), I also agree wholeheartedly with the notion of "doing philosophy of mind with a screwdriver"—testing philosophical ideas with robotics and artificial life in real world laboratories. In an article entitled 'Artificial Life as Philosophy', Dennett (1994) advocates a similar view.

But despite a distinguished tradition of positive and mutually beneficial interaction between philosophy and the sciences, philosophy of mind is, perhaps more than at any time in its history, feeling the heat from scientific areas as those fields 'encroach upon' the study of questions still considered by many to be philosophers' territory. Quite justifiably, methods and strategies from areas like quantum physics, chaos theory, neuroscience, and artificial life have been stirring up the field, eliciting responses from the philosophers' camp ranging from the sort of dismissiveness recalled above to enthusiastic endorsements of method x as that which will at last naturalise the study of mind. Although my own tendencies draw me firmly in the naturalistic direction, explicitly arguing the case for the relevance of scientific fields to 'philosophical' questions really is not my battle. I hope others will simply evaluate my judgement in choosing the tools I have in light of the lines of argument they support, rather than under the shadow of some pre-existing prejudice as to how best to secure the borders of philosophy against infiltration.

2.3 Description Complementarity and Puzzles of Perspective

After the central goal mentioned at the start of the chapter, that of embracing the 'ineffable feel' of phenomenal experience while simultaneously heeding the constraints of physical law, the next most significant thread running through this book concerns the relationships between different levels of description[6] of the same cognitive system and between first and third person perspectives on cognitive systems. I believe both relationships bear crucially on the project of understanding how (or whether) matter can implement a mind.

With respect to the first, a good many problems in the history of philosophy originate, in my opinion, with a failure to appreciate the complementarity between alternative descriptions of the very same thing—a failure which, at times, borders on stubborn-minded silliness. Throughout the explorations of the topic which appear here, I have in mind a background context of working hypotheses inspired by the likes of Davidson (1973) and Hellman and Thompson (1975); I opt for what the latter call 'ontological determination'—the physical is all there is, and everything that happens physically is governed entirely by the low level laws of physics—coupled with 'explanatory anti-reductionism'. In other words, nothing ever happens which is not, at the lowest level, entirely a result of the laws of physics;[7] yet, in giving intelligible *explanations* of processes, we may well have to rely on entities constructed at a higher level of description commensurate with that at which we describe the processes themselves.

For instance, it would be no explanation of how a clock keeps time with its hour and minute hands to describe the interactions, in accordance with the laws of physics, of every single subatomic particle within it. A good explanation would describe instead the interactions of cogs and pendulums or transistors and quartz crystals (depending on the clock) and the relationship of those interactions to the movements of the hands. Yet nothing ever happens to (or between) cogs, pendulums, transistors, quartz crystals, or arms of a clock which doesn't follow straightforwardly from the lowest level laws of physics. The approach to explanation favoured in this book embraces both observations together. The view contrasts starkly with those of Durkheim (1938) or Radcliffe-Brown (1952), for instance, who maintain that good explanations require not only an appropriate level of description, but a new independent level of *things* whose causal properties, significantly, *do not* follow from those of their constituent parts.[8] In the guise of a discussion of evolution and Conway's Game of Life (Poundstone 1985), Dennett (1995c, pp. 166-75) offers the tidiest look at levels of description I have encountered to date.

Interest in the relationships between alternative descriptions of the same thing doesn't end with different levels; a special case of complementary descriptions exists in the distinction between first and third person *perspectives* on cognitive systems. Just as I believe there is a common failure to appreciate the relationships between levels of description, so, too, is there a common failure to appreciate the factors underlying the truism that reasoning *about* a system as a third person differs greatly from *being* that system in the first person. If, for instance, I cannot come to know 'what it is like' for *me* to be in a particular state until I have actually been in that state or one relevantly similar to it, why should it be at all puzzling that I cannot come to know what it is like for someone else to be in such a state? And is there any reason to think I *should* be able to grasp what it is like for me to be in a particular state before having actually been in such a state? Such perspectival issues figure in a wide range of questions about the relationship between mind and the physical world. My direct quarrel with a standard view that problems of perspective support a case against physicalism begins in Chapter 4 and returns in Chapter 6, but a reluctance to accept the standard view underlies much of the rest of the book.

3. COGNITIVE DISSONANCE

It was in one of the first philosophy texts I encountered as an undergraduate, Richard Taylor's (1963) *Metaphysics*, that I vividly recall reading the tale of a poor fellow called Osmo, who perishes while trying to escape a future foretold for him in a special book. With perfect accuracy, the mysterious volume describes every event in Osmo's entire life, from his birth through to the present and on into the future and his eventual death. Like the protagonist in a Greek tragedy, Osmo cannot escape his Fate: eerily, whatever the book foretells *always* comes to pass, and try as he might, his every failed struggle to forge a new and different future for himself only underscores the book's apparent infallibility.

Taylor's point in articulating the story—apart, perhaps, from terrorising impressionable young philosophers—is to suggest that we all should view the future with the eyes of a fatalist. Fatalism should appeal, Taylor argues, for purely logical reasons: since some complete description of all my life's future events is true *right now*, and since that description could *already* have been written down in a book (say, by an omniscient and prolific author), I lack any sort of freedom to change it, and I ought just to stop worrying about it! Despite Taylor's protests to the contrary, the argument for fatalism seems a textbook example of modal fallacy. But nevertheless, I think I can almost grasp what might have led him to suggest the line of thought, despite his fluency with the logic of modal operators. Although most philosophers reject Taylor's reasoning, the years have not dulled my impression that something remains deeply disturbing about the fantasy scenario he describes; a book accurately foretelling my every experience—perhaps even my every thought—for the rest of my life truly would be a frightening prospect.

That peculiar discomfort, I believe, bears a possibly illuminating likeness to the distaste many express for the idea that human cognition and consciousness might one day be explained within a purely materialist framework. I am *not* suggesting that those unhappy about such a possibility are confused about modal logic! Rather, although the analogy is imperfect with respect to logical structure, there seems to me an appealing kind of symmetry between the two cases. On one hand, it is disturbing to think that some book could in principle expose, for all to see, every single event in my entire life—*even though* I know perfectly well that the logical possibility of such a volume, in and of itself, in no way diminishes my freedom in bringing about the events described within it.[9] There might be other reasons why I lack freedom, but the possibility of such a book is certainly not one of them. And on the other hand, some may likewise find it disturbing that a complete materialist theory of consciousness could in principle expose, for all to see, the underlying physical foundation of every single pain or taste or visual impression (or hope or lust or intuition) throughout a subject's entire life. This might seem disturbing *even though* such a prospect would in no way diminish the painfulness of the pain or the lustfulness of the lust; that rich phenomenal experience might have a physical explanation would render it no less rich phenomenal experience.

A usually unarticulated further worry might grow from the notion that anything which admits of a physical explanation is no different in principle from any other physical thing. The possibility of a book explaining my entire conscious existence in the language of physics threatens to reduce that experience itself to nothing more than *mere physics*, siphoning off its *value* down to the level of some least common denominator appropriate for other physical things like bricks and globules of sludge. Probably it is only natural to feel some discomfort at the idea that the very sciences which our own ingenuity created could drain away our fundamental value in this way. On this view, perhaps our value can only be preserved by finding it a nonphysical refuge categorically separate from (and impossible to unify with) the merely mechanical transactions of bricks and balls of sludge.

Needless to say, these are worries I do not share. On the contrary, I feel that such a naturalistic physical explanation could only *add* to the sheer marvellousness of phenomenal experience. There can be little doubt that it would be amazing if it turned out instead that the subject of my phenomenal experience was really an independent, immaterial, ethereal spirit of some sort—that *I* was such a spirit and my consciousness one of that spirit's properties. But how much more truly amazing it would be to discover that the conscious 'I' is instantiated purely physically—that somehow, despite having nothing but that stringy ball of nerve fibres and other organic matter with which to do it, I still manage to enjoy my full remarkable repertoire of rich conscious experience! That matter simply organised in the right pattern and changing in the right ways could actually instantiate *me*, with all my vivid phenomenal experience intact: *that* is a marvel worthy of the name. And, indeed, if what really counts is its rôle in enabling our conscious lives, why should a *pattern* be of any less value than an immaterial spirit?

In the next nine chapters, I set out what I believe are some useful steps toward understanding the sorts of links between mind and matter which might make such a view attractive in its own right and which may allay Osmo-style discomfort. Do I believe I have given a rigorous, complete, and definitive solution to the mind/body problem(s)? Of course not! But I do think a 'solution'—or, rather, a set of solutions to a cluster of related problems—will eventually be found to share the general form, and perhaps some of the details, of what I outline here. Many times in these pages I deliberately reach far out on a limb while constructing some view or other. In so doing, I aim to lay out a sort of tree of possibilities; not all of its branches will ultimately bear fruit, I am sure. But I'm convinced that while some limbs will eventually need cutting out and others will benefit from considerable reshaping, the central thematic trunk is healthy and planted just about where it ought to be. I hope that laying out the tree as I have will provide new opportunities for progress through the process of evaluating and snipping back those bits which don't belong, strengthening those which do, and shaping this nascent theory into something robust and comprehensive.

NOTES

[1] Nowhere do I argue positively for any particular rendition of material monism; I find it hard to grasp what a good argument for material monism would even look like.

[2] Philosophers, understandably concerned to be clear about what it is they're trying to explain before setting out to explain it, frequently adopt the convenient assumption that our first hand, direct experience of consciousness suffices to fix the concept appropriately well. But to paraphrase an example due to Aaron Sloman, the assumption is as unjustified as the claim that even before Einstein's analysis of the concept of simultaneity, we really all knew what it was anyway just through our first hand experience. As for the case of simultaneity, Sloman suggests, it may be that only after constructing some good theories, capable of supporting coherent concepts, will we be able to grasp properly what it was we were trying to explain in the first place!

[3] Since endnotes are easily found at the end of each chapter, when cross-referencing them I usually mention the page number for the discussion which is endnoted rather than giving the page number for the endnote text itself.

[4] Not *all* the excitement is hype. While I am dubious about his tendency to overstate the *ontological* significance of what amount to different dynamical (or non-dynamical) *descriptions* of physical entities—see section 3 of Chapter 9—Tim van Gelder (1995a, b) offers a sober but provocative analysis of each approach. In a forthcoming target article for *Behavioral and Brain Sciences*, he succinctly maps out distinctions between the computational hypothesis—roughly, the view that cognizers are digital computers—and the dynamical hypothesis—roughly, the view that cognizers are dynamical systems.

[5] Very often it is expedient to import science into philosophical discourse—nothing clinches an existence proof like empirical evidence, for instance!—but some of the *really* difficult problems of philosophy might well be those which either cannot be naturalised or which would be very difficult to naturalise. (Ethics comes to mind as a possible example.) I have the utmost respect for these sorts of areas, but my temperament generally tempts me more in the direction of those where some 'easier' progress can be made.

[6] Although I usually refer to 'levels', I mean also to include *parallel* or otherwise complementary descriptions which may not exist in an hierarchical relationship with one another.

[7] In the language of Chapter 6, everything *supervenes* on microphysics; see section 1 of that chapter in particular.

[8] On the last especially, also compare the emergentists such as Alexander (1920), Broad (1925), or Pepper (1926); see McLaughlin (1992) for recent discussion.

[9] Specifically, the purported argument to the contrary requires shifting the scope of a necessity operator from a whole conditional to just the consequent of the conditional. But although it is necessarily true that if such a book correctly proclaims that I will perform action x, then I will perform action x, from this it simply does not follow that such a correct proclamation entails that I necessarily will perform action x.

CHAPTER TWO

Zombies and Their Look-Alikes

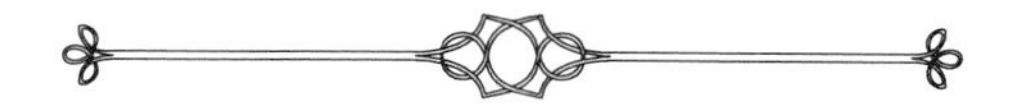

It is an embarrassment to our discipline that what is widely regarded among philosophers as a major theoretical controversy should come down to whether or not...philosophers' zombies...are possible/conceivable. (Dennett 1995b, p. 325)

Zombies of analytic philosophy, unlike the voodoo victims of Haitian folklore, are hypothetical creatures entirely bereft of conscious experience who nonetheless *behave* indistinguishably from the rest of us. Philosophers' zombies walk and talk as if they're conscious, they appear to wake up in the morning, and over breakfast they even speculate on the meaning of dreams they claim to have had. They don't *realise* they're zombies, of course—no feeling of peculiarity spoils the pristine emptiness of their barren phenomenological landscapes—and an enterprising philosopher engaging one in conversation about the topic might well hear all manner of insightful commentary about what the notion of zombies reveals about philosophy of mind.

Daniel Dennett says zombies don't exist. Most people, no doubt, would agree. But more importantly, he suggests they are logically impossible: misguided philosophers who claim they are conceivable merely fail to imagine the *full* repertoire of zombie behaviours. All too often, philosophers define zombies as above and then proceed to argue for some behavioural clue or other which would give them away. But of course there can't be any such clue. "The philosophical tradition of zombies would die overnight", Dennett says (1995b, p. 325; Dennett 1995a is similar), "if philosophers ceased to mis-imagine them". It is unusual that I disagree significantly with Dennett, especially in print (Mulhauser 1997a). But I think his pessimistic prognosis can't be quite right. Dennett's challenge offers a tailor-made warm-up exercise for the rest of this book:

> Show me, please, why the zombie hypothesis deserves to be taken seriously, and I will apologize handsomely for having ridiculed those who think so... If the philosophical concept of zombies is so important, so useful, some philosopher ought to be able to say why in non-question begging terms. I'll be curious to see if anybody can mount such a defence, but I won't be holding my breath. (Dennett 1995b, p. 326)

This chapter takes up the gauntlet, exploring how easily the products of impoverished imaginings may be mistaken for insight into logical necessity. Dennett, unfortunately, is as guilty as those philosophers whom he so expertly (and generally correctly) criticises. Zombies of this sort *are* logically possible, and they underscore the importance of untangling the four-way relationship which obtains between a creature's environment, its behavioural responses to the environment, the internal mechanisms which support that behaviour, and the creature's phenomenal consciousness. (Zombies return briefly in Chapter 6 in a stronger form, a variety which is indistinguishable both in terms of externally observable behaviour *and* in terms of internal construction; such a 'super-zombie', or 'absent qualia' zombie, is incompatible with the theory of consciousness developed later.) A careful assessment of zombies also motivates the next chapter's look at information and its central rôle in making sense of consciousness instantiated by physical structures.

1. AN EMBARRASSMENT TO OUR DISCIPLINE

Philosophical interest in zombies is inextricably bound up with what Alan Turing (1950) called the "imitation game", the notorious Turing Test. Suggested by one of the founding fathers of modern computing theory near the zenith of behaviourism's dominating assault on psychology and philosophy of mind, the 'test' amounts to a question and answer session meant to separate the thinkers from the pretenders on the basis of linguistic behaviour alone. Conceal a computer in one room and a human in a second, Turing suggested, and if the computer can regularly fool a human interlocutor into believing it is the real thinking creature, the computer wins. More often than not, philosophers now refer to consciousness rather than just 'thinking', and the question becomes whether passing the Turing Test merits induction into the privileged club of beings we dub conscious.

Anyone prepared to assert the logical possibility of zombies should of course reject outright the imitation game as a sufficient indicator of consciousness. Zombies *obviously* can pass the Turing Test—they are, after all, behaviourally indistinguishable from the rest of us. During the test, a zombie can tell all about how it feels to be taking the test, and he can tell whether he guessed in advance whether we would ask that question. Surprising him with an accusation of zombie-hood, he can reply with poetic denunciations of modern mechanistic thinking, describe his inner turmoil at the accusation—"That's what my ex-wife always used to say!"—and tailspin into seemingly emotional recollections of early childhood and cogitations about how it all would have been so different if his parents hadn't sent him away to boarding school. Yet, by definition, the zombie isn't conscious. He's just a look-alike. For the believer in zombies as a logical possibility, winning the imitation game cannot be sufficient to guarantee consciousness. Or to anticipate the discussion ahead, the believer in zombies isn't satisfied just with the instantiation of a relationship between environmental stimuli and observable behaviour which is good enough to pass the Turing Test.

The general spirit of the Turing Test suggests a simple formula for explicitly constructing a logically possible zombie, using a real person as a model to be 'zombified'. The zombie constructed here won't be very economical—indeed, we'll spend the provisions of logical possibility with grossly extravagant abandon. It certainly won't yield the sort of thing liable ever to turn up in a real physical universe. But at least a point or two will emerge by the end of the exercise.

2. BUILDING A ZOMBIE

An unnecessary but convenient assumption starts things off: namely, that at any given time, the class of possible sensory inputs which can make any difference to a person's behaviour, while it might be very large, remains finite. The assumption is unremarkable. It is *not* the same as saying that a person can only recognise or experience a finite set of sensations. *Maybe* a person can read any of an infinite class of limericks or mathematical expressions, for instance, or can savour an infinite class of tastes made by combining lime and coriander and baked beans. However, a person could do all these things over a continuous interval of time but with only finitely detailed inputs at any particular moment in time. To be picky about technical justification for the assumption, we could appeal to bounds on the number of distinguishable states of a physical system (see page 212, for instance), but this introduces a presently unnecessary commitment to materialism. In fact, it emerges in due course that the assertion itself is entirely unnecessary anyway, but the exposition is, for now, a little easier for it.

With respect to motor outputs, an even weaker limitation is helpful: we require only that at any given moment, the activations of a person's motor outputs—their various muscular activities—can be specified with finite precision. This assertion hardly needs defending: after all, the motor structures of people really are physical things, regardless of what might be true of the minds which handle their sensing and directing. Motor behaviour demanding infinite precision is physically absurd. In any case, a finite ceiling on the precision of sensory inputs and motor outputs allows us to assign, for each moment in time, a unique number to each member of the set of the zombie's possible sensory inputs and motor outputs. Given some candidate person on whom to base the zombie construction, one way of doing this would be to digitise the activation levels of every single one of the person's sensory input neurons and motor output neurons to some desired level of precision. (Anyone who recoils at the neural emphasis is of course welcome to include something else in our giant digitisation: Quantum microtubules? No problem: add in digital renditions of their wavefunctions. ESP? Include that, too; just add in a new number for every extrasensory perception which makes a difference...and so on.)

No doubt the twisting path toward a new rendition of the venerable old lookup table (akin to Block 1978, pp. 294-295 or Block 1981, for instance) is becoming clear. But, crucially, for the present construction all the looking up is done in advance. We'll see how that's done in a moment. First, a set of numbers quantifying current sensory input and another set quantifying possible motor activations isn't quite enough. Mimicking the behaviour of some candidate person also requires a running record of all the *previous* sets of sensory inputs. So, the first step in the construction is to implement however many different input sensors will cover the same range of sensitivity as a person (perhaps an array of millions of neurons with little DSP chips, or digital signal processors, attached), plus a similar set of motor actuators (perhaps millions of human motor neurons). All these sensors and actuators can be set to update their sensing or their actuating at any desired frequency—say, 44.1 KHz for CD-quality sensing and actuating (although that is almost surely overkill). Now add to the sensors and actuators a running 'life movie' which spools to a large digital memory the new set of input numbers every 44.1 thousand times per second. At each sampling instant, simply concatenating the complete memory contents with the new input number gives a single new unique number. All these possible unique numbers together form a complete list of possible overall inputs, while the numbers indicating different motor activations together form a complete list of possible outputs.

Preparations are almost complete for a rendition of a pre-cooked lookup table. But first, note that so far the construction is entirely possible logically but physically ridiculous. An array of just eighty different input sensors, each digitised with a resolution of ten possible activation levels, already provides an input list of roughly as many possible combinations, 10^{80}, as there are particles in the observable universe. Real humans probably have more than eighty input sensors. But despite Dennett's objection (personal communication) regarding this enterprise that "'logical' possibility isn't always very interesting", an important point is lurking in this bigger than cosmos-sized logically possible but physically ridiculous example.

The zombie recipe calls for a human model: let's use Osmo, the unfortunate Taylor-made anti-fatalist from Chapter 1. Suppose for the sake of the example Osmo is a bona fide conscious person. For every momentary situation in which Osmo might find himself, characterised by a number from the list of possible input/memory combinations, there is some corresponding set of motor outputs, perhaps conditioned in a very complex way by other events in his past and nuances of his own internal musings, with which he might respond—a number from the output list. It's the overall relationship between possible input situations and output responses of a real Osmo which a careful ordering of the input/memory and output lists for the zombie will mimic. Since the construction requires cosmically huge lists of numbers, 'Cosmo' seems a fitting name for the zombie. We run through the matching-up informally first and then more precisely in a possible worlds framework.

Intuitively, the easy way to order the lists is just to match up each possible input/memory combination with an appropriate motor output such that the input/memory and output pairs correspond to one way Osmo might behave if he were in the situations characterised by those inputs and memories. Of course, motor activations of the real Osmo probably depend on all kinds of complex psychological phenomena, including fears, desires, hopes, or superstitions—perhaps even consciousness!—all far beyond the simple sensory 'life movie'. Likewise, the real Osmo almost certainly lacks access to such a finely detailed memory of past sensory experiences. But since the zombie business is all about concentrating solely on behaviour, such details are safely ignored while constructing an ordering of numbers which accurately represents the motor output side (i.e., the behaviour) of the real Osmo for any given circumstances of history and present sensory input. Just to be sure, here's an algorithm for ordering the lists more carefully.

Pick an initial world segment P in which Osmo exists, with the end of P marking the time in Osmo's life which serves as the start of the zombie construction.[1] The set of possible sensory inputs Osmo might experience at the next sampling moment suggests a new set of initial world segments $\{I\}$ (where $P \subset I$ for each I), which can be partitioned according to which set of finitely detailed sensory inputs Osmo was experiencing. For each possible Osmo input, now correlate the matching digitised Cosmo sensory input numbers with digitised versions of Osmo's motor responses. This requires no presuppositions about determinism or free will—or fatalism, for that matter. In response to sets of sensory inputs in each segment I, Osmo might have the option of activating his motor outputs in any of millions of different ways. As shown schematically in Figure 1, just pick one, arbitrarily, and place it in the list next to the appropriate digital rendition of Osmo's sensory inputs after P (i.e., sensory inputs in I). The choices suggest a new set of initial world segments $\{I'\}$ (where $P \subset I \subset I'$), in each of which Osmo has experienced some set of sensory inputs a moment after the end of P and has responded with the motor activation chosen. Now form the set of combinations of possible new inputs in particular I' together with spooled input memories carried over from the previous I. This is just the set of concatenations of stored memory numbers with possible new input numbers. In the list, again correlate each of these concatenations with the numbers representing Cosmo's digital version of any of the different possible motor outputs Osmo prefers in $\{I'\}$. These choices suggest a new set of initial world segments $\{I''\}$, and we proceed as before, discarding any input/memory combinations which are never encountered.[2] This can be carried on for as long as desired—or, at least, until possible world Osmos start to die off—ordering the lists so as to model Osmo's motor behaviour up to any given point in his life.

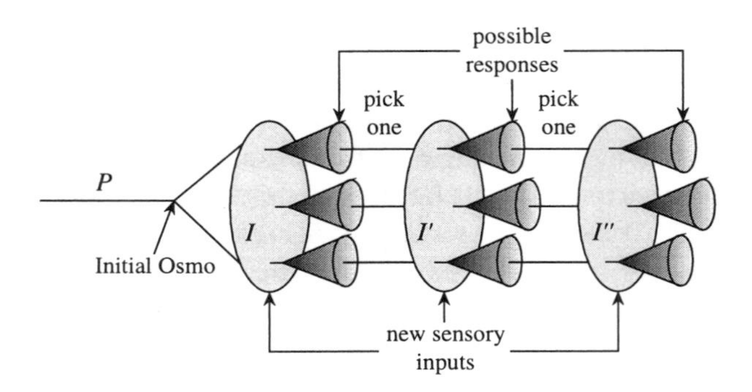

Figure 1. Zombifying Osmo

This creates a list of numbers representing sets of possible input/memory combinations, together with a set of corresponding motor outputs which reflect what at least one real conscious person (namely, Osmo) might do if he found himself in any of those situations indicated by the input/memory combinations.

After going to all the trouble of ordering the lists of numbers into a lookup table, it doesn't actually get used for any looking up. Instead, we equip Cosmo with an appropriate set of input sensors (such as ones for hearing, seeing, smelling, and so on) and motor actuators (such as ones enabling walking and talking and doing back flips) and connect each possible input combination *directly* to the corresponding output actuators. Implementing the input/output values with binary numbers, this arrangement requires only a single AND gate, the computer engineer's electronic version of the familiar logical connective, together with a few inverters, for each connection. An AND gate provides a '1' output if and only if all it's inputs are '1', while an inverter simply turns '1' into '0' and vice versa. As in Figure 2, which for simplicity shows only two bits of input/memory information and omits the actual bit patterns for the output actuators, each input/memory combination activates a single AND gate, providing exactly one '1' signal to the output actuators. This output signal can of course be broken up into a slew of 1s and 0s using inverters. (Inverters appear on the input side of the figure as small circles just before the AND gates, which are themselves notated with half ovals).

Of course this diagram doesn't approach the cosmic scales of a properly wired Cosmo, but the basic idea scales up (very rapidly!) without a hitch. For every possible input/memory combination, associate a single AND gate, and the output of that gate activates the corresponding motor output from the pre-cooked lookup table.[3] This creates a Cosmo for whom appropriate Osmo-imitating response patterns are *pre-wired*. The zombie is complete.

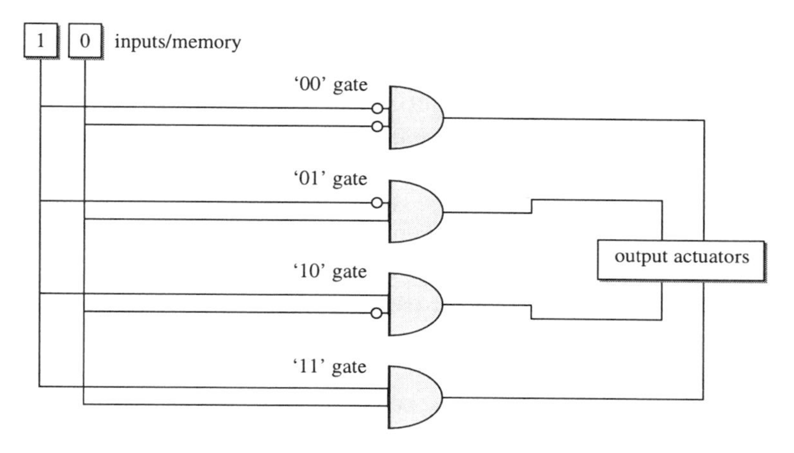

Figure 2. 'One Circuit' Cosmo I/O

Or, at least, I'm inclined to believe Cosmo is a zombie. Dropping him into any possible world where the Osmo from *P* would count as a conscious person, Cosmo follows a sequence of responses to his environment *identical* to a possible sequence of genuine Osmo responses (sampled at 44.1 KHz, anyway). Of course, the algorithm for ordering the lookup table might accidentally allow some sequences in which the real conscious Osmo goes insane or tries to convince people he has no conscious experience or even in which he tries deliberately to fail the Turing Test. But these are all perfectly legitimate sequences of behaviours for a conscious person like Osmo. (If this troubles, we could with puritanical thoroughness rebuild any sequences based on worlds in which Osmo does something untoward, checking every path through the lookup table and selecting alternatives from among Osmo's possible behaviours wherever required.) And although some might dispute the diagnosis, it's difficult to grasp how a creature with *no reflective states at all* could be conscious. While Cosmo applies his motor actuators to insist steadfastly that he has inner feelings—because that is what Osmo would do—he has no inner life at all. While the real Osmo might be thinking of philosophy while chopping garlic in the kitchen, for

Cosmo there is nothing else going on but the activation of a succession of *single* circuits directing garlic chopping and other motor behaviours such as standing upright. (Recall, of course, that a real physical Cosmo of any significant sensitivity and responsiveness would no doubt undergo gravitational collapse and become a black hole before making his mark on either philosophy or *haute cuisine.*)

3. ZOMBIES, FUNCTIONS, AND MATHEMATICAL MYSTICISM

It's time to take stock both of some features of the construction and of how it bears on standard arguments about zombies. First, notice that reducing Osmo's behaviour to a sequence of input numbers and output numbers for the purposes of wiring Cosmo does not require assuming anything about the *computability* of processes in Osmo's brain or even about the computability of any overall input/output relationships he might display. Osmo might actually be an entirely noncomputable Cartesian mind interacting with his body. He might be quantum wave-jumping in the far reaches of his endoplasmic reticulum or tuning a special transworld radio receiver to find the results of computations performed in parallel universes, eventually making nonalgorithmic and very profitable betting decisions at the local race track. But appealing to the possible worlds framework avoids all the details of potentially nonalgorithmic mish mash or even outright sorcery occurring inside Osmo. A finite limit on the precision of inputs and outputs *guarantees* a mathematical step function (for momentary purposes, a lookup table) linking the set of unique input/memory numbers with the set of output numbers. (This point returns in a moment.) But surely, exclaims a worried anti-computationalist, surely if we ask Osmo to calculate a noncomputable value in his head and then use it to answer some question, then he's performing a noncomputable transformation on his inputs...how could Cosmo mimic that? The answer is that Cosmo needn't mimic the noncomputable transformation; he need only take in the input and produce the output. Cosmo may not be able to *compute* a noncomputable transformation, but he doesn't need to: he has all the answers wired in advance. Taking motor outputs from the possible worlds algorithm gives in advance all those answers which Cosmo's wiring requires; 'extracting' those answers requires nothing more than the activation of a single AND gate.

Fans of possible worlds talk will no doubt have noticed long before now what the above paragraph suggests clearly: Cosmo needn't really *respond* to inputs in any way to establish zombies as a logical possibility.[4] Finite precision inputs as well as the pre-cooked lookup table wired by way of AND gates may be discarded. For use in particular possible worlds specified in advance, the construction only requires a Cosmo with appropriately hard wired *outputs* and no inputs at all. Call him *Cosmo Lite*. Each behavioural path picked out by the ordering algorithm from the possible worlds with initial segment *P* features a series of Osmo motor actions which, if wired in advance, would do perfectly well for appearing conscious in all and exactly those possible worlds where Cosmo Lite replaces conscious Osmo but the sensory inputs which the latter would experience remain unchanged. Of course, such a twentieth century appropriation of the old idea of 'pre-established harmony' is so easy as to be trivial, but it isn't very satisfying. Philosophers aren't interested so much in the logical possibility of particular zombies who just happen to make all the right moves to suit a given possible world; they want to know about how a zombie really does respond to stimuli and whether seemingly conscious responses really could occur without consciousness itself. (Thus the reason for taking the trouble to construct a zombie specifically in the input/output spirit.)

The standard reply to accounts inspired by the good old fashioned lookup table—indeed, Dennett's first response (personal communication) to the present zombie construction—concerns *combinatorial explosion*. It emerges in a moment, however, that the problem of combinatorial explosion appears only for the right (or wrong) use of a lookup table. How could we possibly, goes the standard objection, write a computer program, or construct any other device, even to look up the appropriate answers for all the possible questions a person might be asked verbally (let alone all the other possible inputs)? Suppose I'm asked how my dinner tasted last night. Then I'm asked how it felt to be asked how my dinner tasted last night, and would I have preferred stew instead of pasta? Or would I have preferred baked beans with lime and coriander? And how many questions have I been asked, including this one? The sticking point is that I could be asked a vast number of different initial questions, and each of those initial questions could lead to a vast number of other related questions (possibly referring back to earlier questions, thereby keeping sequences unique): the whole affair rapidly blows up into a gigantic tree of input possibilities so vast that any hope of constructing an actual table for *looking up* corresponding outputs evaporates. The usually unstated corollary is that the tree of possibilities actually becomes infinite. The only way to get around the vastness of the tree of possibilities, so the argument goes, is not *simply* to look up an output for each possible input, but instead to simulate the complex introspective psychology of a real conscious person. The old question then returns: is it possible for something to simulate all that complex psychology without actually being conscious?

Virtual reality designers face a similar problem: a subject can't be given free reign to explore a virtual space—analogous to an interlocutor's being given free reign to query our zombie—because of the explosion in information this would require the VR system to keep on hand. Designers must limit the possible avenues of exploration in advance. But, as it might be easier to see from the VR analogue of the standard objection to lookup tables, the combinatorial explosion is actually 'in' the *environment* (the questioner or VR explorer) and needn't be 'in' the candidate zombie.[5] The usual idea of a lookup table is just the wrong kind of table, one which allows combinatorial explosion into the table itself by including, say, complete verbal questions as inputs. Of course no finite computer program could simply implement *this* kind of lookup table, differentiating an infinite set of possible inputs and linking them with their corresponding outputs. But sampling inputs at discrete periods of time instead contains the explosion before it gets started. For any desired sensory resolution and memory structure, we can calculate in advance *exactly* how many AND gates are required for the zombie construction—entirely independently of what language the zombie will speak or how many different questions it might be asked—and that number increases linearly as a function of the span of time for which the zombie operates. The number may be cosmically huge, but it is strictly finite. No explosions here.

If it's so easy to construct a logically possible zombie (or even a zombie lite) which avoids the problem of combinatorial explosion, where could Dennett have gone wrong in crafting his colourful denunciations of the very idea? A clue appears in his application of 'zimboes', introduced first in Dennett (1991) as a special subset of zombies and described recently (Dennett 1995b, p. 322) as "by definition a zombie equipped for higher-order reflective informational states". The point of positing "higher-order reflective informational states" is a little elusive, especially given the tension between such states and the standard notion of a zombie as entirely nonconscious, but the following passage is truly baffling:

> ...zombies behaviourally indistinguishable from us are zimboes, capable of all the higher-order reflections we are capable of, because they are competent, *ex hypothesi*, to execute all the behaviours that, when we perform them, manifestly depend on our higher-order reflections. Only zimboes could pass a demanding Turing Test, for instance, since the judge can ask as many questions as you like about what it was like answering the previous question, what it is like thinking about how to answer this question, and so forth. (Dennett 1995b, p. 323)

Is the dependence of seemingly conscious behaviour on these higher-order reflective states really so manifest? Clearly the behaviour of conscious beings such as most (if not all) readers of this book *does* depend on such reflective states. As far as *that* goes, Dennett is absolutely right. But we've just seen how to construct a logically possible zombie, 'conscious' only in appearance, who doesn't reflect on anything at all. Simply assuming that thoroughly conscious-seeming human-style behaviour requires reflective states just because we do reveals Dennett's unwitting espousal of what might be described as a sort of mathematical mysticism. The idea behind this label will become clear shortly, and it receives an explicit definition below.

As noted on page 22, for any finite set of unique input/output pairs, a step function relating the two is guaranteed. Shown in Figure 3, for instance, is one possible step function relating two sets of numbers, chosen arbitrarily for the example. Mathematically, such a function looks like this:

$$f(x)=\begin{cases} 2.5 \text{ when } x \in [.2,.4] \\ .75 \text{ when } x \in (.4,.75] \\ 1.1 \text{ when } x \in (.75,1.2] \\ 1.4 \text{ when } x \in (1.2,1.6] \\ 1.75 \text{ when } x \in (1.6,1.8] \\ 2.2 \text{ when } x \in (1.8,2] \end{cases}$$

(The square brackets for the first range of x indicate a closed interval, while the combination round and square brackets indicate an open interval carrying on from the previous.)

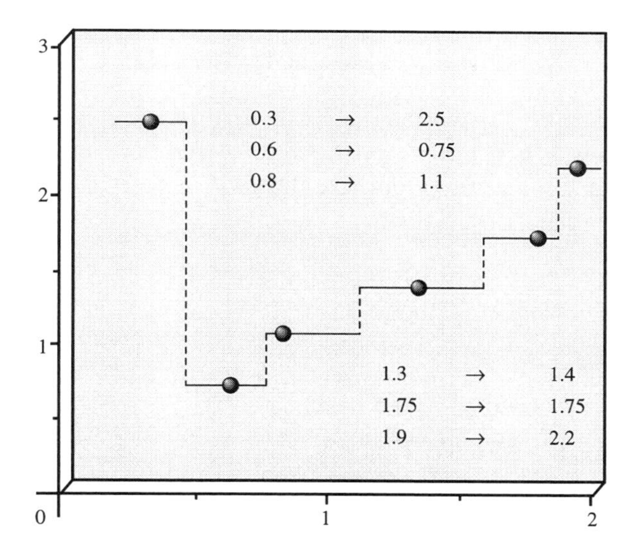

Figure 3. Simple I/O Step Function

It's easy to see how to get a lookup table directly from a step function. The step function may be no more economical than an actual lookup table, since unless adjacent inputs sometimes yield the same outputs, each different input/output pair demands its own entry. Notice that since the widths of the intervals for any such step function can vary, the set of step functions relating any two such finite sets is infinite.

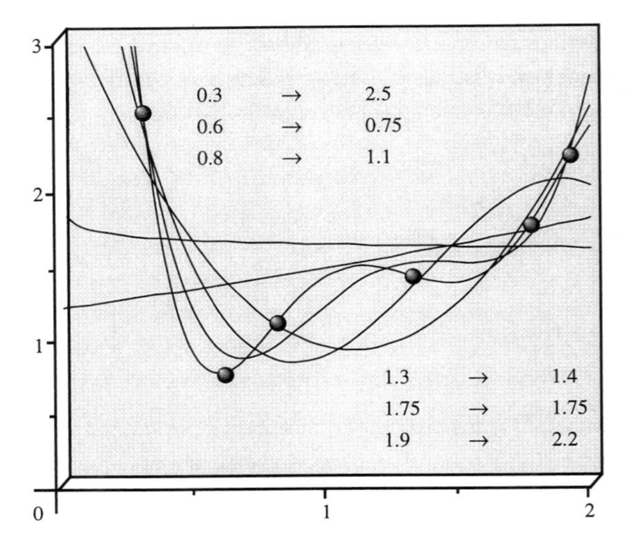

Figure 4. Continuous I/O Functions

Similarly, Figure 4 shows the graphs of six different continuous functions, each of which is a 'best match' for that type of function and the same set of numbers. The actual equations look like this (with each constant rounded to an economical two significant digits):

$$f(x) = 1.2 + 10^{.09x}$$

$$f(x) = 1.6 - .1\log(x)$$

$$f(x) = 3.3 - 4.5x + 2.1x^2$$

$$f(x) = 5.1 - 12x + 10x^2 - 2.4x^3$$

$$f(x) = 8.6 - 32x + 46x^2 - 27x^3 + 5.7x^4$$

$$f(x) = 14 - 69x + 130x^2 - 110x^3 + 47x^4 - 7.2x^5$$

For any matched sets of inputs and outputs, continuous functions like these which approximate the input/output relationship can always, in principle, be found (provided, of course, that any given unique input matches up with one unique output). While some types of functions (exponential, logarithmic, degree 2, etc.) may make for better fits than others, the set of those which fit approximately is obviously infinite.[6] If we're interested in what really goes on *inside* a conscious creature, however, (or inside any other physical system, for that matter), almost always we'll be after a decomposition of some overall behavioural response relationship into a composite set of functions describing the input/output relationships of different mechanisms within the system as a whole. Consider Osmo, for instance. Of course there is an overall step function which describes his conscious-seeming behavioural responses perfectly well, and that overall function can in principle be implemented with a cosmically vast set of parallel AND gates. But a set of equations describing the input/output relationships of a vast set of AND gates doesn't describe what really happens in Osmo—it describes what really happens in Cosmo. There is an infinite set of *other* sets of equations which also describe a behavioural repertoire matching Osmo's, and some of these describe internal mechanisms bearing some resemblance to what really appears inside Osmo. Some of these sets will even match the overall behaviour *and* meet the requirements of the laws of physics.[7] It is these types of functions which we're after for understanding what is really happening: a composite set of functions which describes overall behaviour by means of a set of physically realisable subsystems. For the most part, incidentally, that means attention focuses on *continuous* functions. One interesting feature of continuous functions is that, depending on the complexity of the input/output relationship, often they can be specified considerably more succinctly than a lookup table or step function. Nature is rarely so inefficient as to implement directly something like a lookup table!

Perhaps it's awareness of Nature's adeptness at picking efficient instantiations of overall behavioural response patterns which seduces Dennett into flirtations with mathematical mysticism. The most physically plausible descriptions, whatever they may be, of the overall behavioural relationships of seemingly conscious creatures we observe every day really *are* entire sets of interrelated functions describing complex brains and internal reflective states and so on. The most physically plausible implementations of these overall relationships do not involve lookup tables or step functions. As noted above for Osmo's case, some step function accurately describes his overall behavioural responses and can in principle be instantiated without consciousness—but such a function, implemented theoretically by AND gates in Cosmo, is not the *right* one for understanding the real Osmo from the standpoint of physics. On such a point we need have no quarrel with Dennett. The mathematical mysticism, however, comes in believing, apparently without argument, that of the infinitely many different composite sets of functions which could yield an apparently conscious overall input/output relationship, *all* describe systems which really are conscious (alternatively: "which, if implemented, really would be conscious"). Mathematical mysticism supposes that consciousness is *required* for performing such sophisticated input/output feats as we regularly witness in conscious creatures each day. But what about the infinitely many other systems whose input/output feats are mathematically equivalent yet which have no reflective states whatsoever? In fairness, section 3 of Chapter 3 offers a suggestive, but only suggestive, information theoretic basis for the intuitions of the mathematical mystic.

Whereas the impoverished imaginations of some philosophers mislead them, as Dennett rightly points out, concerning the full range of possible zombie behaviours, Dennett's own imagination apparently misleads him into underestimating the full range of mathematical functions. The passage quoted above, in which Dennett refers to "behaviours that...manifestly depend on our higher-order reflections", exposes the soft underbelly of such an impoverished view exceptionally well: infinitely many functions describe those behaviours and different underlying mechanisms which could be responsible for them, so why should we believe that all of them describe systems with higher-order reflections? *Manifestly*, they do not. Moreover, Dennett is presumably here addressing what is logically possible, not even the lesser physical possibility—although it *still* would smack of mathematical mysticism to suppose without argument that the only functions which could describe seemingly conscious behaviour *and* cohere with the laws of physics are those which correspond to truly conscious creatures.[8] The truth of that view, which we might call 'weak mathematical mysticism' to distinguish it from the stronger version grounded in logical possibility, is an open question.

This and related questions encapsulate all about zombies that is most significant for cognitive science and philosophy of mind. The real question shouldn't be whether zombies are logically possible. It's easy to see they are. The real question should be: which of the infinitely many mathematical descriptions of seemingly conscious input/output relationships are also descriptions of truly conscious creatures? What *sorts* of descriptions are the right ones—and not just mathematically satisfactory input/output look-alikes—for creatures which really are conscious? The corollary, scarcely less important: which of those functions which describe *physically possible* renditions of the right input/output relationships also describe truly conscious creatures?

In fact, I believe the case is strong for the view that pressures of natural selection favour individuals capable of instantiating increasingly complex input/output relationships with their environments. (Something like this view has a long tradition, at least from Spencer 1855 to Godfrey-Smith 1996.) It might be very safe to say that the only kinds of architectures which can actually instantiate complex, seemingly conscious input/output relationships and which can actually have been created by the course of evolutionary history on Earth really are conscious. But on its own, such a view is as speculative as the mathematical mysticism decried above, and I would expect no one to believe it without more argument and more empirical data.

The moral of the story as far as it concerns the project in this book is that it just isn't enough to note that something instantiates a correct input/output relationship, as Cosmo does, or that it is capable of passing the Turing Test. It isn't enough to say that a system 'processes'—or appears to process—some information. Philosophical paydirt comes only after digging far enough into the details of a system to see *how* it instantiates an input/output relationship and *how* it processes information. The job of doing this in a mathematically precise way occupies Chapter 5.

The next chapter nails down a rigorous definition of information and grounds the idea of representation in that definition. This catalyses subsequent observations about the rôle of information in philosophy of mind and makes possible a later formalisation quantifying the complexity of an input/output relationship. Readers who prefer a little less formality might like to read the next two chapters in reverse order, foregoing the details of a mathematical approach to information until seeing some of the use to which it gets put in Chapter 4 (and later in Chapters 6 and 8).

NOTES

[1] As usual when this book glosses a technical detail, the particulars aren't crucial to the main point. However, an initial world segment can be understood here as a way the world might have been up to a particular time, described by a maximal consistent set of non-indexical propositions covering the history of the world up to a particular instant of time relative to Osmo's own inertial reference frame, with simultaneity understood relative to that frame. 'Maximal' indicates a set of propositions to which no other proposition may be added without violating its rules for construction. Trading rigour for simplicity, I use the same letter to denote both a set of propositions and the world or segment to which it refers.

[2] Because part of the construction's sensory inputs are closely coupled to motor activations, arbitrary choices about those motor outputs may pare down the tree of possibilities.

[3] Obviously, this sketch omits some fine details. For instance, we'll want a gate to respond to a particular input/memory pattern only when the total length of the input/memory number is the same as the length of that pattern. (For instance, '11' is distinct from '110000', so gates must be wired appropriately.) But such details only add extra control lines little different from those shown here.

[4] Probably such fans also will have noticed other *minutiae* which complicate the picture without compromising its general flavour: what about those possible worlds where Osmo meets up with Cosmo, for instance? We could never get them even to shake hands properly on the simplified account offered here, but the general point persists.

[5] Dennett, among others, no doubt knows this anyway, given his arguments in the first chapter of Dennett (1991) about freely explorable strong hallucinations and the ridiculously huge informational overhead they would require on account of combinatorial explosion.

[6] Note that *any* computable transformation at all can be encoded as an arithmetical proposition involving only positive integers and the operations of addition, multiplication, and exponentiation. See Davis, *et al.* (1976) for polynomials and Davis, *et al.* (1961) and Jones and Matijasevic (1984) for exponential diophantine equations; also see Chaitin (1987b). The ubiquity

of functions mapping between sets of numbers is responsible for the general problem of underdetermination of scientific theories by data.

[7] The example is slightly muddy here for at least two reasons. First, simple input/output mappings are inappropriate for the lowest level laws of physics, which are probabilistic. Second, the real Osmo is not a system with fixed architecture producing outputs in response to an environment and a perfect internal memory; thus, Osmo's overall behavioural responses are not best described with the same contrived input/output relationship attributed to Cosmo.

[8] The argument against strong mathematical mysticism is independent of whether given physical instantiations of the right input/output relationships are conscious only in some possible worlds (i.e., whether otherwise physically identical systems might be conscious in some possible worlds and not in others). Reasons emerge in Chapter 6 for rejecting such a view, but for now there is no difficulty in accommodating it.

Chapter Three

Information, Complexity, and Representation

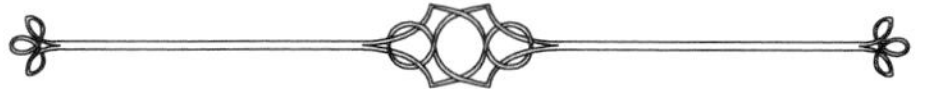

Near where I stayed in Edinburgh while finishing the first draft of this book, a small loch lies nestled in trees at the base of a hill, atop which rests the Royal Observatory. Some evenings I sat by the loch and watched a pair of swans; gentle, quiet, and oblivious to human philosophy, the two helped thoughts to settle and crystallise in ways busy libraries and eager email exchanges did not.

One set of questions catalysed by late night visits to the loch concerned the traditionally problematic idea of *representation*: what made my memory a representation (if it was a representation at all) of this swan over here rather than of that very similar swan over there? Why did I have a mental 'picture' of the moon rather than of the water in which I saw it reflected? And why did a reflection from the surface of the loch suggest a representation of the moon, while the very same batch of electromagnetic radiation reflected from the surface of a swan suggested a representation of a swan? When a passing swan exploded the reflection of a bright white crescent into hundreds of tiny ripples tipped with moonlight, did *it* change what I was representing in that region of space—from a moon to a bunch of wavelets—or did I?

The aim of this chapter isn't to give a philosophically rigorous solution to problems like these, as satisfying as that might be. (Similar questions return in Chapter 8.) Instead, the chapter outlines a formal definition of 'information' which at once suggests some new perspectives on such problems and provides a way of describing any physical system in terms of the information it instantiates. Showing how to quantify the tandem notions of information and complexity, the chapter applies the formal framework to transform the notion of representation into something precise and objective. Where the standard notion leaves considerable room for interpreting what *exactly* it means to say one object represents another, the relacement pins down this relationship clearly and explicitly. Along the way, the chapter provides a brief de-mystifying look at information theoretic versions of Gödel's incompleteness results. These first steps prepare the ground for a progressive effort throughout the remainder of the book to incorporate powerful tools from the field of algorithmic information theory into the standard philosophical repertoire for attacking the mind/body problem.

1. INFORMATION IS PHYSICAL

Stripped of its formal trappings, the gist of the present approach is this: information is and can only be *physical*. Information, on this account, is instantiated by correlations between observable properties of physical systems. When observables[1] of two systems are correlated, they each bear information reciprocally about the other. The physical sciences rely on a physical view of information, largely because it seems the only way of avoiding problems with, among other things, relativity theory, Maxwell's Daemon and the Second Law of Thermodynamics, and bizarre readings of quantum mechanical 'delayed choice' experiments. (Landauer 1991, from which the title of this section is borrowed, makes a very readable introduction to the topic, with plenty of references.)

The account of information is itself more important for present purposes than the rationale from physics, so here is a cursory look at just one example, Scottish physicist James Clerk maxwell's simple 1871 thought experiment which bedevilled scientists for roughly a century (Bennett 1987b, Leff and Rex 1990). 'Maxwell's Daemon' is a hypothetical creature who, by observing individual particles in a container of gas, seems able to separate it into a warmer component and a colder component, violating the Second Law of Thermodynamics. The Second Law stipulates that the overall entropy, or disorder, in the universe may increase or stay the same but may never decrease. Alternatively, it describes the tendency of the universe to fall into increasingly probable configurations: the number of disordered states is vastly larger than the number of ordered states, giving the former a probabilistic advantage over the history of the cosmos. (On the relationship between thermodynamics and Shannon's classical information theory, see Brillouin 1956.) Maxwell's Daemon apparently reverses this trend locally without any global cost to rescue the Second Law.

Consider a box, as in Figure 5, divided in two with a sturdy and non-porous wall and filled with some gas at thermal equilibrium. A small shutter in the wall can open, allowing gas particles to pass through. Since the particles of gas are in thermal motion, some have higher energy than average (indicated in the figure by longer tails), while some have lower energy (indicated by shorter tails). Indeed, gas particles' average speed is proportional to the square root of the temperature measured (in Kelvins, for instance) from absolute zero, or the ground state of matter, and higher temperatures mean greater variation in particle velocity. Maxwell's Daemon watches the zooming particles, and when a higher energy particle coming from one direction (say, the right) is about to hit the shutter, he opens it, allowing it to pass through into the other half of the box. Likewise, the daemon allows lower energy particles to pass through in the opposite direction. At all other times, the shutter remains closed.

After the daemon works for some time, so the story goes, gas particles in the box show a radically different distribution pattern. Now, as in Figure 6, higher energy particles far outnumber those of lower energy in the left compartment, while the reverse is true in the right compartment; the *average* energy, however, remains unchanged. Since the shutter can be very small, requiring arbitrarily minute amounts of energy to operate, whereas the compartments of gas can be huge, it seems the daemon can separate a box of gas at thermal equilibrium into a hot compartment and a cold compartment for arbitrarily little energy expenditure. By simply opening a tiny shutter at the right times, the daemon could heat homes in Siberia by segregating randomly occurring energetic particles in the air. Or, he might maintain artificial ski slopes in the tropics or run a steam engine without fuel.

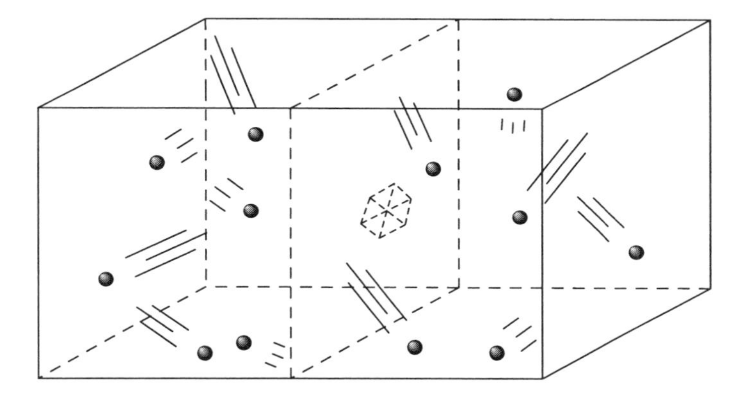

Figure 5. Before Maxwell's Daemon

The daemon's ability to separate the disordered gas at thermal equilibrium into hot and cold apparently violates the Second Law. We can make this a little more precise. It is convenient to measure the entropy of a system as the logarithm of the number of distinct accessible states—frequently, simply 'access states'—in which it could be found. (As Bekenstein 1981a, p. 288 puts it, "entropy is a measure of the phase space available to the system in question".) The number of distinct states varies according to the number of bodies in the system and their degrees of freedom, but for convenience we often consider binary systems. In that case, a system of n binary subsystems has a total of 2^n different possible states; it's entropy is then $\log_2(2^n)$, or simply n. (The logarithmic scale helpfully makes entropy additive.) Thinking of entropy in terms of distinct possible states, it's easy to see how the daemon turns a higher entropy system into a lower one: there are many more possible ways of having a box full of gas where particles of various energies are distributed uniformly than of having a box full of gas where all the higher energy particles are on the left and all the lower are on the right. Alternatively, the daemon's activities can be expressed in terms of heat flow, which transfers an amount of entropy proportional to the quantity of energy flowing divided by the temperature at which that flow takes place. Accordingly, heat flowing from a hot system to a cold one raises the entropy of the cold system more than it lowers that of the hot system—because we divide by a larger number for the hot system than for the cold. Any such flow raises the entropy of the universe. This accounts for the alternative rendition of the Second Law as the maxim that heat differences between different regions of space must decrease overall. But because Maxwell's Daemon arbitrates heat flow in reverse, from a compartment becoming colder to a compartment becoming hotter, entropy apparently decreases.

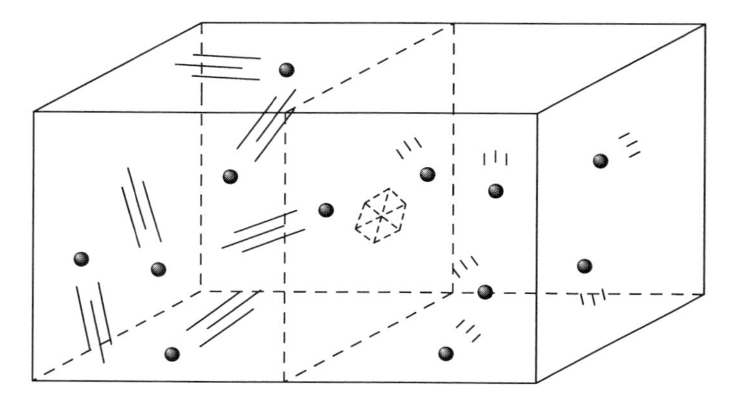

Figure 6. After Maxwell's Daemon

Of course, we all experience maintained differences in temperature like this whenever we open the refrigerator. But that temperature difference is bought with heat dissipated by a grossly inefficient compressor system, and the operation of the compressor system is in turn paid for with new energy entering the system from the electrical grid. Thus, a local temperature imbalance is more than offset by heat dissipated by the compressor, by the plant which generated the electricity and the power lines which deliver it, and so on. Likewise, we humans do a very decent job of maintaining a steady internal temperature in the face of widely variable external temperatures. But here again, that temperature differential is paid for by the body's continual extraction of energy from food. Indeed, were it not for the persistent flow of new energy into Earth's ecosystem—in the form of sunlight—most life on our planet would not last long.

At first glance it is tempting to suspect, following the hunch of physicist Leo Szilard (1929) in the early days of quantum mechanics, that saving the Second

Law from Maxwell's Daemon somehow rests on the daemon's actual gathering of information about the gas particles. Surely, we could be forgiven for thinking, that must cost some energy and dissipate some heat which could help offset the local reversal of the trend towards temperature homogeneity. But work in the theory of reversible computation (Bennett 1973) shows this is wrong: by performing it slowly enough, the energy dissipation of the information transfer itself, the process of observing individual gas particles, can be made arbitrarily small. Contrary to the popular belief for several decades after Szilard published his thoughts in 1929, the quantum theory of radiation doesn't require any minimal energy cost for observing the gas particles. (And the uncertainty principle, which limits the information which can be gained from a measurement, likewise has nothing to say about the energy cost of the measurement itself.)

Curiously, the Second Law of Thermodynamics is rescued not while *gathering* information but while *erasing* it. At some point, the daemon must reset whatever memory storage was used for making an observation in preparation for the next one. Given that memory must be *physically instantiated*, the energy costly step for Maxwell's Daemon is actually the physical erasure of that memory. If there were some way for the daemon to possess information about the gas particles without having to store it physically, energy-free memory updates would allow the Second Law to be violated. One might object that the Daemon could simply keep records of all the information ever gathered, never erasing it. But then, of course, the daemon pays for the decrease in entropy in the box with a corresponding increase in the entropy of his own expanding memory contents.

In general, *any* process which appears locally to buck the universal thermodynamical trend requires either something like this energy costly memory erasure or a continually increasing memory record, offsetting other entropy reductions and allowing the whole process to be reversed. A beautiful example occurs in our own cells while growing a complementary RNA chain from a DNA sequence, the transcription process catalysed by RNA polymerase (Bennett and Landauer 1985; also see Eigen 1992). From a bath of nucleoside triphosphates—i.e., a base of nucleic acid condensed with a molecule of the sugar ribose and a triphosphate group—RNA polymerase selects a complementary base and fixes it to the polymer strand under construction, releasing a free pyrophosphate ion before stepping forward to the next DNA location. By itself, this process is reversible. Near equilibrium, it sometimes goes backward: the RNA polymerase captures a roaming pyrophosphate ion, joins it to a removed base, and backs up one position on the DNA strand. The transcription process near equilibrium slows to a crawl, dissipating very little energy; any increase in strand order can be reversed again. But fortunately for us, other metabolic processes step in to remove pyrophosphate ions and maintain the supply of nucleoside triphosphates, effectively erasing the memory of the transcription process and dramatically increasing the dissipation of energy. Were it not for that memory erasure, fuelled primarily by the Krebs cycle—and thus by the food we eat, and, ultimately, by the sun—the subsequent processes catalysed by reverse transcriptase and so on could not take place: life as we know it could not exist.[2]

Once again, all this rests on a view of information as physically instantiated. Physics just doesn't work properly if disembodied (disinstantiated?) information is allowed to exist. (More precisely, physics doesn't work properly if non-physically instantiated information is allowed to have any effect on the physical world.) That isn't to say any forms of dualism in philosophy of mind ought to be rejected on the basis of these considerations *alone*. At first glance, we could have a heyday with the following set of inconsistent propositions:[3]

1) *There exist minds which contain information.*
2) *There do not exist minds which are physically instantiated.*
3) *All information is physically instantiated.*

But of course the proponent of dualism needn't accept *1)*; she may simply maintain that minds exist and contemplate things but that they don't actually traffic in information as presently understood. Whatever it is they *do* deal with just doesn't count as information as understood here: perhaps 'Cartesian ghosts' or 'Platonic forms' would be more appropriate, but it isn't information. Readers really concerned with this may want to skip ahead briefly to the section 'Platonic Hell' beginning on page 51.

2. MEASURING INFORMATION

Viewing information as physically instantiated leads naturally to a way of measuring it based on descriptions of physical systems, where 'descriptions' means strings of one variety or another correlated in a particular fashion with the physical system whose information content we are measuring. (Strictly speaking, of course, it is the physical instantiation of a string which is correlated with other physical systems.) Rather than English or Swahili or LISP, it is convenient to choose as a language of description programs, in the form of binary strings, for a Universal Turing Machine U.

The Universal Turing Machine, described independently but equivalently by Alan Turing and Emil Post (see Davis 1965[4] or van Heijenoort 1967 for original papers from each), is a general digital computer which can simulate any other digital computer. That is, for any other computer M, a program p for M can be prefixed with a fixed length segment of code σ for U such that σp causes U to perform the same computation as p for M. The fixed length segment σ simply tells U how to simulate the behaviour of M. This is akin to the piece of software which enables a Macintosh personal computer to emulate computers based on an entirely different Intel chip set and thus to run their software unmodified. It is widely—and incorrectly—held that a Turing Machine can simulate any physical process at all. Chapter 9 explores how Turing Machine power may not be *quite* universal, but for present purposes no harm comes from assuming that for a particular finite level of accuracy, some Turing program exists to generate a description of any physical thing. Although the result is irrational (i.e., it can't be expressed as a fraction), some finite string of binary digits causes a Turing Machine to generate successive digits of the ratio of the length of the diagonal of a square to the length of its sides (i.e., $\sqrt{2}$). Another Turing program generates a very detailed, but less than perfectly accurate, description of a Boeing 777 taking off, and there is a Turing program—probably a *very* long one—for playing perfect games of chess (i.e., for winning whenever logically possible, given an opponent's moves, and never losing when a draw is possible).

2.1 Algorithmic Information Content

We define the *algorithmic information content* $I(x)$ of an object (or bit string) x as the length of the minimal self delimiting program(s) for a Universal Turing Machine U which will generate a description of that object for a given level of precision.[5] (Always requiring self-delimiting programs, also known as 'instantaneous code', greatly simplifies the formalism later. Such a program isn't allowed to rely on a blank marker on the Turing Machine tape to signify the end of the program and thus must include within itself an indication of its own length.) An infinite class of different algorithms, or Turing programs, can generate any finite bit string. But the restriction 'minimal', denoting a program which is not itself the output of some still shorter program, narrows interest just to the most concise program. Notice that the choice of a particular U never introduces more than $O(1)$ difference in the numerical value of $I(x)$, since the prefix σ which makes U simulate another computer is of constant length.[6] (The expression $O(f)$, or 'order of f' indicates a function bounded in absolute value by a constant multiple of f.)

The framework of algorithmic information theory was developed independently in the 1960s by R.J. Solomonoff (1964; first discussed in print by Minsky 1962, pp. 41-43) for measuring the simplicity and unifying power of scientific theories; by A.N. Kolmogorov (1965), then of the Soviet Academy of Sciences; and by Gregory J. Chaitin (1966, 1977)[7] while he was an 18-year old undergraduate at the City University of New York. The idea's heritage often earns it the name 'KCS' (for 'Kolmogorov-Chaitin-Solomonoff'), but for this book it's usually 'algorithmic information content', 'algorithmic complexity', or 'algorithmic randomness' (often with the 'algorithmic' prefix omitted). The exposition here primarily follows Chaitin's conventions. Because of the remarkable ease with which seemingly plausible yet entirely erroneous arguments may be concocted with superficial readings of algorithmic information theory—and because many (hopefully not erroneous!) arguments in the rest of this book require it—the remainder of this section explores the approach in some detail.

Consider two strings:

1111111111111111111111111111111111111111

0111011010010100101000110110111010010100

Intuitively, the first string, forty '1' digits, contains much less information than the second, an apparently patternless batch of '1' and '0'. Alternatively, the first seems much less random than the second. The algorithmic information content measure preserves this intuition, since its internal pattern means the first can be specified succinctly with a program instructing a computer just to "print '1' forty times". The complexity of the string is thus bounded from above by the length of the binary sequence required to specify 'forty', together with however much code it takes to tell the computer to print a digit a certain number of times. Equivalently, almost all its information content comes from its length; the actual sequence making up that length, being a simple '1', is comparatively unimportant. In general, since it takes about $\log_2(n)$ binary digits to specify n, the complexity of such a regular string x_n, of length n, satisfies:

$$\begin{aligned} I(x_n) &= I(n) + O(1) \\ &= O(\log_2(n)) \end{aligned}$$

Since $\log_2(n)$ grows much more slowly than n, it's easy to see that programs for generating such regular strings grow much more slowly than the strings themselves. There's comparatively little difference between a program which instructs "print '1' forty times" and another which instructs "print '1' four hundred times". In other words, strings of this first type are highly *compressible*. The underlying pattern—straightforward repetition—makes compressing the string into a shorter space very easy. Of course, many other strings are also highly compressible by virtue of slightly more complex patterns. While the binary expansion of π, for instance, bears little resemblance to the simple string above, its algorithmic information content is low because we can specify a program for generating its digits concisely. The same is true for the many other *cryptoregular* strings such as the binary expansion of $\sqrt{2}$.

But it would seem that the shortest program for calculating the second string above, the apparently patternless set of characters, is one which just prints it out directly:

"Print '0111011010010100101000110110111010010100'."

The absence of easily specified organisation makes the string virtually incompressible. Indeed, *most* strings fall into this category, and in general their information content, again where x_n denotes a string of length n, satisfies:

$$\begin{aligned} I(x_n) &= n + I(n) + O(1) \\ &= n + O(\log_2(n)) \end{aligned}$$

That is, the information content of such an incompressible string comes from its length as well as the actual sequence making up that length. (Recall the requirement that minimal programs be self delimiting, containing within themselves information about their own length.) We say a string is *random*, or *algorithmically complex*, when its information content satisfies the above—i.e., when the minimal program for generating it isn't substantially shorter than the string itself, when its information content approximately equals its length. (This last holds again because $\log_2(n)$ grows much more slowly than n, so the length of x_n comes to dominate $I(x_n)$.) The arbitrariness injected by the word 'approximately' is to be expected. Randomness is normally a vague concept. Algorithmic information theory simply offers a quantitative choice as to when a string should be dubbed random. It's clear, incidentally, that any minimal program is itself random. Consider a minimal program p to generate x. If it weren't the case that p is random, then by definition there would be some substantially shorter program r which generates p. The string x could therefore be generated by a program which says "calculate p from r, and then use p to calculate x". Such a program is only slightly larger than r and is thus shorter than p, so p cannot have been a minimal program.

This notion of randomness applies similarly to infinite strings. An infinite string is random when its finite initial segments x_n of length n are random. More carefully, an infinite string x is random just in case the lengths of minimal programs to generate initial segments of x are bounded from below by a fixed constant subtracted from the lengths of those segments:

$$\exists c \forall n (I(x_n) > n - c)$$

It's easy to see that by the lights of algorithmic information theory, *most* strings have near maximal information content (or, equivalently, most strings are random). Basic cardinality considerations show why. For instance, there are only $2^{n-20} - 2$ programs with length in the range $[1 \ldots n - 20]$, each of which generates at most one string of length n. Thus the proportion of strings of length n with complexity less than $n - 20$ is less than $2^{n-20}/2^n$, or roughly one in a million. That is, fewer than one in a million strings can be compressed by twenty bits or more. In general, the proportion of strings of length n with algorithmic information content less than $n - c$ decays exponentially as c increases. With algorithmically complex strings outnumbering their simpler cousins by such a huge margin, at first glance it seems it should be easy to offer a provably random string of any desired length. However, it is impossible. This fact returns in section 4, but first we explore some of Chaitin's extensions to this basic measurement of information.

2.2 Joint, Relative, and Mutual Information

Given the definition of the information content $I(x)$ of a string x as the length of the minimal program(s) for a Universal Turing Machine U to generate x, the measurement extends easily to relationships between two objects.

First, we define the *joint information* $I(x, y)$ of two objects x and y as the length of the minimal program for computing x and y together. Joint information is subadditive, bounded from above by the sum (plus a constant) of the individual information contents. Similarly, the *conditional*, or *relative* information content $I(x \mid y)$ is the size of the minimal program to calculate x from y. Because a program can always, at the least, simply calculate x while ignoring y, the relative information content of an object x with respect to anything else is bounded from above by its own individual information content $I(x)$. Conditional information content, incidentally, offers another angle on randomness: maximally complex strings have $I(x_n \mid n)$ close to n, while the most ordered strings have $I(x_n \mid n)$ approaching zero. The two measures are related by the fact that joint information content is equivalent to the sum of the lengths of the programs to compute first one string and then the other from it:

$$I(x, y) \quad = \quad I(x) + I(y \mid x) + O(1) \quad = \quad I(y) + I(x \mid y) + O(1)$$

In fact, making this work as above requires a slight cheat. The tidy equality emerges only under the stipulation that rather than $I(x \mid y)$ returning the size of the minimal program to calculate x from y, it returns the size of the minimal program to calculate x from a minimal self-delimiting program for y. Alternatively, as Chaitin notes (1975, p. 339), we can use $I(x \mid y, I(y))$ instead of $I(x \mid y)$. The snag is that without some change, the error bounded above with $O(1)$ becomes unbounded. It's easy to see why. Suppose, as above, that

$$I(x, y) \quad = \quad I(x) + I(y \mid x) + O(1)$$

and, instead of y, ask about the joint information content of a string and that string's own information content:

$$I(x, I(x)) \quad = \quad I(x) + I(I(x) \mid x) + O(1)$$

Notice, however, that

$$I(x, I(x)) \quad = \quad I(x) + O(1)$$

This is obvious because a minimal program for producing x already includes within it a specification of its own length; thus the length of the minimal program which outputs both x and its information content is only slightly longer than a program for producing x alone. Moreover, it is clear that

$$I(I(x) \mid x) \neq O(1)$$

In fact, without the special stipulation, $I(I(x) \mid x)$ is unbounded (Chaitin 1975). That the size of the minimal program to calculate $I(x)$ from x is unbounded is an interesting fact which foreshadows the discussion of incompleteness in section 4, but for now it means the original assumption must be rejected or a way must be found to fix the unboundedness. As above, however, the latter is easy to do, and we use $I(x \mid y)$ always with the understanding that we're provided a minimal program for y.

Returning to the task of extending information measures to relationships between objects, the degree to which $I(x, y) < I(x) + I(y)$ suggests a measure of *mutual information* $I(x : y)$, defined as the extent to which knowledge of one object helps in the calculation of the other. This is equivalent to the extent to which they can be calculated together more succinctly than separately. Since the relationship is symmetrical,

$$\begin{aligned} I(x : y) &= I(x) - I(x \mid y) + O(1) = I(y) - I(y \mid x) + O(1) \\ &= I(x) + I(y) - I(x, y) + O(1) \\ &= I(x) + I(y) - I(y, x) + O(1) \end{aligned}$$

Like many of Chaitin's modern definitions for algorithmic information theory, a slightly different ensemble version of mutual information content appears in Shannon and Weaver's (1949; also see Goldman 1953) original work on efficient information coding and communication through noisy channels.[8] While it doesn't feature in the present discussion, Gács and Körner (1973) also suggest an interesting related measure of *common information*, the largest information content of a string which can be calculated from each of two strings by programs bounded in length by a given value.

Finally, two strings are *algorithmically independent* when their mutual information is as small as possible.[9] It would be handy if these methods of quantifying information could help nail down the slippery notion of representation.

3. GROUNDING REPRESENTATION

In 'grounding' representation, the present aim is to ground the *general* idea of a representation, not particular examples of representations. The latter is the symbol grounding problem—the main source of the questions at the start of this chapter—which reappears in sections 2 and 4.4 of Chapter 8. But making sense of how a particular representational symbol is grounded for an individual cognitive system demands first an account of representation itself. The distinction is roughly just the difference between *being* a representation and being *used as* a representation. Here I suggest a precise and objective definition of the first in terms of algorithmic information theory.

Intuitively, it would be convenient to say one object represents another when it contains information about it that isn't too difficult to extract. In the language of Chaitin's information theory, we might measure the extent to which object y is a representation of object x by asking how much cheaper it is to compute x from y than just to compute x; i.e., how large is the following difference:

$$I(x) - I(x \mid y)$$

But, recalling the identities of section 2, this is just another rendition of the mutual information content $I(x : y)$, revealing that the question about the extent to which object y is a representation of object x is precisely symmetrical with the question of the extent to which x is a representation of y:

$$I(x : y) \quad = \quad I(x) - I(x \mid y) + O(1) \quad = \quad I(y) - I(y \mid x) + O(1)$$

Thus, I suggest the following as a suitably general definition of representation. Two objects *represent* each other to the extent that they are not algorithmically independent—equivalently, to the extent that they have substantial mutual information content. (Because mutual information content may obtain between objects and *phase trajectories* as well, this definition also applies without difficulty to the relationship between one object and another's phase trajectory, or between two phase trajectories.) The vagueness inherent in 'substantial' is analogous to the vagueness in the definition of randomness, and it is to be expected. After all, many objects may represent a given one to a degree, but some will simply be *better* representations. Mutual information content precisely quantifies that notion of 'better' in terms of the extra number of bits needed to make up for a bad representation or the number of bits saved by a good representation. (Readers concerned that a naïve correspondence theory has just been brought on board may allay their fears by skipping briefly to section 2.3 of Chapter 8, beginning on page 185; the definition given below of functional representation may also calm some worries.)

Incidentally, I am the first to admit that remaining within the Turing Machine framework slightly constrains the progress which can be made with this definition toward understanding the kinds of questions with which the chapter began. Although it can be useful to step out of the Turing Machine framework and consider the effectiveness of representations directly with respect to other systems such as brains, however, the difficulties actually are not so enormous under the particular approach adopted later in Chapter 8. Indeed, disadvantages of the Turing framework are offset by a powerful advantage: uncontroversial logical scrutability, limited only by incompleteness results. Brains and similar cognitive substrates, for now anyway, lack that advantage. Thus we keep the general definition of representation couched in terms of the rigorous and only step into less well understood territory when the need becomes pressing.

Where before it was easy to see that algorithmically random strings dominate the space of all strings, pairs which represent each other are comparatively rare. Almost all strings are also random relative to each other: for most pairs, it is no cheaper to compute them together than to compute them separately. While information itself is abundant, good representations are hard to find. Of course, section 3 of Chapter 2 made much of the fact that infinitely many functions relate any two matched sets of numbers. The catch, however, is that the vast majority of these functions require more information to perform their feats than is contained in the numbers themselves. This uncovers some rationale behind Dennett's mathematical mysticism: where representations *seem* to be involved—recall those "manifestly" required "higher-order reflective states"—their most economical instantiation *is* a genuine representation on the current definition. Whether real conscious entities are instantiated economically, and what that has to do with consciousness, however, are separate questions altogether.

Finally, we extend the definition with a notion of *functional* representation which, like ordinary representation, plays a leading rôle in the remainder of this book. Intuitively, this should capture the idea of a representation which includes information about both an object and some of its possible transformations.[10] For instance, a representation of a cube which included information about how that cube should be translated or rotated in space under the action of a vector field would be a functional one.

Consider a range $R = \{x^t\}$ consisting of possible transformations of an object x. These $\{x^t\}$ result from corresponding initial conditions—i.e., specifications for particular transformations—in some domain $D = \{i^t\}$. (This broad construction of initial conditions allows other information into the condition apart from just states of x.) Thus there is a family of minimal programs $\{a_m\}$ which map initial conditions $\{i^t\}$ in the domain to transformations $\{x^t\}$ of x in the range, $a: D \rightarrow R$.

At first it might seem the simplest formalisation of a functional representation ignores the domain altogether and just measures the extent to which it is cheaper to calculate corresponding members $\{x^t\}$ of R using some representation y than it is to calculate the transformations from x itself. This captures the idea of including information about the look of x under transformations; perhaps we might look for a representation y where (with positive c):

$$\exists c \forall i(I(x^i \mid y) \leq I(x^i \mid \mathrm{x}) - c)$$

In other words, for all the transformed $\{x^i\}$ in range R, it would always be cheaper (to an extent quantified by c, a measure of the representation's 'efficiency') to calculate the transformation from the representation than from x itself. But one easy way for y to be a functional representation is of course to include, in addition to x, a representation of an axiom system which governs the behaviour of x.[11] Then the minimal program to generate a transformation of x from y simply picks the relevant transformational rules from within y and applies them to x. This reveals the flaw in this first go at a definition: when x itself contains information about its behaviour under transformations (i.e., some information about a member of $\{a_m\}$), it might be the case that over the set of all possible y, *none* satisfies the above. That is,

$$\sim\exists y \exists c \forall i(I(x^i \mid y) \leq I(x^i \mid \mathrm{x}) - c) \quad = \quad \forall y, c \exists i(I(x^i \mid \mathrm{x}) < I(x^i \mid y) + c)$$

The solution to the difficulty is a simpler definition of functional representation: an object y ***functionally represents*** an object x with respect to a domain of conditions D when y represents x and y represents some member of $\{a_m\}$. That is, $I(x : y)$ is substantial and, for some m, $I(a_m : y)$ is substantial. A relationship of functional representation may obtain between objects, phase trajectories, or mixes of the two. As before, the vagueness of 'substantial' reflects intuition: just as some representations are better than others, so, too, do some representations better reflect the behaviour of that which they represent.[12]

Much more could be said about these rudimentary applications of formal information measures to the slippery concept of representation, and much more could be said about the information theoretic framework itself. Chapter 5 extends the framework to include measures of logical depth and functional logical depth, but this so far suffices for a little while. The next section turns to some basic considerations about the computability of the information measures and limits to decidability in general before some closing thoughts on the place of information in philosophy.

4. GÖDEL DE-MYSTIFIED

Gödel's theorem...is now considered easy to prove and almost obvious. (Chaitin 1982, p. 942)

The general problem of incompleteness, of which Gödel's (1931) famous theorem and Turing's (1936) halting problem are symptoms, is this: within a formal system described by axioms with a given algorithmic information content, we cannot prove any theorem whose information content exceeds that of the axiom system by more than a fixed number of bits. That's it. Gödel derived his original incompleteness results only by way of a long and ingenious proof phrased in the language of number theory. At the time, mathematics lacked many of the tools today taken for granted—probably the most significant of which is the general computing framework of Turing and Post—and as a result Gödel's proof was naturally arduous. But modern developments, especially algorithmic information theory, render its more general form transparent. We'll have still a little more to say about it, but if ever confusion looms on the horizon, it can all be summed up with this basic idea. As Chaitin renders it, "if one has ten pounds of axioms and a twenty-pound theorem, then that theorem cannot be derived from those axioms" (Chaitin 1982, p. 942).

John von Neumann (1963) said the "conceptual shock" induced by Gödel's incompleteness theorem was comparable only to two other scientific developments of this century: quantum mechanics and relativity theory. Gödel's paper spawned reams of valuable contributions in artificial intelligence and philosophy of mind (not to mention quite a few nearly worthless obfuscations), and to this day some people (Lucas 1961, Penrose 1989, Bringsjord 1992[13]) still seem to find appealing the view that incompleteness means human persons can be neither simulated by nor equivalent to formal computational systems. (For a peculiar move in the opposite direction, see Kim and Kim 1995.) There can hardly be a researcher in the whole field of philosophy of mind who hasn't heard something about what Gödel did or didn't show. But much of the philosophical furore over Gödel's theorem rests on confusion about what exactly it is. The purpose of this section is two-fold: first, to explain how easy it is to understand incompleteness and second, to recognise some limitations it places on algorithmic information theory.

4.1 Decidability and the Halting Problem

The 'incompleteness' moniker relates closely to the paradox of the liar (see Quine 1962): "This statement is false", a statement which apparently is true if and only if it is false. Gödel's rendition for formalisations of number theory is a statement which says of itself that it is *unprovable*. If such a statement in a given formalisation were provable, then *any* statement in that system would be provable, because the system would be inconsistent. Alternatively, if the statement really were unprovable, then that formalisation of number theory would include a true statement which had no proof: if would be incomplete. Thus, every consistent formalisation of number theory is incomplete.

The more general form of this result is easier to understand as a consequence of Turing's halting problem (which, historically, came second). Given a numbered[14] class of all computer programs which output sets of natural numbers, consider the set of numbers of programs which do not include their own numbers in their output sets. Clearly this set cannot be listed by any computer program, because that program's number must be included if it isn't and mustn't if it is. (This is reminiscent of Bertrand Russell's 1908 paradoxical barber who shaves all and exactly those in his town who do not shave themselves; he must shave himself if he doesn't and mustn't if he does.) The solution to the paradox is simply that there can be no *effective procedure*—i.e., no Turing Machine algorithm, no finite program, which is guaranteed to produce the desired result—for deciding whether an arbitrary program ever outputs a specific number. Given an arbitrary Turing Machine program, we cannot show in advance whether it outputs a specific number or, equivalently, whether it halts (as opposed to looping or otherwise executing instructions forever). If any particular program does halt, we can find out eventually just by running it for long enough, but in the general case we can't show that programs *fail* to halt.

To see it from another direction, number as above the programs which output sets of natural numbers, and suppose the halting problem is solvable. That is, suppose there is an effective procedure for deciding whether any given program halts. Then we could embed that procedure in a larger program which determines whether the Nth program outputs an Nth digit and constructs a number D which is different at each location N to the Nth digit output by the Nth program (say, with binary outputs, by NOT-ing the Nth digit). But no such program could output D, since it must itself have a number N, and by definition the Nth digit of D is different from the Nth digit output by the program numbered N. The original assumption, that an effective procedure exists which could be used to determine whether the Nth program outputs an Nth digit and thus to construct D, must be incorrect. There can be no effective procedure to decide whether an arbitrary program halts (and thus whether it ever outputs an Nth digit).

Gödel's incompleteness conclusions follow immediately from the unsolvability of the halting problem. Suppose it were the case that every true statement were provable. Then every true statement of the form "program N halts" is provable. And, indeed, any program which *does* halt can be proven (eventually) to halt simply by running it. This much is fine. Supposing every true statement is provable, however, also means every true statement of the form "program N does not halt" can be proven. But the main requirement for a formal axiomatic system is that there be an effective procedure for deciding whether a particular proof is valid. (Thus an alternative rendition of a formal axiomatic system is simply as a Turing algorithm for enumerating a set of theorems, generating possible proofs and testing them directly for validity.) So for an arbitrary Turing Machine program N, we can simply generate all the valid proofs in search of one that proves the halting status of N, and on the assumption that all true statements are provable, we're guaranteed to find one, eventually. This contradicts Turing's theorem that there can be no effective procedure for determining the halting status of an arbitrary N. Therefore, the original assumption that every true statement is provable must be wrong.

In fact, the set of numbers of *halting* programs can be generated by a Turing program (although, of course, we never know when we're finished), but the *complement* of that set, the set of numbers of programs which don't halt, cannot. Thus, not all true statements of the form "program N does not halt" are theorems with associated proofs. That is, not all such true statements are provable. Sets like the set of numbers of halting programs, those which can be generated by a Turing program, are called *recursively enumerable*, or 'r.e.'; when, as in the present example, there is no effective test for set membership—equivalently, when the complement of the set is not recursively enumerable—it is called a *recursively enumerable nonrecursive* set. The members of an r.e. set which is nonrecursive can be generated by a Turing algorithm, but *not in order*. (For an early formulation of such sets in this way, see Post's elegant 1944 paper; also see Kleene 1952 and Rogers 1967).

But all this about earlier forms of incompleteness theorems is more difficult than the compact general statement with which this section began: within a given formal system, we cannot prove any theorem whose information content exceeds that of the system's axioms by more than a fixed number of bits. It's easy to see the rationale behind this general statement. Consider a program for a given axiom system whose job it is to find the first theorem (say, in order of length of its proof) with an information content greater than that of the axiom system and the program. Would such a program ever find such a theorem—i.e., would it ever find a theorem which it is impossible to generate with a program and axiom system its size? Of course not! This is reminiscent of G.G. Berry's paradox, first noted by Bertrand Russell (1908): Find the smallest positive integer which to be specified requires more characters than there are in this sentence.

A corollary of this limitation on the power of formal systems is that N bits of the right axioms can allow us to establish the halting status of all programs shorter than N but not of any with length N or above. For instance, an N-bit number describing the proportion of those programs shorter than N which halt, together with a generous supply of patience, would allow us to determine exactly which programs do halt: simply start testing programs shorter than N bits, and when we've found enough that halt to satisfy the N-bit axiom, we're finished; the procedure itself is guaranteed to halt. The first N bits of Chaitin's (1975, 1987b; also see Gardner 1979) remarkable random number Ω would also do the trick. (Ω is a maximally complex infinite string defined as the probability that Turing Machine U will halt if successive bits of its program are generated by tossing an unbiased coin. In a sense, Ω expresses all constructive mathematical truth as concisely as possible.) Similarly, it is not possible to prove with N bits of axioms any string's information content that is greater than a fixed number of bits above N. In fact, proving complexities and establishing halting status for given lengths turn out to be equivalent.

Interestingly, it is one of Nature's (logic's, God's, whatever's...) little jokes that any formal system in which it is possible to determine the complexity of strings with information content less than N either has relatively few bits of axioms and requires some very large proofs, or it has concise proofs and very many bits of 'axioms' which aren't algorithmically independent. Why? Consider the function $a(n)$, defined as the greatest natural number with algorithmic information content less than or equal to n. That is, $a(n)$ returns the largest natural number which can be specified with an n bit program. (This is the information theoretic equivalent of the Busy Beaver function $\Sigma(n)$, defined as the greatest number which can be computed by an n state Turing Machine.) But now consider a function which can be specified briefly but grows quickly, such as $f(n) = n!!!!!!!!!!!$, where the single factorial is defined as

$$n! = \prod_{i=0}^{n-1} (n-i)$$

Multiple factorials grow rather rapidly. Most hand calculators, for instance, balk even at the double factorial $n!!$ above $n = 4$ (which, to eighteen significant digits, weighs in somewhere in the neighbourhood of 620,448,401,733,239,439, 000,000). Yet, using recursive programming[15], a general function which returns factorials of arbitrary depth, taking a number like n and the desired depth as arguments, is specified very concisely in pseudo-code:

```
FUNCTION BIGFAC (OPERAND, DEPTH)
    IF DEPTH > 1 THEN BIGFAC:= BIGFAC (BIGFAC (OPERAND, DEPTH - 1), 1)
    ELSE IF OPERAND > 1 THEN BIGFAC:= OPERAND * BIGFAC (OPERAND - 1, 1)
    ELSE BIGFAC:= 1
END BIGFAC
```

This function returns `OPERAND`!...! for the number of '!' specified by '`DEPTH`'. The length of the function itself remains fixed regardless of the depth of factorial it is asked to compute. The information content of a number which can be calculated by such a function is thus bounded from above by this small constant overhead plus the number of digits required to specify the two arguments '`OPERAND`' and '`DEPTH`' (which, for economy, might of course be the same number). As this example function suggests, some strings with comparatively little information content are *extremely* long. Indeed, $a(n)$ increases more quickly than any computable function of n, and it is $a(n)$ which governs the relationship between length of proof and bits of axioms (Chaitin 1974; for early results on the curious relationship between logic size and length of proof, see Gödel 1936). With only N bits of axioms, calculating the set of all strings of complexity less than N bits requires *extremely* long proofs—on the order of $a(N)$. The alternative extreme is to assume all that we want to prove, which, not surprisingly, gives more concise proofs, but the number of bits of axioms which must be assumed can easily outstrip the number of particles in the cosmos. (Of course, such 'axioms' are not algorithmically independent.)

The upshot of all this is that the framework of algorithmic information content offers an extremely powerful *theoretical* setting for understanding information and relationships between information bearing objects, but in the general case it does not provide a direct recipe for calculating complexity. Indeed, it is difficult to imagine *any* objective measure of information or complexity which is simultaneously both decidable *and* free of arbitrary manipulation as a result of changes in coding or representation. Any candidate measure would need to be entirely logically independent (and algorithmically independent, for that matter) of algorithmic information content: otherwise, the latter's decidability limitations could be carried back to it. Alternatively, no such candidate measure could itself be computably related to algorithmic information content, since it would then provide an indirect avenue for computing noncomputable values! Insofar as algorithmic information content preserves many normal intuitions about complexity, an algorithmically independent rival would thus clash with the same. This limitation shouldn't cause any great concern. As noted above on page 42, it can sometimes be useful to step out of the rigorous but undecidable Turing Machine framework—but any hopes of grabbing something concrete directly with respect to brains or other cognitive substrates must be tempered with the awareness that doing so invites arbitrarily large variation in the complexity of whatever we might grab. Such excursions are best understood as approaches *inspired* by formal information theory but devoid of the latter's underlying rigour.

4.2 Incompleteness and the Limits of Human Cognizers

Having found ways through some of the mystique commonly thought to shroud Gödel's incompleteness results, it's interesting to consider what sort of argument might be formulated to the effect that incompleteness implies human persons can't be simulated with a formal system (or that they can't be computationally equivalent to such a system). To my knowledge, no one has ever advanced a respectable argument to this effect.

One argument would exhibit a proposition p which we know to be true but which provably cannot be proven within any formal system which could be included in a human mind. (Obviously it is of no help just to exhibit a true proposition which *hasn't been* proven, since that by itself illustrates nothing about whether it *can be* proven.) Regardless of whether it used the actual language of algorithmic information theory, a good argument would show that p has greater information content than any formal system a human could use. Obviously, a good argument would also assume that p is consistent with whatever we *can* generate with whatever formal systems we do use. In other words, p isn't a proposition we know to be true but which contradicts conclusions of, say, formal logic. That would be an interesting illustration that we are not formal systems, but it has nothing to do with incompleteness. Finally, a good argument would show that the reason we know p to be true is not merely that we've assumed it as an axiom (say, after witnessing a physical example of its truth). From a host of problems with this approach, two immediately leap into focus.

First, there are of course infinitely many formal systems in which p can be proven, and merely *understanding* the proposition p illustrates that we can include at least $I(p)$ information in whatever formal system takes our fancy. (The degenerate case is of course just assuming p as an axiom in some formal system such as standard logic, but vastly many other axiom systems with about the same information content would also do the trick.) At a long shot, such a primitive argument might work against the position that humans are rigid formal systems which never adopt new axioms (and which are thus limited to whatever information content they have when their lives begin)—but as far as I am aware no one seriously maintains that view.

More importantly, even showing that the information content of proposition p exceeds that of a given rigid formal system which never adds new axioms is equivalent to determining whether the theorem-generating program for the given formal system ever produces a particular proposition. Since the propositions proven by the theorem-generator obviously form a recursively enumerable nonrecursive set, however, this is a tall order. The fact that we can exhibit propositions which can't be proven within *particular* formal systems, incidentally, shows only that we are not equivalent to *those* formal systems: a metatheorem about a particular formal system clearly is not crafted in the framework of that particular system.

All in all, I'm not the best one to provide an example of a respectable argument of this type, since I confess complete bafflement at the task of ridding hypothetical arguments of incoherence. For readers interested in such ramifications of incompleteness for philosophy of mind, it would be better to consult some of the other literature on the topic, such as that cited above—although armed with the material of this chapter I suspect most will find it challenging to locate a satisfying argument.[16] For entirely different reasons why human beings may in fact have no more computational power than finite automata, see the discussion on the physical limits of computation starting on page 212 in Chapter 9.

5. PLATONIC HELL

It's easy to see how a great mathematician, for instance, accustomed as she might be to daily feats of sophisticated mental gymnastics and no doubt very impressed with the aptitudes of the human intellect, could be misled by the thought that Gödel's theorem categorically separates her mathematical thought processes from merely *formal* systems. Continually experiencing as she does her seemingly unbounded capacity to analyse hypothetical axiomatic systems apparently without having her thoughts suffer any of their formal limitations—and her capacity to sniff out truth about these systems in great leaps of mathematical insight—she might easily overlook the fact that she hadn't ever actually produced a single example of a mathematical truth which couldn't have been generated by some formal system or other.

It's also easy to see how a great philosopher, for instance, accustomed as she might be to daily feats of sophisticated mental gymnastics and no doubt very impressed with the aptitudes of the human intellect, could be misled by the thought that her keen grasp of logic categorically separates her philosophical thought processes from merely *physical* systems. Continually experiencing as she does her seemingly unbounded capacity to analyse hypothetical physical systems apparently without having her thoughts suffer any of their physical limitations—and her capacity to sniff out truth about these systems in great leaps of philosophical insight—she might easily overlook the fact that she hadn't ever actually produced a single example of a logical truth whose information content wasn't instantiated in some physical system or other.

It's easy to see how a philosopher could emphasise the powers of logic at the expense of the realities of the physical instantiation of information. A philosopher might easily forget that while pondering an infinite random number like Ω, she's not *really* thinking a thought which embodies all constructive mathematical truth in its most compact form: no physical part of her suddenly acquires infinite algorithmic complexity just by virtue of pondering Ω. Likewise for any other logical construction. I suggest that any thought whatsoever, whether it 'refers' to something real or imaginary, is limited by its physical instantiation and that its information content (ignoring the oddity of extending a formal measure to something like a *thought* for the sake of putting the point loosely) is bounded by the information content of its instantiating structures. Elevating logical constructions to some sort of Platonic Heaven where they can exist independently of real physical instantiation—and, worse, positing some direct apprehension of that realm by human minds—requires a corresponding demotion of the coherence of discourse about philosophy of mind. Later there is much more to say about the sense in which logical constructions might be conditioned upon empirical facts about physics. (In particular, see section 4 of the next chapter on logic as used by cognizers, section 4 of Chapter 6 on supervenience, and section 3 of Chapter 7 on logic and physics.)

But an opposite tack also leads somewhere: while we ignore the boundary conditions set by physics at our peril, it is no less hazardous to ignore the *additions* of information made possible by physics, over and above logic. The forgetful philosopher might easily overlook the fact that, while reading a book, the information transfers occurring as a result of the laws of physics far exceed the mere logical implications of what she is reading. Even supposing she studies physics in great depth and acquires a clear understanding of all the physical processes of information transfer occurring while she is reading a book, she might easily neglect a key point of the next chapter: *using physics to deduce the consequences of all those physical processes does not bring about the same information transfers as actually engaging in the act of reading the book.* The next chapter explores both these points as they bear on philosophy of mind and applies other observations about physically instantiated information in the first steps of a quest to bring to the light of day a view of mind which might otherwise remain doomed to the wrong end of a Platonic see-saw.

NOTES

[1] For an introduction to the language of *observables* in quantum mechanics, check Chapter 7, especially section 2. Note that the use of 'correlated' here need *not* refer to quantum mechanical entanglement, however. One job of section 2 of the present chapter is to quantify precisely the notion of information content under a wider sense of 'correlated'.

[2] Not all life on Earth ultimately relies on energy from the sun. Geochemical energy released from deep-sea vents will also do.

[3] Of course, very much rides on the first proposition and the word 'contain'; I've taken no great care to sort this out, as I don't intend the suggested inconsistencies to be taken very seriously for the moment. I do believe they are responsible for some of the confusions discussed in section 5, however.

[4] Davis (1965) is a favourite for original papers from Gödel, Church, Turing, Rosser, Kleene, and Post.

[5] For present purposes, when speaking of an 'object', we might just as easily speak of a 'bit string describing an object'; the two are used interchangeably here. At the lowest level of description, a single bit of information may be associated with the 'yes/no' answer as to whether a system's eigenvalue with respect to a particular quantum observable lies within a specified interval.

[6] It is this choice of a U, incidentally, which makes precise the particular way in which descriptions correlate with objects.

[7] Chaitin (1987a) is an indispensable, if somewhat repetitive, collection which includes most of his papers cited in this book. For an extensive collection of papers, tutorials (including PostScript and PDF versions of the entire book *The Limits of Mathematics: A Course On Information Theory & the Limits of Formal Reasoning*), and other links on the topic, see Chaitin's web page at http://www.cs.auckland.ac.nz/CDMTCS/chaitin.

[8] There, information itself is understood in terms of reduction of uncertainty in a set of choices within a given context. For applications of this classic version of information theory in a psychological framework, see Baars (1988, especially Chapter 5) and, for a simpler approach, Godfrey-Smith (1996); also see Miller (1953) and, of course, Dretske (1981). For critiques of the last, see Coulter (1995) and the commentaries accompanying Dretske (1983).

[9] It might instead be said "is zero", but sometimes—as in the case of two strings each of length n—residual mutual information can't be removed, since knowing one does tell something about the other (in this case, its length). Likewise, for physical objects, the level of precision (page 1) bears on the mutual information content contributed by atomic and molecular similarity and so on. In general, however, concern centres on *relative* comparisons of mutual information content, where such details are washed out.

[10] By 'transformation', I specifically include not only translation, reflection, and so on, but also *deformation* (such as conformal or quasiconformal mappings, etc.). All mention of transformation in the context of functional representation should be understood in this broad sense.

[11] Obviously, such representation needn't involve *explicit* representation of axioms. The reflector of an automobile headlight, for instance, contains information about the transformation axioms governing the free flight of a piano launched from a window—since both the reflector and the piano's trajectory are parabolic in cross section—yet nowhere is such information represented explicitly.

[12] Note that functional representation subsumes the notion of representation by second order isomorphism advanced by Shepard (1968). (Also see Shepard and Chipman 1970, Palmer 1978, Holland *et al.* 1986, and Gallistel 1990.) Second order isomorphism—between, on one hand, *relations* between external objects and, on the other, *relations* between their corresponding internal representations (rather than first order isomorphism between external objects and their internal representations)—boils down to the representation of both objects and the transformations which they may undergo.

[13] See Mulhauser (1995d) for a rather negative review.

[14] The easy way to number computer programs is simply to represent their instructions with numbers—say, binary strings—and then associate with each the full natural number resulting from the concatenation of all the instructions. Contrariwise, each natural number has associated its own program, and we can put them all in numerical order, allowing us to speak of the Nth computer program in the list.

[15] Here, 'recursive' denotes a programming technique in which a subroutine—in this case, a function—calls itself in part of its calculation. The example function checks first to see if it is being asked for a single depth factorial, and if it isn't, it calculates the factorial of the result of calling on itself to compute the factorial of the operand at one lower depth. If it is instead just calculating a single depth factorial of an operand larger than 1, it multiplies the operand by the result of calling on itself to calculate the factorial for the next smaller number. Finally, when it is asked merely to calculate 1!, it returns the value 1 and stops. I just made up the function for the present example; no doubt it can be done even more economically.

[16] Incompleteness does, however, have some bearing on how supervenience relations should be understood; the topic returns briefly in section 4.2 of Chapter 6.

CHAPTER FOUR

A Rose, By Any Other Name (Or Description)

Nothing you could tell of a physical sort captures the smell of a rose... Therefore, Physicalism is false. (Jackson 1982, p. 127)

Many philosophers, including myself, find the above verdict of self-ascribed 'qualia freak' Frank Jackson extremely appealing. Consider an alternative premise: 'Nothing you could tell of a physical sort captures the mass of a rose'. That clearly is false. We *define* mass as a physical property, and 'capturing' it demands nothing more than specifying some subset of the rose's complete physical description. A physical description of rose mass misses out nothing, because there is nothing else to it but that physical description.

The smell of a rose, on the other hand, seems to elude this sort of description. If I've *already* smelled a particular rose, someone might capture that smell in a 'physical' description just by referring to 'the odour you sensed just now when you smelled that red object in the garden'. But this differs only mildly from something like 'the secret I told you last weekend': each description merely acts as a placeholder, and for someone who hasn't actually smelled the rose in the garden or heard the secret last weekend, this sort of description captures nothing. Whereas describing rose mass in physical terms conveys all the relevant information from one person to the next, describing rose smell in physical terms seems to convey only a tiny portion of what is relevant for grasping rose smell. A physical description seems to miss out exactly how it is the rose actually smells.

One rendering of this view is the maxim that information about what an object is like solely in terms of physical properties does not logically entail corresponding information about what it is like as an object of phenomenal experience. Indeed, Jackson and other 'qualia freaks' might deny there can even *be* information, on the strictly physical definition from the last chapter, about phenomenal experience. (Jackson himself implicitly denies—1982, p. 127—that all information is physical.) On that view, physical descriptions and phenomenal descriptions are simply incommensurable.

In keeping with the account of information advanced in the previous chapter, the task here is to extricate ourselves from some difficulties caused by focusing solely on the logical implications between objective physical descriptions and subjective phenomenal ones (or phenomenal experience itself). Exploring what *information* may be transferred by means of a description suggests that the two—logical implication and physical information transfer—are far from equivalent. Neither denying incommensurability between the physical and phenomenal nor asserting that the former can *somehow* yield the latter is equivalent to asserting that someone could logically *derive* the latter from the former. Indeed, observing this inequivalence turns out to render many philosophical problems with the perspectival nature of experience far less baffling. An information theoretic perspective also sheds some light on the *third person problem*, the job—analogous to the 'problem of other minds'—of evaluating another person's (or machine's, etc.) claims to be having some conscious sensation or other experience.

1. SENSATIONS AND DESCRIPTIONS—PRELIMINARIES

Instead of asking the peculiar question of whether a physical description could logically imply a rose smell[1], consider a different one: Can the process of reading a physical description of a rose achieve exactly the same transfer of information as directly smelling a rose? It's very nearly obvious that it cannot in the general case, if for no other reason than that reading a description transfers some information by way of visual perception—or tactile perception, for a description written in Braille, for instance—which shoving a rose under a nose and inhaling does not. That is, the process of reading causally instantiates information relationships between the reader and the read which are not the same as relationships causally instantiated between the smeller and the smelled during the process of smelling. Indeed, it seems the information transfer caused by reading a description could be the same as that caused by some given direct experience only in the following convoluted circumstance: when the description being read is actually *of* the direct experience of reading a description of the direct experience!

But below it emerges that, with a few *provisos*, the answer is affirmative for a slightly different question: Can the process of reading a physical description of a rose achieve a transfer of information which is properly a *superset* of information transferred by directly smelling a rose? Far from missing out some important information, it is possible not only to learn how a rose smells from a physical description, but it is possible to learn *more* that way than by just smelling it directly.

First, it is worth noting that in pondering physical descriptions and whether they miss out important information about smells or other sensations, descriptions ought to be understood as read (heard, touched, etc.) by systems which are at least capable both of reading the descriptions and of experiencing the relevant sensations. Suppose there is some description written in English which *does* capture the smell of a rose. Perhaps it's found that if only we set the same old descriptions in iambic pentameter, they suddenly come alive with vivid rose smells for all who read them. (Alternatively, perhaps intoning some alchemical spells or sprinkling the pages with a little eye of newt just before reading brings similar consequences.) Would it be at all relevant that such new fortified descriptions, when scanned by computerised character recognition systems or when perused by monolingual speakers of Swahili, have no such effect? It hardly seems it could matter; for all those readers capable *both* of reading English *and* of experiencing a rose smell, the new descriptions still work just fine, and their failure for others is no fault of the descriptions themselves. It can be no argument against physicalism to point out that a physical description fails to capture the smell of a rose *for a toadstool*, for instance. To be convincing, any alleged 'missing out' ought to result from some connection, or lack thereof, between descriptions and smells—*not* from some inherent difficulty either in reading descriptions or in experiencing smells.

2. GETTING SENSATIONS FROM DESCRIPTIONS

With these observations to hand, this section explores how a physical description may 'capture' the smell of a rose. In particular, we're after a description which enables a capable reader to learn how a rose smells phenomenally, such that after really smelling an example rose, the reader/smeller learns nothing new about the actual smell. (Obviously, some new things might be learned through a subsequent smelling experience such as what it is like to smell a particular rose at a particular time or what it is like to experience rose smell for only the second time ever, but for the sake of tidiness I assume such things don't add to the smeller's grasp of the smell itself.)

A description which somehow actually enables a reader to *experience* a rose smell would be especially convenient. Maybe other lesser descriptions could also do the trick, but surely a description which does make possible an actual phenomenal experience of rose smell won't have missed the mark. The present aim in producing such a description is not (as stressed in Chapter 1) to argue *for* physicalism, since the discussion below appeals to some physicalist assumptions; instead, the story teases out the source of mystery in Jackson's pronouncement and shows why it fails as an argument *against* physicalism. In particular, the present story is formulated in terms of actual experience so as to avoid allowing anything significant to hinge on the troublesome word 'knowledge'. Jackson's original example relies heavily on the oddities of knowledge and the difficulty in accounting for how someone could know all the physical facts about an experience without knowing what it was like to have the experience. If we manage to tell a parallel but non-mysterious story while keeping 'knowledge' off centre stage, this suggests the mystery is in the *words*, not the world. (And as in Chapter 1, our priority is 'world first'!) The upshot is a deflationary reply to the following question: how *could* physicalism be true when it seems clear that no physical description captures the smell of a rose? In fact, as the discussion below takes shape, it becomes clear that physicalism very nearly *entails* that Jackson's problem should arise.

The example builds on Jackson's Mary, the neurophysiologist who studies the world from inside a black and white room through a black and white television monitor and who has never actually seen a red object. (Commonly, extra stipulations ensure that she doesn't get to see red by, say, accidentally slicing her finger or rubbing her eyes. See Thompson 1992.) Without leaving her greyscale exile, Mary learns all there is to know about the physical aspects of seeing the colour red; she grasps in every detail the precise neurophysiological processes which would occur if she herself were to have the experience. Yet even with all this information, she still learns something *new*, so the story goes, as soon as she goes outside her room or switches to a colour television and, for the first time in her life, actually sets eyes upon a red object like a rose. Thus since, *ex hypothesi*, she could already have learnt *all* the physical information about the process of seeing red before actually having seen something red herself, the story apparently shows that not all the information about seeing red is physical.

Consider Mary's associate Margaret, a botanist who learns in great detail all the physical properties of roses[2] but who has never actually smelled one directly. Maggie has at her disposal a vast quantity of information about roses. She knows all about their lowest level cellular chemistry, she knows their average mass, and she knows how the ballistic coefficient of a fresh bud compares with that of an open bloom. Unlike Jackson's Mary, she has even seen (but not smelled) a red rose. And, since borrowing some of Mary's textbooks to brush up on her neurophysiology, she knows the full effects on human nerve cells of certain particles released by roses. All this is 'physical information' on the present (and probably almost any other) definition of the term.

Perhaps Maggie has amassed a comprehensive library of notes she has taken during her years of rose studies, and she wonders what to do with it all. One thing she might do is rewrite it all in Braille; then people could take in all that information through their fingertips. She might also record lectures of her notes on audio tape; then people could take in all that information through their ears. Or, more interestingly for the moment, she might use selected portions of her library to construct an artificial stimulator for nerve cells in the human nose. Then people could take in some of that information through their noses. Either feeling Braille books or listening to audio tapes is bound to be more pleasant than putting electrodes in one's nose, but the artificial nerve stimulator has an important advantage which might make the experience worthwhile: it makes it possible to use the *same information* in a modality which happens to accommodate smell sensations.

Of course, these different versions of Maggie's vast library don't all include *exactly* the same information: ordinary books written in English have mutual information content with dictionaries which artificial nerve cell stimulators lack, and Braille texts have mutual information content with Braille printing presses which ordinary books do not. (For those reading chapters out of order, see section 2.2 of the last chapter for mutual information content.) But the mutual information content between certain physical features of roses and the relevant bits of Maggie's library in the form of ordinary books, Braille books, or artificial nerve stimulators remains roughly the same. (Consequently, the mutual information content of the nerve stimulator with the relevant portion of a Braille book or ordinary book is close to maximal, given the peculiarities of the individual instantiations: it is approximately the same as the individual information content of any single form on its own, with the peculiar features of the bumps, the typeface, and so forth subtracted out.)

For now, we allow the extravagance of supposing without argument that stimulation of appropriate nerves in a subject's nose initiates some process which ultimately results in the subject's experiencing—if awake, not drugged, and so on—a smell sensation. It doesn't matter for now what the connection might be between nose stimulation and smell sensation; it matters only that there is one. On the assumption of such a connection, it seems at least one of Margaret's possible uses for her information *does* enable her to learn exactly how a rose smells even if she lacks one to sniff directly. There appears to be no difference relevant to Maggie's smell sensation whether her nose is stimulated directly by some rose particles or artificially in the way it *would be* stimulated by a rose directly.

This might not be convincing for someone who, for example, believes that an extra property of roses, apart from their direct effect on nerve cells in a subject's nose, somehow gets passed along when they are smelled. Maybe some unique 'rose essence' makes a direct connection from bloom to brain, for instance—or from bloom to Cartesian spirit. Maybe then a special alchemical incantation or some eye of newt *could* help 'capture' that distinctive essence, but clearly no physical description, whether instantiated in ordinary books, bumpy Braille books, or artificial nerve stimulators, would do the trick. Such a possibility of something else important *beyond* the nose stimulation we'll unceremoniously ignore, because it misses the point both of Jackson's argument and of the reply formulated here. The present aim is simply to tell a physicalist story according to which Jackson's argument loses its punch, and such a 'rose essence' possibility is irrelevant to that project; in any case, 'qualia freaks' likely will have a different sort of objection.

That objection comes to this: we're cheating! The point of the argument is that no physical description could, *by itself*, capture the smell of a rose. Using a human being's nose and brain and mind to get something *extra* out of the description just concedes that all the relevant information wasn't there in the description in the first place: it was, if anywhere, in the human. There is still no way for a normal subject to *derive*, logically, an understanding of his own phenomenal experience of a rose's smell solely on the basis of its physical description.

On the last point, I have no quarrel. But drawing out what is wrong with the rest of the objection, by way of another example, is not too difficult.

3. 'MACMARY'

Where before attention centred on Mary's colleague Margaret, who, like Mary, was a fellow human being possessed of huge quantities of physical information but deprived of some particular direct experience, now consider MacMary, Mary's little Scottish computer counterpart. (No one in Scotland is really called 'MacMary', but there are plenty of computers called 'Mac-this' or 'Mac-that'...)

Suppose I've typed into MacMary's word processor a book all about the way its disk drive works. The book might include detailed specifications of the electrical signals which activate the disk drive to perform each of its functions; in particular, I might give examples of exactly the signals which would be sent inside the machine to make it format a disk. Yet no matter how eloquently I express the information about the computer's internal signals, in iambic pentameter or otherwise, that information *as a book* cannot by itself make MacMary format a disk. Typing into MacMary a book's worth of information about chip-level disk interface signals has no effect whatsoever on its own actual chip-level disk interface signals. The information about disk interface signals has an effect only on MacMary's chip-level *memory* signals.

Yet, using the *same information* in a different way, I can cause MacMary's disk drive to format a disk without any difficulty simply by getting into the machine's insides, finding the appropriate control lines, and imposing my own signals on them. That I cannot impose those signals on the relevant control lines simply by typing into MacMary's word processor information about what I would like them to be is purely an (intentional!) accident of modern computer design. The software with which I compose the book about disk drives has no direct access to any of the computer's chip-level signals, and indeed neither does most of the computer's own operating system. Even the computer's lowest level machine language likewise lacks direct access to chip-level signals *qua* chip level signals in all but a few special cases, simply because machine language instructions refer to higher level things like register contents and operations on them. Fortunately for computer designers, this means register contents and operations on them can be physically implemented in any number of different ways, with vastly different patterns and voltages at the chip level, without any effect whatsoever on the relationship between codes in the computer's instruction set or on the operation of its programs. (Of course, all this holds independently of the banal observation that executing machine language instructions or commands at the operating system level—or even typing books into word processors—always does cause changes in chip-level signals.)

But none of these impediments to MacMary's formatting a disk when all the relevant information about chip-level signals has been typed into it as a book are by any means *necessary* characteristics of computer design. Although the programming languages and physical computer layout feature in entirely different levels of description and remain neatly segregated for the most part, nearly all computers do allow limited direct access—what programmers sometimes call 'writing on the metal'—for a few special cases.[3] Consider a computer which allows it for some large subgroup of its chip set. Maybe it allows, for example, direct modification of chip-level signals in accordance with information typed into a word processing file. For such a computer, unlike MacMary, the task of formatting a disk just by typing portions of a book about how disk drives work might be perfectly straightforward. Whether or not it is possible for MacMary to have a 'disk formatting experience' as a result of having a book typed into it all about how disk drives work clearly has no bearing whatsoever on the question of whether the book captures all the physical information about how disk drives work. Either MacMary or the imaginary computer outfitted for extensive writing on the metal might be equipped with the *same* disk drive, and the *same* book still accurately describes all there is about the physical operation of the drive.

The MacMary example helps uncover the flaw in the objection at the end of the last section, and it suggests a significant additional observation about both Mary and Margaret. First, the line of argument above which suggested Maggie was cheating by using an artificial nerve stimulator to 'derive' (empirically) a smell sensation from physical information about a rose is roughly analogous to an argument which suggests it's cheating to open up MacMary and manually impose chip-level disk controller signals. But such a 'cheat' in the computer example applies no novel information beyond what was already there in the book about disk drives, and the corresponding 'cheat' in Maggie's case applies no novel information beyond what was already there in the books about physical rose properties. That MacMary doesn't format a disk even when it has a complete book all about disk drive operation and that Maggie doesn't smell a rose even when she has a complete library all about physical rose properties says nothing at all about whether the respective book or complete library 'misses out' something important about disk drives or about roses.[4]

What both MacMary's failure to format a disk and Maggie's ignorance about how a rose actually smells *do* reveal is simply that there is more than one way a computer or a human might make use of the same physical information. Alternatively, there is more than one way a computer or a human might instantiate the same physical information. What is *different* when MacMary does format a disk or when Maggie learns how a rose smells—for clearly something *is* different—is not a matter of information, but a matter of *state*. Maggie in a state of really understanding how a rose smells (or in a state of simply smelling one) need instantiate no more information about roses than Maggie in a state of understanding all the relevant physical information while remaining ignorant of how a rose smells. But the two, while perhaps equivalent with respect to their mutual information content with particular physical information about roses, are obviously distinct states. (This appeals to the assumption, acceptable presumably even to most dualists, that two cognitive states of the same person being distinct entails their corresponding physical states are also distinct. See Chapter 6 for expansions on this line of thought and supervenience.) That is, a state of experiencing a rose smell or formatting a disk is simply *different* from a state of understanding a description of the same.[5]

MacMary, in possession of its complete word processor file all about disk drives, is in a state which, among other things, instantiates all the physical information about how its disk drive operates; it is *not* in a state of MacMary actually formatting a disk. But appropriate application of the information could get MacMary to that state. Maggie, in possession of an entire library's worth of physical rose property descriptions, is in a state which, among other things, instantiates all the physical information about roses; she is *not* in a state of Maggie actually smelling a rose. But appropriate application of the information could get Maggie to that state. This distinction between different physical states instantiating the same information in different ways is, I suggest, the source of all the mystery in thought experiments about the likes of Mary, MacMary, or Margaret.

Consider now a Maggie or a Mary modified in line with the hypothetical computer allowing extensive direct access to its chip-level signals. Suppose Maggie could stimulate her own nose nerves *directly*, without the external assistance of an artificial nerve stimulator. That is, suppose Maggie had an extra set of nerves which could provide some desired signal to her nose nerves, in a way such that their efferent signals sent on to the olfactory cortex match the signals they *would* deliver if they were directly stimulated by, say, a rose. (Alternatively, suppose she just sends the appropriate signals directly to her olfactory cortex, missing out the nose altogether.) Under the assumption above that interaction with nose nerves initiates some process, however complex or even other-worldly it might be, which eventually results in smell sensations, there seems no reason to think the new Maggie oughtn't be able to learn exactly how a rose smells just by reading a relevant physical description and sending the right signals to her nose. Just as a computer might be designed in such a way that it could place some part of itself in a particular state according to a typed description, there seems no reason to think a person couldn't be 'designed' in such a way that she could put her nose or olfactory cortex in a particular state according to a written description. It is, however, just as much an accident in Maggie's case as in MacMary's—an accident which for Maggie no doubt has very good evolutionary rationale—that she cannot 'write on the brain cells' in this way.

The conclusion seems difficult to avoid: a physical description of a rose, or of the colour red or of the operation of a disk drive, needn't 'miss out' anything. Moreover, it is actually possible to learn *more* from appropriate use of a complete physical description of an experience than from direct experience alone: after all, the former can yield both the experience itself *and* an understanding of the physical processes underlying it, whereas the latter offers only the direct experience. At least in this context, anyway, the physical and phenomenal are not incommensurable. Yet all this coheres perfectly well with the observation that a subject cannot logically derive how a rose will smell from a complete physical description of one. Starting with a complete physical description of a rose, Maggie cannot logically derive herself into a state of understanding how a rose smells. Appropriate use of the very same physical information, however, can causally drive her into a state of understanding exactly that. The information theoretic perspective exposes any 'missing out' or 'incommensurability' as a matter of physiology and not of logic.

4. A TASTY SIDE NOTE

The fact that a subjective phenomenal understanding of an object of sensation cannot be had as a direct logical consequence of a complete physical description of that object—but could be had as a physical, causal consequence of the same information—suggests some interesting observations and questions about logic itself.

First, it underscores the point of section 5 of the last chapter, which warned against elevating logic at the expense of basic physical boundary conditions. Restricting attention to the logical consequences of a complete physical description, for example, obscures some rather simple facts about how that description might actually be used by a real cognizer. No one would believe that physicalism is false just because a book about disk drive operations doesn't immediately make MacMary format a disk; yet, in the absence of physical considerations to temper logic, the parallel fact that complete physical information about roses doesn't immediately make Maggie smell a rose (or enable her to derive an understanding of rose smell) apparently tempts many to that very conclusion.

More interestingly, recall the modified Maggie who tickles her own nose nerves directly, without the assistance of an external artificial stimulator. What if Maggie and everyone else were capable not only of this useful feat of 'writing on the brain cells', but also of doing it *automatically*, without giving it a thought—their noses automatically responding to descriptions they read? If everyone really did experience this kind of high level linguistic to low level olfactory 'short circuit', if humans automatically associated smells with physical descriptions of objects like roses, might they come to believe that the smell of a rose was actually *logically entailed* by physical descriptions of roses? Understanding logical entailment with respect to the ordinary logic we all know and love, the answer is almost certainly negative. But I wonder: would such people understand logic in the ordinary way?

It's interesting to consider the degree to which the natural understanding of *P* from (*P* & *Q*), for instance, might depend on the right 'short circuits' being in place in normal subjects' brains. If it were possible for some brain defect to deprive subjects of the obvious recognition of *P* in (*P* & *Q*), for instance, clearly for those with that defect *P* would *not* follow obviously from (*P* & *Q*). Of course someone might simply say that such people don't properly understand the *meaning* of a sentence like *P* or a sentence like (*P* & *Q*). But a plausible response is then that the reason everyone else *does* understand such sentences is that they have 'normal' short circuits which the others lack. (Indeed, for an argument against the claim that such 'obvious' inferences are constitutive of understanding, see Cherniak 1981.) After all, on a physicalist view, deducing one proposition from another, much like coming to understand how a rose smells, is a matter of entering a new physical state.[6]

Rather than pondering humans who might fail to recognise something as simple as the *P* in a (*P* & *Q*), consider instead some whose associative connections are a *superset* rather than a subset of these kinds of logical connectives. For about ten adults in one million, sensory input in one modality regularly, uniformly, and automatically evokes sensation in a second modality. For someone with *synaesthesia* (Cytowic 1989, 1993), the word 'logic' might be yellow, while for another, a lemon might taste triangular. More than seven decades ago, researchers in the US explored fascinating puzzles about synaesthesia and blind subjects' understanding of colour (Wheeler 1920, Wheeler and Cutsforth 1922), and more recently the phenomenon has proven richly suggestive in work on creativity and perception of metaphor (O'Malley 1964; Ralston 1976; Marks, *et al* 1987).

Suppose now that synaesthetes were the norm and 'normal' people the exception: then the colour orange might be just as obvious an association—and perhaps even an *implication*[7]—of (P & Q) as most of us consider P to be, and those who couldn't 'see' the orange in (P & Q) would understandably be viewed as just as impoverished or noncomprehending as we consider those who don't see the P. (In fact, while individual synaesthetes report remarkably self consistent associations, those reported by *different* synaesthetic subjects rarely agree; thus, the example requires a small cheat to ensure the hypothetical synaesthetic norm is truly normative.) Following the 'short circuit' approach from above, we might think of the P from (P & Q) connection as a kind of 'single modality synaesthesia' with which almost everyone is affected.

The idea suggests a radically empirical approach, conditioning the meanings not only of words but even of the basic logical connectives upon physical facts about cognizers. The prospect of this sort of 'bio-Kantianism', as one reviewer of an earlier version of this book called it, is intriguing, but here is not the place either to develop it further or to defend it against a host of difficulties, two tasks which would lead well beyond the present scope. (Readers interested in this general area will find the discussion of metaphysical supervenience with *a posteriori* necessity in Chapter 6 relevant.) None of this book's later projects rest on any particular position with respect to this question, but one clarification as well as a related objection nonetheless merit remarks.

First, the clarification: suggesting an empirical approach to the use and understanding of logic by cognizers certainly does *not* require suggesting that logic understood classically as a body of truth-preserving transformations of propositions is somehow merely 'in the mind' or determined by inter-subjective agreement. Even less does it require any stance on the sorts of issues kicked off by Quine's (1951) famous paper[8] and later debated in Putnam's (1969) and Dummett's (1976) identically titled papers 'Is Logic Empirical?': I am concerned here not with the revisability of logic in the face of empirical evidence (Quine and Putnam) or new accounts of meaning (Dummett), but with the physical preconditions which allow a cognizer to grasp logic in the first place. Were this the place for exploring the question more deeply, a good place to begin would instead be the suggestion that human beings acquire through the course of evolution the capacity to discern forms of logic appropriate for reasoning about an environment governed by particular laws of physics. On this view, normal understanding of logic truly does reflect objective relationships between real things in the world—that is why we use it successfully to reason about the world in a truth-preserving fashion. But the reason for this normal understanding of logic is *not* that minds have some direct insight into logic or rationality or truth or an *a priori* stipulation that having a mind at all presupposes rational behaviour[9]; instead, it is the mundane fact that minds are instantiated by brains which have evolved in part to incorporate *information* about how objects in their environment behave under the laws of physics. (See section 3 of Chapter 7 on the logic of physics and especially David Deutsch's observation on page 211 in Chapter 9 about the reason why logic is logical.) The point of these observations about synaesthesia and the normal grasp of logic is merely that altering minds' material substrates may alter their understanding of what is logical. Score one for a view counterbalancing that from section 5 of the last chapter: here logic keeps its privileged status, but we note the physical boundary conditions on the cognizers trying to grasp it.

Second, the objection: Jim Edwards (personal communication) expresses deep misgivings about theories of mind which encourage such a treatment of logic by virtue of an information theoretic view of representation; I take his particular concern to apply equally to the present hint of an empirical view of logic and to the approach to consciousness which occupies much of the remainder of this book. Specifically, Edwards notes the parallel between representational theories of this sort and dispositional theories of meaning in the Wittgenstein (1953) tradition. Saul Kripke's (1982) well known critique is widely taken to have demolished all forms of dispositionalism, on the grounds that finite physical things are not the right sorts of things to embody the meanings of terms whose correct rules of usage may be defined over infinite domains. (Also see McCulloch's tidy 1995 discussion, pp. 100-106.)

Following Horwich (1995; also see 1990), however, I have reservations about Kripke's argument against dispositionalism. Horwich argues persuasively that the only move which salvages Kripke's basic argument from a fallacy of equivocation does so at the cost of rendering it unsound. Horwich's attack requires the premise that the disquotational approach to truth or a similar 'deflationary' account is correct and that alternative 'inflationary' accounts are false; thus, philosophers who find inflationary accounts appealing are liable to remain unconvinced. However, I expect those sympathetic to inflationary accounts of truth will level other complaints, well beyond Kripke's critique, against both the theory of consciousness outlined later and this present business about logic. Again, the issue stays almost entirely untouched; I mention it here only to note that dispositionalism is far from dead (also see Peacocke 1992 and Dummett 1978) and that whatever objections some might raise against present projects by taking Kripke's argument as a point of departure remain far from conclusive.

5. THE THIRD PERSON PROBLEM

The line of thought developed in the last three sections suggests an interesting corollary: there can be no 'objective' description of a rose's smell in terms of physical information about the rose. The reason is clear. Rose information *by itself* is not sufficient to determine the physical state of a particular cognizer smelling one; that state depends on information not only about the rose but also about the cognizer, information which might vary widely between individuals. (Of course, neither all of that information nor all of that state is liable to be *relevant* to an individual's sensation of smell; the point appears again below and in Chapters 5 and 6.) Just as some colour blind subjects cannot discriminate between, say, red and green, probably not everyone distinguishes rose smell from banana smell—and for such people, their experience of roses, bananas, or both is surely different from my own. Indeed, for a synaesthete, a rose might even smell *square*. On the other hand, for a toadstool, a rose presumably hasn't any smell at all. Evidently such differences in rose smell are due not to the roses but to the smellers.

That this is true prompts a curious related question. What is to be made of the claim of another cognizer to be having a particular sensation of some object? (With 'claim' I mean to suggest Dennett-style heterophenomenological[10] detachment, where narratives about internal experience are evaluated from a third person perspective.) Would having complete information about both the object of sensation and the cognizer (or even just the latter) help us understand the cognizer's sensation? Would it enable us to understand what it is like *for the other person* to experience the sensation? Making sense of the claims of a third party with respect to their own sensations, a special case of the 'problem of other minds', I refer to here as the *third person problem*. An information theoretic approach to the problem motivates the exploration of functional relevance in the next chapter and sets the stage for the full-blown theory of consciousness which occupies Chapter 6.

5.1 What is it Like to be Nagel?

One aspect of the third person problem relates closely to Nagel's famous question (1974) about what it is like to be a bat. The preceding discussion observed that understanding a particular description is not the same as being in a state described by that description; one way around this physical (as opposed to logical) barrier to learning a rose's smell from a description of its effects on noses, for instance, is simply to use the information directly to simulate sensations in an appropriate modality, such as olfaction. But when trying to grasp what it is like *for another person* to smell a rose, new problems intrude, analogous to difficulties with understanding, for instance, what it is like *for a bat* to hang upside down. It is one thing to place ourselves in a state very similar to the state we would be in if real rose particles were stimulating our noses. But it is an altogether different thing to place ourselves in a state very similar to the state *someone else* would be in if rose particles were stimulating his or her nose.

Nagel distinguishes between *perceptual imagination*, the capacity to place oneself in a state resembling the state one is in when one actually perceives a thing, and *sympathetic imagination*, the capacity to imagine being in the conscious state of another being. There is little doubt that while I might perceptually imagine myself experiencing hanging upside down, I could not sympathetically imagine myself *being a bat* hanging upside down. Nagel's preferred explanation for this apparent obstruction includes at least two components. First, so the story goes, I cannot even *conceive* of bat phenomenology because it is too alien to anything I have ever actually experienced. (How we're to know this in advance of actually experiencing it, I'm not sure...) Moreover, I apparently cannot surmount the difficulty by understanding any physical description of bat physiology or behaviour or anything else because such a description necessarily misses out what it is I'm actually after: the 'subjective character' of what a given experience is like for the bat. But an explanation along these lines is tantalizingly unsatisfying.

The matter of missing out the subjective character should be the first to trigger alarm bells. Consider an alternative: 'What is it like to be a human being hanging upside down?' This question clearly yields to the kind of strategy applied earlier to rose smells. On this view, the matter of missing out subjective qualities depends on the relationship between the subject reading the description and that which is described. To the extent that I cannot apply a physical description of a bat's experience of hanging upside down to place myself in roughly the same state as the bat, Nagel is right that something about the bat's experience gets lost. However, if it were possible to translate a physical description of a bat's experience of hanging upside down into bat-readable form, I imagine it might be just as comprehensible for a bat as a human-readable description of a human's experience of hanging upside down is for me. The difference has to do not with the quality of the description but with the differences between a bat and a human. Of course, the difficulty really grows out of something like the first part of Nagel's explanation: the reason a physical description doesn't capture, *for me*, what it is like for a bat to hang upside down is not merely that bat *phenomenology* is too alien to what I can experience, but that bat *physiology* is too different from what I can instantiate. These two components of Nagel's explanation are like bits of an iceberg floating above the water, two ideas unified by the larger underlying framework of an information theoretic approach.[11]

On the surface, it seems at least some instances of the third person problem, where attention rests on similar organisms rather than drastically different creatures like bats, ought to be free of such considerations. For example, suppose someone *very* physically similar to me (perhaps my physical near duplicate) claims to be smelling a rose, and suppose I know that when I am in the same physical state, I smell a rose. Surely then, we might think, I am justified in believing they're experiencing a rose smell. But should it be so easy? Just on the basis of what has come so far, it isn't; fortunately, though, the exploration isn't finished yet.

5.2 Explanation—Good Arguments and Bad

A traditional argument attacks the inference from the existence of uniform correlations between my own physical states and my phenomenal experience on the one hand to the existence of similar correlations between someone else's physical states and their phenomenal experience on the other. The view supposedly takes support from the maxim that proper application of scientific explanations never permits extrapolation from characteristics of a single unique phenomenon to other, as yet unobserved, phenomena. Only after building a catalogue of many similar observations, received wisdom dictates, can we predict anything about other relevantly similar cases—or even understand what would make other cases 'relevantly similar'. Along these lines, perhaps my observing *my* correlations between phenomenal experience and physical states—the only such correlations which I can experience first hand—affords me no justification whatsoever for inferring the existence of similar correlations in other people. Almost exactly this argument is taught with a straight face when introducing undergraduates to the 'problem of other minds', and often it seems the standard preference of physicalist philosophers of mind or cognitive scientists is to sidestep it rather than to address it directly.

It is a poor argument which mysteriously elicits more respect than it should. Here's another maxim about scientific explanation: we expect coherence between whatever explanation correctly applies to any single occurrence of an event and whatever explanation correctly applies to any other previously observed events, regardless of whether these past events are directly related to the one in question. For instance, to the extent that we are confident in the relative 'correctness' of quantum electrodynamics—i.e., in the congruity between predictions of QED and subsequent observations—we do not expect a good explanation of gravity or a good explanation of snowfall in the Swiss Alps to contradict it. Moreover, the subtle reciprocal relationship between explanation and theoretical unification[12] of similar events suggests we ought to expect the *same* explanation to apply equally to one single event and to other relevantly similar events. We don't expect that one theory explains all experiments with water, *except* when they involve some otherwise indistinguishable water available only from a Fountain of Youth somewhere in South America, which requires a different theory; explanatory theories should be both time-independent and location-independent (except, obviously, insofar as time or physical location figure in a theory as variables).

So suppose some class of theories explains why my sensations correlate[13] in the way they do, *ex hypothesi*, with my physical states. (Chapter 6 explores the supervenience relation such a theory might involve and addresses more critically several issues of explanation within different supervenience frameworks.) Far from resigning myself to scientific impotence on the grounds that I've really only one experimental situation (or *ex post facto* body of data), I should happily conclude *at a minimum* that at least some of these theories apply similarly well to other physical systems relevantly similar to me. Just as with any scientific endeavour, we should expect that the correct explanation of a single observed phenomenon, whatever form it might take, does its job for more than just the single observed phenomenon at hand. Equivalently, when extrapolating accounts of unobserved systems on the basis of one or more observed ones, we expect continuity over at least some surface in the variable space of the type of system under consideration.

To see the point from another direction, suppose some class of theories accounts for all there is to know about all the *other* brains and minds, if any of the latter exist, in the entire cosmos except for mine, about which we remain for the moment agnostic. Of course, someone always might argue that the class of theories accounting for all the other brains and minds is empty or that all the theories account for minds as entirely separate entities unrelated to brains or some such. But such a position amounts to *assuming* that we won't find what we're after anyway, and thus it isn't much of an *argument* to that effect. So assume for the sake of argument that there *is* some non-empty class of theories specifically relating mind and brain. Then to the extent that *my* brain is similar to the other brains, we should expect that I also have (or lack) a mind in accordance with the same explanatory theory which accounts for the features of all the other organisms with brains. Since the other organisms' lacking a mind would obviously imply *my* lacking a mind, and since I know that I don't lack a mind, the inference is immediate that the other creatures possessed of similar brains also don't lack minds.

Notice, incidentally, that all these characteristics of explanatory extrapolation in scientific investigation remain independent of whether in the end our preferred explanation of the first person correlations comes down to Cartesian dualism or whatever. Requiring only that there be *some* theory uniformly relating brain and mind demands no particular allegiance to material monism: if we are prepared to admit Cartesian dualism or any other *ad hoc* premises into our understanding of the relationship between neural activity and sensation in the first person (or the third), then we ought to be similarly prepared to extrapolate it to our understanding of that relationship in the third person (or the first). Provided that *there is some theory*, whether or not we know what it is or even how it might look, the extrapolation ought to be uncontroversial.

It remains to see what 'relevant similarity' might mean. It would be convenient if an answer could be found specifically in information theoretic properties of cognizers, rather than just directly in their physical states. An answer to the question in information theoretic terms would be nice: 'What makes *my* rose smelling state like (or unlike) that of someone else?'. An opportune answer would be that information transformation at some particular level (or levels) of description holds the key for deciding whether two cognizers are relevantly similar and for addressing the third person problem of evaluating another cognizer's claims of conscious sensation. But in any given system as complex as a human being, vast quantities of information are created, transformed, and destroyed every second; surely not all of it can be relevant to conscious sensation! The next chapter sets out to describe the level (or levels) at which cognizers can fruitfully be understood in functional and information theoretic terms; this provides the first of the tools to begin building a theory of consciousness.

As before, readers who prefer a little less formality may wish to read the following two chapters in reverse order. While Chapter 6 will be less comprehensible without the technical details of Chapter 5, seeing a little of the general gist of the later chapter may make the time required to understand the earlier a more attractive investment.

NOTES

[1] Jackson's 'knowledge argument' is often taken as an argument against the logical supervenience of facts about consciousness on physical facts, but it is nothing of the sort. Supervenience returns in the case made in Chapter 6 for understanding consciousness on a cognitive basis.

[2] Expansion of the set of propositions describing rose properties through arbitrary conjunction ('some roses are red *and* Rome is in Italy') presents some difficulties for simple renderings of this idea. If required, which it is not for this particular example, 'all physical information' for a given physical system might be grounded in the system's complete wavefunction and its implications under the laws of quantum mechanics. (See Chapter 7.)

[3] These are the special cases for which software 'drivers' are written, acting as intermediaries to allow other software programs to communicate with particular pieces of hardware.

[4] I've taken no great care to craft an absolutely watertight analogy between Maggie and MacMary. A better analogy might instead employ a book all about the physical processes which occur when a user selects an operating system command for formatting a disk. This is closer to the book all about the physical processes which occur when a botanist selects a rose and puts it under her nose. Likewise, direct stimulation of nerves or of chip control lines does involve a *small* amount of new information describing, for instance, where the nose nerves and control lines are physically located. But the important feature of the analogy remains the lack of communication between the book typed into the word processor and the chip-level disk controller signals on the one hand and the lack of communication between the parts of Margaret for reading books and her parts for smelling roses on the other.

[5] Nothing too significant should be read into this loose equation of experience with particular states. Section 3 of Chapter 6 suggests it is nonsense to suppose it is like anything at all to

be in a particular instantaneous physical state and explores why conscious sensation is more appropriately linked with *changes* in state.

[6] As an aside, note that on a physicalist view, experiencing normal visual sensation of some object also requires a transition to a new state—*as does blindsight experience*—but the two rather obviously require *different* underlying state changes.

[7] For synaesthetes, it is the words *themselves* (rather than sentences) which have smells or tastes or textures, but the important point is just the automatic cross modal evocation of a sensation. If synaesthesia is possible, then it ought to be possible for *combinations* of words (such as descriptions) to correlate systematically with induced sensations matching those which would be caused by the object of description.

[8] As Dummett (1976, p. 270) observes, incidentally, Quine's radical critique of the *a priori* in 'Two Dogmas' was largely reversed two decades later (Quine 1970).

[9] For the view that we *must* assume persons are rational to attribute to them beliefs, desires, and other intentional states, see Dennett (1978, 1981). Here, I prefer the perspective of Stich (1994), who suggests that attribution of intentional state requires presupposing only that the persons under consideration are *similar to ourselves*.

[10] At ten syllables, 'heterophenomenological' ties with 'electroencephalographical' for first place on my list of most long-winded but reasonably useful cognitive science terms!

[11] Without the information theory, I first explored this question in a very primitive paper from which the title of this subsection is stolen (Mulhauser 1993). Understanding points of view is not *quite* the same as solving the third person problem, but the two relate very closely, and progress on the second may shed at least a little light on the first. These issues receive more thorough treatment in section 4 of Chapter 6.

[12] Recall that quantifying the unifying capability of scientific theories in terms of the compression of observational data they make possible was the original purpose of Solomonoff's work in algorithmic information theory; see page 1. See Kitcher (1989), for instance—especially pp. 477-505—on the view that explanation *is* unification.

[13] Section 1 of Chapter 5 explores the slipperiness of mere *correlations*; it would be better to phrase the argument in terms of supervenience, but for now the point can be made loosely.

CHAPTER FIVE

Functional Systems

The previous chapter suggested some basic ways information theory might augment the standard logical context for understanding problems in philosophy of mind and cognitive science. The processes of information transfer made possible by physically instantiated descriptions of objects, for instance, turn out to be more important for some purposes than the logical implications of those descriptions. The discussion indicated how, in the first person case, it is possible to understand what something is like as an object of sensation on the basis of its physical description. The problem of extrapolating that understanding to the third person case and of making sense of third person claims about conscious experiences, however, raised the question of what properties make cognizers relevantly similar. How do *I* know that *Mary* is sufficiently similar to me in enough of the right ways that I should believe her experience of red is anything like mine—or like anything at all?

To understand relevant similarity, this chapter adopts a roughly functionalist stance, taking up the problematic question of what features justify labelling two systems functionally equivalent. After surveying briefly some well known problems with ordinary functionalism, we set out some goals and requirements for a new approach. With these goals in mind, section 2 extends the complexity measures of Chapter 3, and section 3 describes a precise and objective theoretical framework in which to understand functional modules and their corresponding levels of description. Finally, section 4 comments briefly on issues including the debate between connectionists and symbolicists, natural kinds, and the 'language of thought' hypothesis. An information theoretic solution to the question about Mary's experience of red and an explanation of why it feels like *anything at all* to be particular functional systems await in the next chapter.

1. LIBERAL FUNCTIONALISM

By most lights, functionalism of one flavour or another commands by far the widest audience of any basic approach to philosophy of mind or cognitive science. Developed initially largely in response to the shortcomings of undiluted behaviourism, functionalism might be thought of as behaviourism extended to internal psychological states. Originally championed, and later rejected, by Hilary Putnam (1960), early functionalists took internal psychological states as direct analogues of the internal states of a Turing Machine. A machine table indicates, for any particular input and present internal state, an output and a transition to a new internal state, thereby fully specifying the behaviour of a Turing Machine. Likewise for internal psychological states: all that was thought important about a psychological state was entirely captured by the machine table analogy. Their significance had nothing to do with how they felt, or 'what it was like' to possess them, but was instead wrapped up with their rôle in determining behavioural output and governing transitions to new internal states.

Later, functionalism grew much more sophisticated. Probabilistic automata replaced Turing Machines, and the idea of simply correlating single monolithic psychological states with single internal machine states subsided while the notion persisted that a psychological state's functional rôle in ultimately determining behaviour remained its single most significant feature. Contemporary renditions of functionalism typically add some notion of *structure* to the basic state transition framework—accommodating composite, or vector, states which may describe causally interacting internal components or subsystems. Complex cognitive tasks and behaviours are understood to arise through the interactions of these more fundamental modules, and complete decomposition into a set of straightforward, easily analysed, and thoroughly non-sentient modules is the eventual explanatory aim.

A side note: the question of relating 'mental states' to functional states gets put off until the next chapter, and for now the distinction between functional *analysis* as an approach to psychological *explanation* and functional*ism* as a view of what mental states *are* is not an issue. To make any sense of this chapter, it is crucial to focus on the task of describing systems in functional terms specifically *without* any prior conviction as to how a 'mental state' should be understood. (Mental states stage a comeback beginning on page 109, in section 3 of the next chapter.)

1.1 General Trivialities

The attraction of explaining complex cognition in terms of an array of simple modules, each of which admits, in principle, of full description with physics or neuroscience or whatever is powerful. But well known difficulties threaten any simple account of what exactly a functional (or computational) system is or of how to understand what functional system any particular physical entity might instantiate. In different ways, Block (1978)[1], Paul Churchland (1981), and Putnam (1988)—the latter now having entered a post-functionalist phase—have each drawn attention to the vacuity of simple functional claims, and from Kripke (1982) it is clear that even under very stringent interpretational constraints, we still may not be able to determine the class of functions a real finite physical system computes. More recently, Searle (1990) has suggested his wall could be interpreted as implementing a word processing program.

The gist of the criticisms arises from shortcomings in the naïve notion that a physical system instantiates a particular functional or computational system when component states of the physical system can be correlated with component states of a functional system (either real or abstract) in such a way that causal transitions between states of the physical system mirror functional or computational transitions in the correlated system.[2] The appeal here is *not* to the quantum mechanical notion of states of two physical systems correlated by virtue of an interaction; 'correlation' in the present context means nothing more than *correspondence*: some function maps states of a physical system to states of another physical or abstract system. The mapping is a sort of 'translator' which permits interpretation of states of one system in terms of the other. We might be forgiven for thinking that when we can reliably interpret states of a physical object as states of some abstract functional system, that physical object is (or implements) just such a functional system—even though the details of 'translating' might make it difficult on first glance to pick out the system as a functional one.

But the trouble is that such translation functions are very thick on the ground. The function might be extraordinarily complicated. In fact, the information content of some might exceed that of either the system in question or the candidate functional system. But for almost any two systems there will be one. Recall the observations from the section 'Zombies, Functions, and Mathematical Mysticism' in Chapter 2 and especially the easy availability of infinitely many functions relating suitable sets of numbers. In the present context, given two finite numbers characterising the states of two systems, there always exists a function mapping one to the other. *Obviously*, such a function might be *gerrymandered*, to use a term common in the philosophy of science literature, but that doesn't mean there won't be one. The very easiest of examples appears schematically in Figure 7, showing the evolution along vertical time axes of two different systems each with two degrees of freedom; mappings between the two phase trajectories are easy to come by.

Since trajectories of each system intersect a given constant-time plane at a unique point, establishing a function between the two is a 'simple' matter of mapping those intersections for one to intersections for the other. The continuous time axis for one system may be dispensed with if we're interested, for instance, in a correlation with an abstract computing device defined with a *merely ordered* time index. In that case, the above type of mapping (which would no longer be bijective) simply gets replaced with a one-to-one mapping between states of the computing device and whole *classes* of states of the physical system. The existence of a translation function in no way depends on details of this simplistic example, and the problems don't disappear if we appeal to counterfactual information about how the temporal evolution of a physical object *would be* different under other conditions (alternatively, how its trajectory looks under perturbation). As it happens, appeals to counterfactuals open a Pandora's Box of other troubles even more difficult than that of characterising functional systems. This point returns below.

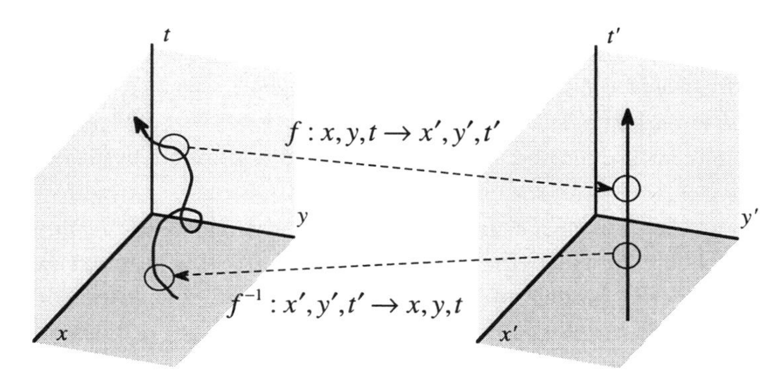

Figure 7. 'Translation' Functions are Cheap

Notice that the existence of a translation function remains entirely independent of the fact that the mutual information content of two systems may be essentially nil. That is, it might always cost just as much to compute the behaviour of each system individually than to compute one given a description of the other (or to compute the two together). For two arbitrarily 'translatable' systems A and B, it may well be that

$$I(A) \quad \approx \quad I(A \mid B)$$

Whereas, anticipating the ideas of the next section, we may reasonably expect that 'good' translation should be available only between two systems which aren't algorithmically independent, so that

$$I(A') \quad > \quad I(A' \mid B')$$

to some significant degree.

In any case, all these complications surface straightforwardly as symptoms of decades-old problems with understanding the character of natural laws and our descriptions both of the laws themselves and of the behaviour of physical systems in accordance with them. Why should a gerrymandered translation function arbitrarily relating a physical system to an abstract computational one be any less appropriate an explanation or description of the system than some simpler one (perhaps which fails to relate it to computation at all)? These difficulties have lingered in various forms at least since the inception of the central 'covering law' model in Hempel and Oppenheim's (1948) seminal paper—although really they may be traced back to David Hume's scepticism about 'necessary connexions', first aired more than two centuries ago in section VII of his 1748 *Enquiry Concerning Human Understanding*. (On the Humean background, also see Scheffler 1963.) Ernst Nagel (1961) explores the issue in minute detail but offers no definitive solution, while Rescher (1970) takes the enticing view that natural laws are mind-dependent and do not reflect objective features of nature.[3] Skyrms (1980) opts for what has become one of the most prevalent approaches among functionalists who care about such things: pragmatics. The remarkable lack of cross-fertilisation between philosophy of science on the one hand and cognitive science or philosophy of mind on the other, however, never fails to amaze; many in the latter fields (including myself, I hasten to add) go about their business with only minimal appreciation of developments in the former. One result is that the same non-solutions to this problem are offered, sometimes with a good deal of fanfare, over and over again. Another result is that many theorists merrily carry on with their work seemingly unaware that any difficulties even exist.

1.2 Particular Trivialities

An example of the latter is David Chalmers, who blithely appeals to counterfactual conditions in asserting that functional organisation cannot arise in an arbitrary system like a pail of water except by hugely improbable coincidence (Chalmers 1996b, p. 252)—an assertion he intends to be underwritten by his formal framework for implementing a combinatorial state automaton, or CSA (p. 318). Unfortunately, the CSA criterion, which demands one version of the usual mapping from physical states, physical I/O, and causal transitions on the one hand to formal states, formal I/O, and formal transitions on the other, is trivially satisfied by vastly more physical systems than Chalmers seems to realise, despite his stipulation that the mapping hold counterfactually.

The most glaring problem is that Chalmers requires only that *if* the physical system is in a certain state (to be matched with an analogous formal state), *and* it receives a certain input (to be matched with an analogous formal input), *then* this causes it to enter a new state in a transition which mirrors that of the formal system when it receives the analogous formal input in the analogous formal state. The immediate difficulty is that nothing requires that the real physical system ever *actually* is in a certain state which maps to a particular formal state or that it ever *actually* receives a relevant input. For the case of a bucket of water, for instance, there is nothing at all incoherent in supposing that water molecules *could be* arranged in such a way that, in response to some particular perturbation interpreted as the input 'what is your name?', they interact so as to produce an output (say, in patterns of ripples) interpreted as 'my name is Bucky'. But surely this counterfactual oughtn't be relevant to whether the bucket of water actually *is* a functional 'Bucky'-answerer. Maybe this is easier to see when considering a bucket of brains instead: given that *my* brain is arranged to produce sensible outputs when it receives 'what is your name?' inputs while in particular states, surely a *bucket* of well mixed brain bits could also be so arranged, even if it never *in fact* gets so arranged. *Obviously*, such an actual arrangement of molecules might be very unlikely to emerge spontaneously (at least, on a naïve interpretation of how states and so on are to be defined, but see below). But Chalmers's CSA criterion requires only that *if* the system is in a particular state mapping to a particular formal state, *then* it reliably behaves in a way mimicking the CSA, a conditional which is satisfied vacuously (i.e., in virtue of a false antecedent) by numerous physical systems for vastly many CSAs. And the problem clearly cannot be discharged in the general case merely by stipulating that a system is, at some particular time, *in fact* in some physical state which maps to a particular formal state.[4] Strangely, Chalmers acknowledges this shortcoming neither in the book cited nor in his attack (Chalmers 1996a) on Putnam's (1988, pp. 120-125) proof that every open system realises every abstract finite state automaton.

The CSA criterion itself is infested with other problems less relevant to immediate purposes but interesting nonetheless. For instance, Chalmers's appeal to counterfactuals in constraining CSA implementation ignores sticky muddles first exposed half a century ago by Chisholm (1946), Goodman (1947a, b, and 1955), Kneale (1950), and subsequent authors. Chalmers professes not to see any problems for his CSA counterfactuals deriving from indexicals and echoes Carnap (1947a) in his debate with Goodman (continued in Carnap 1947b), asserting (personal communication) that he just doesn't allow such indexicals in his descriptions of physical states. Chalmers also ignores arguments by the likes of Maudlin (1989) against involving counterfactuals in such contexts in the first place. (As Maudlin might protest, why should counterfactual capabilities of a cognizer, never displayed in the actual world, bear in any way on our understanding of its experience or cognitive structure in the actual world?)

Finally, Chalmers leaves unaddressed the possibility of vacuous satisfaction of his CSA implementation criterion by way of physical system decompositions chosen so poorly that it turns out the physical system simply cannot receive the relevant input when in the relevant state. As above, this is consistent with the requirement that *if* it could receive such input, it would behave accordingly. And of course, simply *asserting* that there are no such troublesome decompositions for a particular CSA specification amounts to an assertion about the infinite space[5] of all decompositions. (In fact, see page 97 for a note on potential formal decidability problems with Chalmers's criterion. See Mulhauser 1996[6] for notes on trivial CSAs and other problems with the book cited.)

In a recent paper, Hardcastle (1995) seems more aware of the subtleties of the standard problems with functionalism, and her exposition is well informed by the relevant literature. She concedes that "there (may) be no principled way of deciding which natural number recursive function best captures the activity of some system" (p. 306), justifiably rejecting Stabler's (1987) suggestion that the functional status of a system be interpreted against a backdrop of 'normal conditions' on the grounds that apparently no principled strategy exists for determining the class of 'normal conditions'. In the end, she adopts a pragmatic approach, pointing to "rough agreement that we need a pragmatics of explanation" (p. 307) and citing Hempel (1966), Kitcher (1989), Railton (1981), and van Fraassen (1980) for support. Curiously, however, Hempel (1965, p. 338) is very clear on the limitations of pragmatics, while the Kitcher essay Hardcastle cites actually includes a section (pp. 415-417) called 'Why Pragmatics Is Not Enough'. Elsewhere, Kitcher and Salmon (1987) argue convincingly that van Fraassen's pragmatic method is trivial. Unfortunately, the approach Hardcastle ultimately recommends in terms of an 'interpretation function' and two narrowing conditions—one of which, ironically, is that we assign the computational function to a system in order to *explain* its behaviour—suffers all the same traditional difficulties. The approach leads nowhere that philosophy of science hasn't already been and found lacking.[7]

1.3 Teleofunctionalism

One increasingly fashionable reply to the 'troubles with functionalism' (borrowing Block's 1978 title) is *teleofunctionalism*, a view which shifts the very meaning of the word 'function' so as to underwrite a contrast between a component's historically relevant biological function and its merely accidental side effects. Debates about function may then be adjudicated on the basis of an organism's phylogenetic history and the teleological—or, more properly, *teleonomic* (see Pittendrigh 1958 and Williams 1966)—rôle of a module in that history. In other words, the approach recommends decomposing an organism into modules and determining the precise functions of those modules by appeal to an organism's evolutionary history.[8] Up to a point, the strategy makes terrific sense. From a practical perspective, where we are able to identify modules common to a line of organisms, it's at least plausible that *those* modules should figure in a functional account. For instance, it seems a safe bet on this view that the human optic nerve, as an abstraction from the particular details of neural wiring in individual humans, is an appropriate module to include in some variety of functionalist account of vision. However, neither my *particular* optic nerve details, nor optic nerve-plus-brown hair, nor optic nerve-before-midnight, nor optic nerve-of-exactly-10-centimetres, should figure in a functionalist account, because none of them played the right rôle (or *any* rôle, probably) in the history of natural selection. Likewise, on this approach, the function of my optic nerve is to conduct electrochemical signals produced by my retina and not, say, to consume oxygen.[9]

But teleofunctionalism amounts primarily to a methodology for exploring questions about what some component of an organism is *for*—i.e., why it is there, or why it was selected during the course of phylogenetic history—not questions about how such a component *works*. Teleofunctionalism is an extremely valuable tool for any project to naturalise the study of mind, but the teleonomic or etiological notion of function to which it appeals addresses a very different sort of question than those discussed earlier in this section. Teleofunctionalism does not in any way solve the problems with functionalism; it simply disengages them by shifting to a different (complementary) underlying definition. In other words, if we are interested in understanding how an organism (or a mind) *works*, teleofunctionalism is the right answer to the wrong set of questions. Attempting to overcome the standard troubles with functionalism by appealing straightforwardly to teleofunctionalism, as is occasionally done—"but I'm a *teleo*functionalist, not *just* a functionalist!"—invites immediate difficulties not only with recasting the questions in a suitable form but also with the inherently temporal nature of teleofunctionalist analysis.

The latter is the great weakness of attempts to recruit the teleofunctional approach for purposes for which it wasn't (or shouldn't have been) intended: it grounds attributions of functionality in facts *historically removed* from the actual system and its contemporaneous environment. Consider some hypothetical system which, for whatever reason, has *no evolutionary history*, not even by virtue of being designed by an organism (such as a human) with a real evolutionary history to call its own. The teleofunctional approach can tell nothing at all about such a system's functional components—except perhaps that it has no functional components. (It's components aren't *for* anything, in an etiological sense.) But there certainly oughtn't be any *logical* requirement that all functional systems or even all conscious systems are products of natural selection (even though the latter, at least, might well be an empirical fact). There is no *logical* reason why my physical duplicate (or Millikan's 'swamp man') couldn't pop into existence tomorrow as a result of some bizarre wanderings of atoms. If teleofunctionalism really were all we had to go on, such a physical duplicate would not be a functional system at all, and, presumably, he wouldn't think—or, if he did, he wouldn't think using the same modules as I do. Clearly there is more to being a functional system than displaying the right evolutionary heritage; systems may be understood both in terms of etiological function and in terms of how they work. It is the latter project which figures in this chapter. It would be convenient to understand functional decomposition, to be understood in the sense of a decomposition which tells us *how something works*, in a way which is at once objective and independent of facts about historical events taking place before a system even existed.

(The two methods do, of course, meet up. I describe in this chapter an objective and 'temporally blind' method for decomposing a system into functional components. It may then be appropriate to ask what, if anything, these functional components are for in the teleonomic sense: in virtue of what behavioural capacity supported by a given functional module did the presence of that module become a selectively relevant factor in phylogenetic progression?)

In addition to objectivity and temporal sensibility, several goals motivate the approach advanced here. The primary aim is to narrow, in a mathematically principled way, the class of functional systems which given physical systems are understood to implement.[10] The method should yield an equivalence relation between systems, such that sense can be made of functional equivalence (or similarity) between distinct systems. It should underwrite useful distinctions between the functional capacities of, say, a bucket of water, a human brain, and a book describing every causal relationship within a human brain. It would also be nice to link functional descriptions to particular kinds of environmental stimuli and motor behaviours. Human beings are almost certainly *not* functional systems with respect to each and every kind of environmental input: the functional breakdown useful for understanding responses to visual stimuli, for instance, is very likely irrelevant for understanding responses to cosmic rays. Likewise, *as conductors of heat*, humans and cows might be functionally equivalent, while *as reflectors of sound waves*, grapefruits might be functionally equivalent to people. Yet, *as responders to the impact of flying hailstones*, all three might function differently. Finally, as a corollary to the goal of locating functional bits of a system, the strategy should suggest an appropriate level (or levels) of description useful for understanding and explaining the behaviour of cognitive systems. The strategy should be flexible enough to accommodate functional modules picked out at *different* levels of description. The exploration begins with some extensions to algorithmic information theory.

2. LOGICAL DEPTH MEASURES

Recall from section 2 of Chapter 3 the definition of an object's or bit string's algorithmic complexity as the length, in bits, of the shortest Turing program which generates it. Here three extensions to that fundamental measure of complexity expand the utility of information theory. The first, called *logical depth*, is due to Charles Bennett, Chaitin's colleague at IBM's Thomas J. Watson Research Center. (The measure was originally discussed in print by Chaitin 1977; also see Bennett 1982, 1987a, 1990.) Where algorithmic complexity is the length of the shortest program for describing an object, logical depth is the *exe-*

cution time of the shortest program.[11] Bennett originally intended that "a string's logical depth should reflect the amount of computational work required to expose its buried redundancy" (Chaitin 1977, p. 358, quoting an unpublished manuscript), but this description is misleading: the concern is not the computational effort of *finding* buried redundancy in a string, but the computational effort of *producing* that redundancy from its most concise description.

In other words, the central idea of the measure is that the structure of a logically deep string should reveal its roots in a long computation or dynamical process. Logical depth is the first of the complexity measures motivated partly by interest in what kinds of *processes* might yield an object. Restricting attention as before to self delimiting programs, recall (from page 39) that the information content for most strings, where x_n denotes a string of length n, satisfies:

$$\begin{aligned} I(x_n) &= n + I(n) + O(1) \\ &= n + O(\log_2(n)) \end{aligned}$$

Thus, the shortest program for producing most strings, the algorithmically random ones, is one which just prints them out. And the execution time of such a program, for large n, is dominated by the task of simply outputting those n bits already included in the program itself. Logical depth thus takes minimum values when the algorithmic information content is either maximal for a given string length—when the program includes many bits but very simple output tasks—or minimal for that length—such as when the program includes only a few bits simply to be repeated with no extra computations. The *variation* in logical depth, however, is also greatest for the class of strings with minimal information content for their length. Figure 8 shows schematically a very roughly estimated relationship (which obviously cannot be computed directly) between logical depth, plotted vertically on a logarithmic scale, and the difference between a string's length and its information content.[12] The curve for maximal logical depth is *very* steep indeed, since it is closely related to the function $a(n)$ mentioned on page 48, defined as the greatest natural number with information content less than or equal to n (and which grows more rapidly than any computable function). As noted before, an n-bit program may produce a truly huge number after an even larger number of computational steps; as a result, logical depth skyrockets as x_n approaches the value of $a(n)$.

Thus, except for maximally random strings, logical depth usefully distinguishes between individual strings all of which may have the same algorithmic information content. (The only catch, of course, is that most strings *are* maximally random.) Deutsch (1985) justifies a slight modification to logical depth with the observation that in Nature, random states are produced not by 'long programs'—i.e., by processes which actually include descriptions of their random output—but by short programs exploiting nondeterministic hardware. Deutsch suggests a measure called quantum logical depth, or Q-logical depth, keyed to the execution times of the shortest programs for his own Universal Quantum Computer. The quantum analogue of logical depth reflects the fact that very short programs for the quantum mechanical version of the Turing Machine generate random sequences very quickly by exploiting probabilistic features of quantum mechanics. In this regard, however, Q-logical depth provides no significant advantage over ordinary logical depth measured, for instance, with respect to a classical Bernoulli-Turing Machine—a Universal Turing Machine equipped with a random number generator. After all, the bulk of computing time for large n is still taken up with the details of actually outputting the desired string. Allowing for the option of emending logical depth to reflect execution time on a Bernoulli-Turing Machine, differences between it and Q-logical depth are irrelevant for this context.

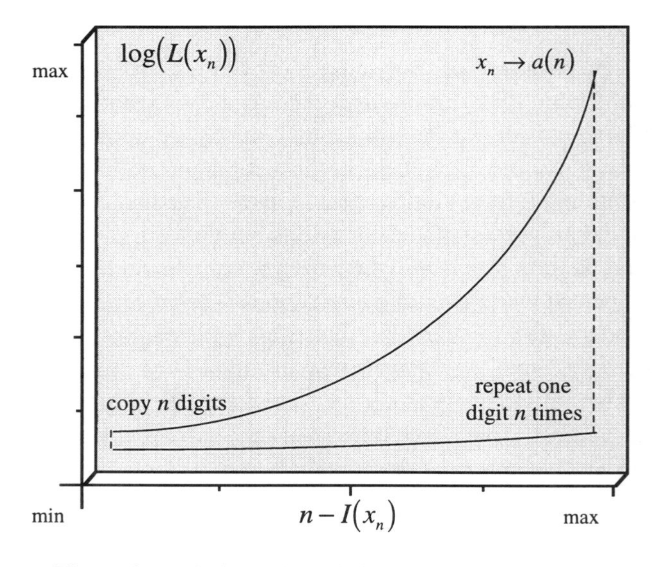

Figure 8. Logical Depth Variation vs. Extra String Length

For present goals, the greatest limitation of both logical depth measures (and of ordinary information content) is their focus on individual bit strings. It would be helpful to consider more than single strings; for purposes of analysing cognitive systems, which may produce countless new outputs every second (or continuously variable outputs), it would be nice to quantify the complexity of the actual *processes* yielding those outputs. At first glance it might seem that a good way to measure the complexity of such a process would be to examine the algorithmic information content of the shortest description of its input/output relationship. But of course this measures the complexity of the relationship interpreted *as a string*, not *as a process*. That is, it reflects only the complexity of a description of a process, not of the process itself. Of all the minimal n-bit descriptions, for instance, some describe actual processes which demand huge resources, such as computing deeply nested factorials using an input number as a seed, while others describe very simple processes which are comparatively cheap computationally, such as adding a particular almost-n-bit random number to an input number. Of course, measuring the logical depth of the shortest description of a relationship doesn't help either, since all n-bit minimal descriptions have almost exactly the same logical depth. (This is because any n-bit minimal program, itself being a random number, can be produced by a slightly longer program which simply prints it out.)

Instead, we measure the *functional logical depth* of a relationship or process as the mean execution time, over some set of inputs, of the minimal self delimiting Bernoulli-Turing Machine program which actually produces the outputs of that process, to within some specified degree of precision. (The measure first appeared accessibly, complete with many errors, in Mulhauser 1995b; the modified treatment here is preferable.) When a relationship is instantiated physically by some system, this means it is the average over a set of inputs of the length of time taken by the shortest program (or the harmonic mean of the average times taken by all such programs) to produce outputs roughly identical to that which the system in question produces in response to the same inputs. For example, consider a system yielding outputs in some range when given inputs in a domain specified in x and y, together with a Bernoulli-Turing program $S([x], [y])$—where $[x]$ denotes a minimal description of x—which reflects its overall input/output behaviour. S reproduces, at some level of precision[13] ε, appropriate strings from the system's output range when given strings from its domain. (Note that different ε might demand different S, since the shortest program to generate very low precision outputs, for instance, might differ markedly from the shortest program for generating very precise outputs.) We call $E(S, \varepsilon)$ a 'pseudo-function' which returns the execution time for S working to precision ε. Then, stipulating a simply shaped rectangular input domain and pretending the values returned by E form an easy manifold, the functional logical depth, or F-logical depth, of the system whose input/output relationship for that level of precision is given by S might be phrased as:

$$F(S,\varepsilon) = \frac{\int_{y_0}^{y_b}\int_{x_0}^{x_a} E\big(S([x],[y]),\varepsilon\big)dxdy}{(x_a - x_0)(y_b - y_0)}$$

Of course, it's a little silly to express it this way, since we have no idea what the execution time manifold will look like, and since we're usually interested in discrete domains and summation rather than continuous ones and integration, but it helps illustrate the central idea. The equation indicates that we integrate over the input domain to find the volume under the execution time manifold and then divide by the area of the domain to find average execution time. Graphically, as in Figure 9, this is a process of scanning across an (x, y) plane and plotting a height along an execution time axis corresponding to the time taken to produce outputs for each point in the domain. The two integrals give the volume under this manifold, from which we derive the average height by dividing by the area of the domain (given in the denominator).

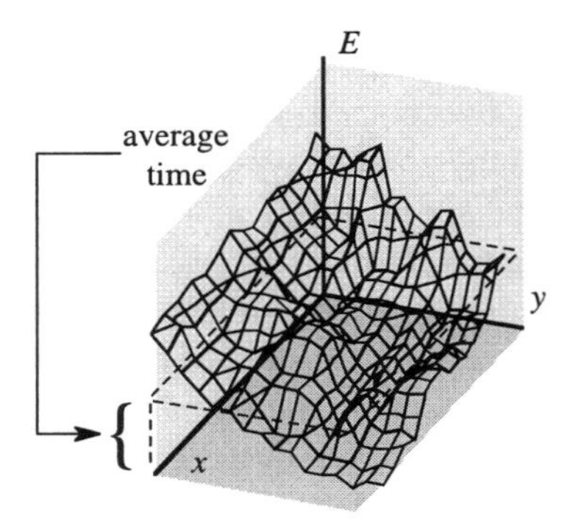

Figure 9. Execution Time Manifold—Not Really

A few minor points deserve attention before some more significant observations about F-logical depth and its applications. First, for nondeterministic systems, it is more useful to require not that S produce individual outputs identical to that of the real system, but merely that its probability density function matches to within some level of precision related to ε. It is clear, incidentally, that the utility of functional logical depth distinctions for all systems, deterministic or otherwise, depends strongly on the value of ε. At low precision, not surprisingly, all processes look F-logically shallow. Second, notice that comparisons of individual systems ignore the time it might take otherwise output-identical systems to produce their outputs, since F-logical depth measures the complexity of the relationship itself and not whether one system is quicker or slower at instantiating that relationship. (Of course, if temporal features are specifically of interest, we may simply include a temporal dimension in the input domain, the output range, or both.) Thus, F-logical depth is a measure behaviourists would like: it doesn't necessarily reflect the *internal* complexity really included in a particular physical process, but only the complexity of the *overall* input/output relationship (or 'behaviour') of that process. So, any two processes which instantiate the same relationship at a given level of precision are, at that level and with respect to the given domain and range, equivalent in terms of functional logical depth, regardless of the peculiarities of what actually occurs in them internally. Finally, perhaps the clumsiness of the pseudo-calculus above is to be excused in light of the fact that, in requiring *minimal* Bernoulli-Turing programs, functional logical depth is no more a computable value in the general case than algorithmic information content, logical depth, or Q-logical depth.

Several other features of the functional logical depth measure appear from the formalism quite straightforwardly. For instance, the measure returns minimal values when no useful function reliably relates a unique input to a unique output—such as in cases where outputs in response to that input are not unique—and when there is, moreover, no useful probability relation between the two. (Here 'useful' describes a function which can be specified succinctly enough to be included in the shortest description of the overall input/output relationship for a given level of precision.) In such a case, S can simply offer appropriately scaled values from its random number generator and be done with it. Similarly, F-logical depth takes minimal values for trivial relationships; the shortest program to mimic the relationship might simply output the input number again or some such, according to the details of the trivial system in question. More interestingly, other things being equal, functional logical depth is affected by the degree of mutual information content between correlated input and output values.[14] When relatively few input/output pairs are algorithmically independent (i.e., when most have significant mutual information content), the shortest program for instantiating the overall input/output relationship may use the input values for help in producing the output values, resulting in a corresponding increase in execution time relative to a program which simply produces outputs by, say, looking them up in a table. This coheres with the intuitive requirement that those systems which actually act on their inputs in order to produce outputs are more complex, as compared to others which simply use them to select, without much calculation, some output from a set of options. This observation also hints at the fact that questions about functional logical depth are more meaningfully asked with respect to finely grained or very large input spaces. Small or coarsely grained input spaces limit the number of corresponding output classes—which are of course single member classes in the deterministic case—with the result that it may be shorter and quicker to exploit a simple lookup table than to undertake real calculations. (For instance, consider a process which computes the nth bit to the right of the point in the binary expansion of π. The most concise program reproducing this relationship with respect to a domain consisting only of $\{n = 8\}$ simply outputs '1' rather than actually calculating the expansion and returning that digit.) This is unsurprising: with respect to just a few inputs and outputs, even the world's largest supercomputers instantiate only the simplest of overall relationships.

At the opposite end of the complexity spectrum are found those kinds of deep processes the products of which something like logical depth might have been suited to pick out were it not for the fact that the shortest program for generating most strings is still one which just (relatively quickly) prints it out. Defining F-logical depth over an entire input domain circumvents a difficulty which arises for ordinary logical depth: returning low complexity for those strings which truly were created by long and computationally intensive processes but whose initial conditions in terms of those processes, together with a description of the processes themselves, require very many bits to specify. The definition of F-logical depth takes as *givens* a minimal specification of the process and a particular input domain; thus it doesn't share this susceptibility to variation arising just from the algorithmic information content of inputs. Likewise, it ignores the sort of deviation which occurs when, for instance, offering typewriters to a million generations of a million chimpanzees really does result in the creation of a highly ordered United States Tax Code (or alternative varieties of mindless bureaucracy). The measure simply indicates the process *itself* is of low depth and makes no claims about the complexity of individual outputs. Individual outputs are better analysed within the framework of algorithmic information theory or ordinary logical depth; the three measures are best seen as complementary approaches each primarily suited to slightly different jobs.

In any case, this exploration of depth measures in general and functional logical depth in particular provides a firm mathematical foundation on which to base a strategy for understanding functional modules in cognitive systems (or in any other physical system, for that matter).

3. CHOOSING MODULES BY MINIMISING COMPLEXITY

As I suggested back in section 2.3 of Chapter 1, a major theme of this book concerns the different levels relevant for describing cognitive systems and the relationships between phenomena occurring at those different levels. The problem of actually deciding on a good level of description for explaining particular phenomena, however, relates closely to the difficulties associated with making sense of functional descriptions. I suggest that the best explanations of at least *some* kinds of phenomena, including cognition—and, in a roundabout way, consciousness—are functional ones (where, again, function is understood in the 'engineering' sense of section 1.3 rather than in an etiological sense). That is, the best explanations of cognition refer to functional modules and the interactions between them in the way that a good explanation of how a clock keeps time refers to cogs and pendulums or transistors and quartz crystals (rather than atoms and molecules). The snag is to decide how exactly to justify the emphasis on functional modules like cogs and pendulums.

I take it that choosing the right level of description for picking out functional modules is not *simply* a matter of coarse graining the lowest level descriptions of particles until we've discarded enough detail to arrive at some higher level where we describe arbitrary collections of such particles. Levels of description and functional or organisational features maintain a subtle reciprocal relationship. Very roughly stated, the present strategy for attacking the problem is to choose the simplest functional modules which can be interconnected in the simplest ways to give a system whose overall input/output behaviour matches that which we're describing. This strategy *begins* with a particular physical system and objectively decomposes it into functional modules, yielding the functional system which it implements; this differs greatly to the method of all other strategies of which I am aware, which begin with both a particular physical system and a candidate formal system and then attempt to adjudicate on whether one implements the other. Fleshing out this general strategy and its consequences is the main aim of this section.

3.1 Minimising Functional Logical Depth

The first step in the strategy is to choose an input domain and corresponding output range with respect to which a given system or class of systems is to be evaluated. Clearly, evaluating systems with respect to different kinds of inputs ought to yield different functional descriptions. As noted in section 1, a human might be one sort of functional system with respect to certain interactions with visible light, for instance, but an entirely different functional system—or not a functional system at all—with respect to interactions with cosmic rays. (Determining functional modules over specified domains and ranges does not, incidentally, make the strategy any less *objective*; it simply makes functional decomposition *relative*.)

Given an overall system (or class of systems) to analyse, together with an input domain and an output range, we choose components of the system to act as functional modules in accordance with the goal of minimising simultaneously the complexity of three relationships. First, we want modules which can be extracted from the entire system cheaply: this requires minimising the functional logical depth of the relationship which takes as input a minimal description of the system and produces a partition of that system into modules, together with associated domains and ranges for each. Next, we minimise the functional logical depth of the input/output relationships for each of the modules selected. Finally, we minimise the complexity of a relationship which as input takes minimal descriptions of the input/output relationships instantiated by the functional modules and yields as output an arrangement of those subsystems together whose overall input/output relationship matches that of the system or systems being considered.[15] That's all there is to the complexity minimisation approach to functional decomposition, but a few things need clarification before continuing.

First, selecting parts of the system as modules amounts simply to partitioning the set of particles in a system into subsets, together with a description for each specifying which states of which particles will be treated as inputs and outputs.[16] Of course, nothing stops our taking entire modules themselves as both inputs and outputs, such as in the case of a module to be treated as a whole coherent dynamical system, and nothing stops our taking inputs as *perturbations* to such a system. It is allowed that modules or their input or output sets may be defined fuzzily or vaguely; there might be no matter of fact, for instance, as to whether a particular atom near the surface of a neuron is part of the neuron or part of the surrounding fluid. (Shortly it becomes clear that such minor details, as long as they are not crucial to the overall operation of the system, get cancelled out in the overall procedure.) The logical depth of this partitioning process is one quantity to minimise. Since selection of modules amounts to a partitioning of the whole physical system, modules automatically acquire some degree of physical plausibility in the sense that they may be constructed only of particles within the system, and they may operate only on inputs which are effected by states, or relationships between states, of particles within them. For instance, no module in my brain can be composed of neurons and moondust, since there is no moondust in my brain; and no modules in my brain can take as inputs radio transmissions from Mars—unless there are particular particles in my brain affected by radio transmissions from Mars, perhaps with the help of radio receivers and speakers and my ears.

Fixing the partition scheme naturally fixes *total* physical interactions between the modules within a given system as well. However, the strategy requires choosing particular input domains and output ranges which contribute to minimising the functional logical depth of the input/output relationships instantiated by the modules—without compromising either the simplicity of the process for specifying them or the eventual reassembly of the individual input/output relationships into an overall arrangement whose input/output relationship matches that of the system as a whole.

Finally, it's important that in considering the complexity of arranging the subsystems back into an overall system, we begin with (minimal) descriptions of the input/output relationships of the modules, not with descriptions of the actual physical entities themselves. The program performing the arranging need only produce from these some matching up of modules' respective inputs and outputs which enables the complete arrangement to reproduce the overall input/output relationship of the system in question. One way of achieving this arrangement might be to specify links between individual Bernoulli-Turing Machines, each of which runs a program for a particular module, while another might be to specify one large program which uses the programs describing particular modules as sub-routines. The only restriction on this overall arrangement is again just that it makes physical sense with respect to the system we're considering: while they may be described at whatever level is appropriate for the modules concerned, 'links' between outputs of one module and inputs of another must be *instantiated* by direct physical correlation—not mathematical correlation—(i.e., by direct causal interaction) between particles in one module and those in another. Thus, it's no good specifying an arrangement which links the retina and primary visual cortex, except by mediation of the optic nerve, optic chiasm, lateral geniculate nucleus, and so forth (or superior colliculus and pulvinar). Likewise, links which reflect information transferred by, say, electric or magnetic fields, are understood in terms of the effects of those fields on the states of particles affected by them.

An intuitive feel for some consequences of this strategy can be had by considering two extreme examples of 'bad' choices of functional modules for some structured object like a human being, understood as a system displaying some range of behaviours over some interesting domain of stimuli. First, consider as modules individual particles in the body, assuming for the example the framework of classical physics. This probably yields modules of very low complexity, insofar as the input/output relationships instantiated by individual particles (taking their states themselves as both input and output) amount to straightforward and comparatively cheap calculations from the laws of physics. Likewise, specifying the individual 'modules' is not too difficult, since the specification program is already given a minimal description of the entire system. However, creating an appropriate arrangement of modules then becomes a daunting task, worse than just describing the entire system again: describing the overall relationship amounts to specifying interactions between trillions upon trillions of programs, reflecting the interactions which may take place between hordes of particles within even a very small volume of the body. At the other extreme, consider the entire body itself as a module. In that case, the overall arrangement, consisting of only one module, is easy. But, as luck would have it, the functional logical depth of that single module then presumably becomes rather large.

What characterises 'good' choices of modules, choices which achieve the minimum possible F-logical depth for the functional modules, their interconnections, and the task of choosing modules in the first place? It's not too difficult to reason about good module choices and about what features of a real system might be preserved or ignored in describing the modules and their interconnections. Perhaps the most important observation is that the strategy automatically ignores any extra information which doesn't play a rôle in instantiating a system's overall response to inputs, whether that information appears in inputs or is generated internally. Other things being equal, the shortest Bernoulli-Turing programs for producing a given module's outputs are those which ignore irrelevant input information. Consequently, although it might sometimes be cheaper to specify inputs on the basis of supersets of the relevant particles, any irrelevancies so introduced will be ignored both by the minimal descriptions of the modules' input/output relationships and by the program for linking them all together. In ignoring the irrelevancies, the strategy ought to select functional modules and connections between them at one—or, more likely, *several*—intermediate levels of description.

3.2 Functional Relevance, Equivalence, and Similarity

This strategy of minimising functional logical depth, then, somewhat formalises the standard goal of explaining a system's overall behaviour on the basis of all and only those internal features which play a functionally relevant rôle in shaping that behaviour. More importantly, it suggests grounds on which particular subsets of the overall system may be treated as functional modules, maximising the simplicity of those modules to the greatest extent possible without compromising the simplicity of their interconnections. The input/output relationships of the chosen modules, together with the interconnections between programs describing those input/output relationships, figure as the theoretical entities in the explanation of the system's behaviour *as a functional system*. We dub *functionally relevant* exactly those modules and interconnections which the complexity minimisation strategy returns for a given choice of input domain and output range. (Notice, incidentally, that since domains and ranges may include temporal dimensions, any module which contributes to a system's overall input/output relationship at one particular time is functionally relevant, even if it plays no rôle in shaping outputs at *some other* particular time.[17]) Constraining the methods for choosing functional modules and interconnections only with the weak stipulation that choices make physical sense divorces a system's functional decomposition, seen as the basis of an explanation of its overall behaviour, from its implementational details—while keeping both that functional decomposition and associated explanation firmly grounded in what the system can physically instantiate. In other words, a functional explanation comes in terms of input/output relationships—mathematical abstractions—but nonetheless the choice of those particular mathematical abstractions is constrained by the capacity of the particular physical system to instantiate them.

When any two systems admit of the same functional decomposition with respect to particular input domains and output ranges—that is, when the complexity minimisation strategy applied to each yields the same *programs* describing modules' input/output relationships and the same overall interconnection between them—they are *functionally equivalent*. Again, emphasising the relationships themselves (properly speaking, emphasising the Bernoulli-Turing programs for matching modules' input/output relationships as well as for stitching them all together) permits abstracting away from the details of particular physical instantiations. Notice that this definition does *not* require that the relationships picking out the modules and input/output sets from physical objects be the same. Thus the actual physical structures of two systems functionally equivalent over some input domain might differ enormously, yet functional equivalence means the computationally cheapest way of reproducing the behaviour of either is in terms of modules performing the same functional jobs in each. Functional equivalence understood in this way is clearly an equivalence relation[18] with respect to functional decomposition. Whether a given system ought properly to be called 'functional' is of no concern here. The important questions about functionalism centre on functional modules and their rôles in producing a system's behaviour, not on stale linguistic issues surrounding what kinds of functional decompositions really deserve the name 'functional'.

Finally, *functional similarity*, in the absence of outright equivalence, is probably best understood in terms of the mutual information content between the complete overall arrangements yielded by the complexity minimisation strategy. Thus, given the definitions of section 3 of Chapter 3, since overall arrangements contain within themselves information specifying their own operation, functionally similar systems functionally represent (and thus also represent) each other. These features of functional similarity account for the intuitively appealing conditional information relationships given above on page 75, contrasting systems which can be sensibly 'translated' and those where translation requires gerrymandering.

The complexity minimisation strategy offers fertile ground for exploring a host of questions in the cognitive sciences and philosophy of science, but not all of its ramifications concern the present enterprise. Here we just take stock of a few of the most salient features and of how they bear on the goals set out at the end of section 1; broader concerns return in section 4.

3.3 Observations and Applications

First, the complexity minimisation framework is both objective and temporally sensible. The strategy requires no reference whatsoever to the evolutionary history of a system or organism; interpretation of a system in functional terms does not rest on any facts about such a history. I and my 'physical duplicate' with no evolutionary background instantiate exactly the same functional system. Functional modules themselves are determined by their contribution within a system to that system's behaviour over a particular input domain. The framework clearly permits functional modules appearing at different levels of description, and the strategy is objective in the sense that it fixes the class of functional systems to which a given system is equivalent in a mathematically precise way; yet, it also accommodates the fact that questions about functional decomposition may demand different answers with respect to different input domains. And while the number of ways of *instantiating* a given functional system might well be infinite, the framework also pares down the class of mathematical descriptions appropriate for understanding a given particular physical system in functional terms from an infinite number of possibilities to only one or very few.[19] The approach requires no extra information about counterfactual behaviour, and gerrymandered decompositions are barred by basic properties of the information theoretic framework. (On the present strategy, *time* replaces the counterfactual; while questions may meaningfully be asked about how a system evolves into other states over time, given its current state, questions about counterfactual current states have no place.)

Section 1 suggested that a good framework should underwrite useful distinctions between the functional capacities of a bucket of water, a human brain, and a book listing all the causal relationships within a human brain. Famously, we *could* in principle come up with a translation function between activity of molecules in a pail of water—or fluctuations in the Bolivian economy (as in Block 1978, p. 315)—and activity of molecules, neurons, or whatever in a human brain. Such a translation function could even correlate particular arrangements of water molecules with motor outputs of a human being, such that water arrangements could be translated into speech acts. But consider some vast collection of water molecules picked out to perform the job of, say, human primary visual cortex. While the input/output relationship instantiated by that collection of water molecules, *as* a mathematical abstraction, might be just the same as that of visual cortex, the job of actually specifying that module and its input and output sets would be *enormous*. Specifying such modules and connection points for a real human brain, while still very difficult, is not so daunting on account of structural redundancies in the actual wetware which allow compression of information about modules and about which particles or neurons (or whatever) transmit information. Equivalently, a complete description of the human brain is nowhere near maximally random. A complete description of water in a pail, on the other hand, is almost guaranteed to be nearly maximally random[20]; a program for specifying water particles to act as visual cortex or, that specified, to accept input from a watery lateral geniculate nucleus could thus do little better than naming every single one by precise location. (This observation relates closely to a lemma given in section 7 of Chaitin 1979 to the effect that easily described large subspaces of random spaces must themselves be random; where the subspaces are not random, they cannot be easily described.) Consequently, while some function does, in principle, translate pail of water activity into something like brain activity, that *that* functional breakdown would not emerge as the preferred one from the complexity minimisation strategy. More likely, depending on the level of detail specified in the domain and range, the best functional decomposition of a pail of water would be in terms of one large module yielding statistical correlations, given by fluid dynamics, between inputs and outputs. A more finely detailed domain and range would simply shift attention down to the level of interactions between molecules, because there doesn't appear to be any intermediate level of structural redundancy between that of individual molecules and that of statistically described eddies and currents.

The story is similarly deflating for the case of a book describing in English all the causal relationships within a brain. On its own, with respect to normal human-style inputs and outputs, the book is hardly a functional system, since in the absence of an elaborate translation function it doesn't instantiate any relevantly human-style input/output relationship at all. (And with a gerrymandered translation function, the example is no different from a bucket of water.) But suppose it were used in an arrangement akin to Searle's (1980) Chinese Room. Perhaps it could list causal relationships in a way which would make it convenient to look up overall outputs in response to any particular inputs. Then, with the addition of some hardware for actually doing the looking up and spitting out the given answers as outputs, we could ask about the functional status of the book together with its input and output systems. Here we avoid the water bucket problem of random particle arrangements, since at an appropriate level the description in the book clearly does manifest structural redundancy (in virtue of the structural redundancies in English, in the alphabet, and in the underlying properties of the brain which the book describes). But the trouble instead is that the only functional modules causally communicating information are the input apparatus, the output apparatus, and the book itself. A book which *describes* causal interactions between the lateral geniculate nucleus and primary visual cortex, for instance, doesn't actually *instantiate* any physical modules which play their rôle. Under the right translation functions, the bucket of water, the human brain, and the book system could all instantiate the same input/output relationship and even evince the same gerrymandered functional modules, but the particular functions and modules which matter to the complexity minimisation strategy are not the same at all: even without a handy functional logical depth calculator, it's easy to see that with respect to functional decomposition, all three are entirely different.

Section 4 continues with some broader observations about the view of functional decomposition outlined here. First, however, I summarise briefly an alternative approach to understanding modularity.

3.4 Chaitin's Mathematical Approach to 'Life'

With some embarrassment, I must admit that when framing the original primitive version of functional logical depth and the complexity minimisation strategy (Mulhauser 1995b), I had not yet discovered Greg Chaitin's work on a similar problem some two decades earlier (Chaitin 1970, 1979). Inspired by John von Neumann's interest in mathematical accounts of life and self-reproducing systems (von Neumann 1966), Chaitin considers the degree of organisation or structure of an entity situated in a discrete model universe. The ultimate goal is to ground the notion of 'organism' or 'life' in some mathematically rigorous definition of structure. While I think meeting anything like that goal would require much more work, the basic idea is useful.

Chaitin defines the *d*-diameter complexity $I_d(X)$ of an object X as the size of the minimal program for constructing X from separate and independently described parts, each with diameter bounded from above by d. That is, consider all the ways of partitioning X into parts of diameter bounded by d, asking how difficult it is to reassemble those parts back into X with a program α. $I_d(X)$ reflects the cheapest decomposition and reassembly method for the given d:

$$I_d(X) = \min\left(I(\alpha) + \sum_{i<k} I(X_i) \right)$$

Here we sum over the parts of X given by a particular partition for diameter d and minimise with respect to all the possible partitions for that d. It's easy to see that for an essentially random object, one with little or no structural redundancy, $I_d(X)$ stays very close to ordinary algorithmic information content as d decreases. Obviously, $I_d(X)$ is bounded from below by $I(X)$: the sum of the information contents for independently described parts of X, together with that of a program for reassembling them into X, cannot be less than the length of a minimal program for generating X directly. When d is larger than X, the partition can simply pick out X itself, and the two measures coincide.

When an object does have some structure at a particular partition level, however, $I_d(X)$ grows rapidly as we decrease d below that level and we can no longer exploit structural patterns appearing for larger d either for reconstructing X or for describing individual elements of the partition. Chaitin suggests that $I_d(X)$ as a function of d is like a Fourier transform of X; $I_d(X)$ increases each time d passes through a diameter where X displays significant patterns. The rate of increase in the difference between $I_d(X)$ and $I(X)$ as d decreases through these zones reflects the degree of structure within X.

Similarities between this approach to structural organisation and the complexity minimisation strategy are striking—especially since both seek to minimise some quantity related to the complexity of decomposing and then reassembling an object from parts—but so are the differences. The most significant difference is that Chaitin's approach tells something about how an object *looks* as a static entity, whereas the present method of functional decomposition tells something about how it *works* as a dynamic entity. To the notion of substrate neutral functional equivalence, there would correspond under Chaitin's framework a notion of substrate neutral equivalence with respect to structural complexity: where the *d*-diameter complexity 'spectra' of two objects match, we might say the two objects display equivalent structural complexity. Understood in this way, however, functional equivalence clearly does not imply equivalent structural complexity, since (among other reasons) the very same system may admit of different functional decompositions depending on the choice of input domain. Under this definition of equivalence for structural complexity, the converse is true as well; after all, two distinct objects could be built using the same intermediate structures in different ways. In any such case where the best partitions are thus the same and where the minimal reassembly programs happen to be the same length, *d*-diameter complexity returns the same values for every choice of *d*, yet the overall objects might *function* entirely differently. Tightening the definition of equivalence for structural complexity by requiring identity both between partition schemes and between reassembly programs degenerates the definition to simple identity between descriptions; *that* sort of equivalence unsurprisingly does imply functional equivalence with respect to any chosen domain, but the definition renders the actual *d*-diameter complexity framework superfluous, and the notion of 'structure' disappears.

4. BROADER CONCERNS ABOUT FUNCTIONAL DECOMPOSITION

Having covered above only some of the most immediately relevant consequences of the complexity minimisation approach to functional decomposition, some broader considerations wait in the wings. My first concern is to set aside a worry which may have troubled some since the first mention of another formal measure of complexity: decidability.

4.1 Decidability and Functional Decomposition

Defining functional logical depth in terms of execution times for *minimal* Bernoulli-Turing programs garners for the measure at least as many decidability problems as ordinary logical depth or algorithmic information content. To the extent that it calls for simultaneous minimisation of functional logical depth for three different relationships, perhaps the strategy outlined above embodies even more. But the point has never been to provide an actual recipe for *producing* a particular best description or functional decomposition of a given system. (To my knowledge, no one has yet offered *any* direct recipe for drawing up a functional description of something like a brain, let alone a computationally tractable one.) It has been merely to show, in a way which is neither trivial nor vacuous, that *there is one*—and, moreover, that it is essentially unique. Actually finding such a best description is, in the general case, an *empirical* matter. Suppose on the contrary that some effective procedure yielded a functional decomposition of any arbitrary system. Then, along the lines of the earlier comment on page 49 that decidable alternatives to algorithmic information content must be entirely logically independent of it, no such 'recipe' could, in the general case, yield a decomposition bearing any computable relation to the products of the present strategy. (If it did, then as before the procedure would offer an indirect method of computing noncomputable values, yielding a contradiction.[21]) In other words, no such 'recipe' can provably have the benefits of complexity minimisation with respect to computational resources, precision, and so on.

This shouldn't be a great concern, although it serves as a reminder of the limits of pure reason and might be taken as a comment about the right way to do cognitive science and philosophy of mind. Recall from the 'Gödel De-Mystified' discussion in Chapter 3 that problems of incompleteness limit what is *provable*, not what is *true*: Kurt Gödel, Alan Turing, and (most transparently) Greg Chaitin have shown that we get no more information from formal systems of reasoning than we put in. (It is helpful in this respect to think of the Turing Machine by analogy as 'logic automated'.) With the help of a 200-page, 17,000-variable equation encoding the number Ω—see page 47—Chaitin (1987b) has even demonstrated rather spectacularly the existence of truths absolutely impenetrable to reason alone, and far beyond the self-referential statements of Gödel, in the so-called 'queen' of mathematics, elementary number theory. At the other extreme, comparatively short and otherwise unsuspicious truths can be altogether out of reach of logic; see Jones (1978, 1982) or Paris and Harrington (1977), for instance. For this reason, Chaitin has long advocated approaching mathematics with the empirical spirit of the physicist. Such thinking extends well beyond isolated enclaves of mathematicians intimate with the absolute limits dictated by decidability. Cherniak, for instance, in discussing the minimum requirements for an agent to be rational in the light of both absolute decidability and computational tractability, notes that "not only is acceptance of a metatheoretically adequate deductive system not transcendentally indispensable for an agent's rationality, but in important cases it is inadvisable and perhaps even incompatible with that rationality" (1984, p. 755).

Likewise for the strategy for functional decomposition advanced in this chapter: the above has shown that, with respect to the stated constraints, there is a 'best' functional decomposition. But *finding* that best decomposition, just like finding mathematical truths with information content greater than that of a chosen axiomatic system, remains an empirical goal[22] which cannot be reached by armchair theorising. Such an empirical search might well take guidance from something like the complexity minimisation strategy applied to known programs running on existing computers, but a very basic observation about incompleteness is that we can never (except by inflating the information content of our reasoning system by adopting new assumptions) *prove* that a particular empirically suggested decomposition is best. An attractive analogy with this relationship between finding functional decompositions and showing they exist comes in the next chapter, where it turns out that showing consciousness supervenes on the physical world is a slightly different project to discovering an actual logical story linking the two.

Finally, it is interesting to note that even if the many problems of Chalmers's criterion for implementing a combinatorial state automaton (see section 1.2) and, by extension, for implementing a functional system, could somehow be fixed up, it, too, *might* encounter problems of decidability.[23] In particular, although Chalmers (1996b, p. 320) is keen to point out that for every given CSA, there is a fact of the matter as to whether a particular physical system implements it, it comes as no surprise that he offers no decision procedure for answering the question. It is difficult to speculate, since the criterion is so fraught with difficulties that fixing it up might involve changing it in ways I haven't considered; but assuming it could be done somehow and further assuming that real physical systems can be perturbed in infinitely many distinct ways with infinitely many distinct effects (consider a translational nudge, for instance), formal undecidability seems a likely side effect. Why? If we happen to know that a particular decomposition satisfies his requirements, verifying the fact is straightforward. Likewise, if we happen to know that, for instance, a particular physical system cannot display enough distinct states to match the distinct states of the CSA, showing there can be no decomposition is similarly straightforward. But unless we know there is some such logical inconsistency in the very idea of a satisfactory decomposition, we cannot *in the general case* prove that no decomposition satisfies the criterion. Doing so would amount to proving that no algorithm ever produces a particular output—namely, one which would describe a satisfactory decomposition. This all is exactly analogous to the fact that any halting Turing program can be proven to halt (eventually), some runaway programs can be proven not to halt, but *in general* the halting status of Turing programs cannot be determined. (The 'Gödel De-Mystified' discussion will clear this up for readers taking chapters out of order.)

4.2 Natural Kinds

The strategy outlined here suggests some curiosities about another issue in the philosophy of science. What results from applying the complexity minimisation strategy recursively to a system and its functional components? That is, consider the strategy applied to a system, then to the modules picked out, then to modules picked out of those modules, and so on. Successive applications return finer and finer decompositions, until the different levels of computationally useful structure are exhausted. (This is akin to the 'Fourier transform' interpretation of *d*-diameter complexity described above in 3.4.) I suggest that the functional modules picked out on each successive application are none other than *natural kinds* with respect to the sort of system in question. To appropriate Plato's phrase, the selected modules 'carve the system at its joints'.

In the philosophical literature, natural kinds are often described as the entities which figure in natural laws and explanations. For instance, the laws of physics apply to natural kinds like protons and gravity and the strong nuclear force. Nonnatural kinds, by contrast, are meant to be things like the class of (physical) *objects the same age as the Eiffel Tower*. This example is Sterelny's (1990), who goes on to say, "There are no laws that apply to all and only the age cohorts of the Eiffel Tower" (p. 42). But of course, *contra* Sterelny, there *is* some (probably very large) 'gerrymandered' description of the behaviour of precisely that class.[24] Restricting attention to finite spacetime manifolds and thus to finite classes of age cohorts of the Eiffel Tower, it is even the case that such a description is finite. But surely it is liable to be *extraordinarily* large. The reason should by now be clear: such a description cannot exploit much redundancy in the physical properties of such objects. The class might include the Eiffel Tower, some bats, a few flakes of rust, a rose or two, and so on. A description which applied to behaviour of all and only those things likely would have to include information distinguishing in great detail nearly every single member of the class. The quantity of information required for such an endeavour is apt to be vast. But scientists naturally prefer much shorter descriptions, ones which compress huge amounts of information into a compact rule. Functional decomposition simply quantifies that notion of compactness in terms of length and speed of application. (Recall Solomonoff's original purpose for algorithmic complexity; see page 37.) I believe the complexity minimisation strategy captures almost exactly what many people have in mind when considering natural kinds. Probably there is very much more to be said about the topic; I mention it here only to draw attention to the obvious superficial connection without developing it more deeply.

4.3 Levels, Languages, and Symbols

At first glance, it might seem that the strategy outlined in this chapter either presupposes or at least counts strongly in favour of a symbolic approach to cognition. After all, it describes cognitive systems in terms of Turing Machine submodules whose interconnections are dictated by another Turing Machine.

Surely, it might be thought, manipulation of symbols on a machine tape comes about as close as possible to the hard-core symbolic framework sometimes called 'strong AI'. Yet, as pointed out in note 15 on page 88, the assumption at the moment is only that all physical processes can be satisfactorily described at some level by a Turing program: this is an assumption about how an overall input/output relationship of a physical process may be *described*, not about whether it really *is* a process of manipulating symbols. (A quick visit to Chapter 8 reveals the real cognitive processing bias of this book, and it is not symbolic!)

For example, consider a tornado in Kansas. Assuming the laws of physics apply universally, there is at least one level at which this system can be described symbolically: the symbols come from mathematics, and the rules of use come from physical law. Some very large Turing program could describe the complete temporal evolution of the tornado and the precise trajectories of Dorothy, Toto, and Auntie Em within it. Yet, is a tornado in Kansas a symbolic system simply by virtue of the fact that it can be described symbolically? Of course it isn't.

Likewise, the strategy of *describing* cognitive systems with Turing programs says nothing at all about whether cognition is best understood in terms of classical symbolic artificial intelligence, connectionism, or little green men shuffling information in ways hitherto never even considered. The real debate between connectionists and symbolicists, sparked famously[25] in a special issue of *Cognition*, comes down to problems with finding good levels of description for giving explanations. No one disputes the fact that when we peek inside a human skull, we really do find, among many other things, networks of neurons—and some of the modules picked out by the complexity minimisation strategy almost certainly are networks of neurons. The more important concern centres on the capabilities of networks *as* networks and whether cognition can best be explained in terms of them. For instance, can networks generalise logical relationships between linguistic elements without actually *implementing*, at some level of description, a symbolic system which directly represents those linguistic elements? Are the abilities of cognizers who can generalise logical relationships best explained in terms of network dynamics or in terms of symbol manipulations? For present purposes, the substance of the debate is entirely irrelevant, and the complexity minimisation strategy should not be taken as an endorsement of one side or the other. (Indeed, I am inclined to think the substance of the debate is entirely irrelevant to most purposes, but that is an axe better ground another time.)

While the complexity minimisation strategy is silent on the question of whether symbolicism works to fix explanatory levels of description, it does seem an unlikely place to look for support of Fodor's (1975) 'language of thought' hypothesis. (I take it that a Fodor-style language of thought implies symbolicism but not *vice versa*; thus, while advocating the language of thought would count in favour of symbolicism, a negative comment on the former is consistent with agnosticism on the latter.) Without rehearsing specific arguments here, it seems improbable that the approach to functional decomposition would return modules whose communications with each other could so handily be interpreted at a single level of description so as to resemble anything at all like a formal language. Functional modules, at least when described in the computational terms of the complexity minimisation strategy, likely operate on input and output sets chosen at widely varying levels of description, such that coherent language-like relationships between inputs and outputs across the entire system are an unlikely prospect. In the absence of independent reasons to think input and output sets appear at a single level (or at least at very strongly constrained multiple levels) of description, it seems far more plausible that the processes of natural selection which created real cognizers will have exploited functional relationships at whatever widely varying levels happened to be available during particular stages of phylogenetic development. (Of course, this isn't to say that an artificially created cognizer mightn't be built around such a homogenised architecture.) Input and output sets selected at perhaps widely varying levels by the functional decomposition approach are not, if I understand his approach correctly, anything at all like the kinds of language elements Fodor is after. The symbolic view of mind comes in for additional friendly bashing in section 4.4 of Chapter 8.

In any case, having seen in this chapter how to make precise the problematic notion of a functional system, the set of abstract functional systems to which it may be said a given physical system is equivalent can be pared down. But the chapter has said little about the question which prompted the exploration in the first place: How do *I* know that *Mary* is sufficiently similar to me in enough of the right ways that I should believe her experience of red is anything like mine? It is that question—and the question of why Mary's (or my) experience should be like *anything at all*—which occupies the next chapter.

NOTES

[1] Block's excellent article—source of the label 'liberal functionalism'—remains, nearly twenty years on, one of the most insightful critiques of simple versions of functionalism and should be required reading for every course in philosophy of mind.

[2] Despite its mathematical triviality, Sterelny (1990, pp. 17-18) notes that something resembling this view remains quite prevalent in cognitive science, figuring in Haugeland (1985) and Newell (1980), for instance.

[3] For more on the relationship between laws and objective features of nature, see Cartwright's (1983) view of an inverse relationship between a theory's simplicity (and its capacity to explain) and its *truth*; she argues that laws of physics don't really represent the facts, they merely

classify them conveniently. Elsewhere, Cartwright (1989) describes laws as abstract claims which deliberately *ignore* most messy details of the real world and which thus cannot really be approximating actual systems, many features of which they don't even mention.

[4] Why won't the easy fix work? Well, it just *might*, if we were only interested in systems which can evolve easily into any given state from any initial state. (Meeting the CSA criterion and existing in one of the privileged physical states guarantees *only* that the physical system implements the right transitions, etc. for those states accessible from that initial state.) Unfortunately, the bulk of interesting systems do not satisfy the requirement—as Chalmers himself observes (1996a, section 3).

[5] Why *infinite*? Decompositions, on Chalmers's view, are not merely partitionings of the set of particles constituting a system; they include specifications of different inputs (physical perturbations) and outputs.

[6] Regrettably, our published exchange (continued in Chalmers 1996c) and PSYCHE-D discussion list comments (see http://psyche.cs.monash.edu.au) were less positive than they might have been, with my own too judgmental criticisms met by Chalmers questioning my professional motives, accusing me of "cheap point-scoring and unargued rhetoric", and labelling the entire existing literature on functionalist formalisms (and their problems) "very superficial".

[7] This is not to condemn all views which appeal to explanatory salience, such as, in different ways, Wright (1973) and Cummins (1975)—or, following the latter, Neander (1991) and Amundson and Lauder (1994)—it is simply to say that such approaches fail to solve the problem of assigning a unique computational description.

[8] Variations on this theme appear in Dennett (1975), Bogen (1981), Lycan (1981, 1987), Millikan (1984), Papineau (1987), Sterelny (1990), and Mundale and Bechtel (1996).

[9] Even this much is by no means straightforward and may even be pathologically tricky in this context. Philosophy of science strikes again: natural selection no more offers up easy solutions to problems about what is relevant in an explanation (Salmon 1970, 1984; Humphreys 1989)—i.e., what *exactly* was selectively relevant—than any other branch of the natural sciences.

[10] Note that the approach advanced here is specifically not intended as a tractable *recipe* for formulating that class; instead, it provides an objective foundation for talk about that class which grounds claims about functional organisation more substantially than other mathematically trivial alternatives.

[11] More precisely, it is better understood as the harmonic mean of all such programs, which we may think of as a weighted average strongly favouring the short programs.

[12] The comparatively small but non-zero variation in logical depth for strings with maximal algorithmic information content comes from differences in overhead for processing the length of the string to be output (the bits which are reflected in the $O(\log_2(n))$ expression).

[13] The level of precision might be given in terms of significant digits in the outputs or, where outputs are of a constant length, in terms of the distance between outputs of S and those of the system in question. Distance in this case is usefully measured with respect to the *Hamming metric*, defined as the number of bit locations at which two strings differ.

[14] This suggests a different measure of the complexity of an input/output relationship: average mutual information content between correlated input/output pairs.

[15] Incidentally, the following discussion assumes that all functional modules instantiate computable input/output relationships; Chapter 9 outlines some reasons for questioning that assumption.

[16] A minor technicality: the method requires that temporal dimensions of inputs or outputs, when included, be reflected directly in the temporal relationships between such particle states.

[17] Since, in evaluating functional relevance with respect to temporally dimensioned inputs, we might be asking for an overall arrangement of modules which corresponds to a system which learns or adapts, modules which play a rôle in shaping learning or adaptation over time are always functionally relevant. Specifically incorporating temporally dimensioned inputs implies that usually the overall Turing Machine arrangement yielded by a particular functional decomposition is not itself functionally equivalent—on the definition of the next paragraph—to that which was decomposed. (The Turing Machine arrangement treats temporal data like any other data rather than manifesting it directly in the timing of its own internal operation.) In other words, the overall Turing Machine arrangement may constitute a *simulation* of the functional system but not a *realisation* of it.

[18] Proving symmetry, reflexivity, and transitivity are, as they say, left as an exercise.

[19] The class could have more than one member in the exceptional case that two distinct, minimal, and appropriately accurate Bernoulli-Turing programs which might figure in the functional description each require precisely the same execution time.

[20] In the absence of independent reasons to believe something is *not* maximally random, the simple cardinality considerations of section 2 of Chapter 3 give every reason to believe it *is*. For the case of any string which we know we *can* compress, however—i.e., in which we have established the existence of structural redundancies—the degree of compression yields an effective upper bound on its information content. Structural redundancies in something like a human brain are visible to the naked eye.

[21] In general, this is an interesting strategy. When trying to quantify something, such as the best functional description of a given physical object, which seems almost impervious to logical analysis, try instead to quantify it in such a way that if algorithmic information theory were decidable, then a specific answer could be had. If we succeed in finding such an information theoretic route to the goal, and if knowing a specific answer would allow an inference backwards to undecidable facts about information theory, then it is immediately clear that there can be no path to the goal which *doesn't* pass through undecidable territory.

[22] Chaitin (1996, p. 30) quotes a 1951 manuscript in which Gödel, a Platonist, expresses the idea with ironic clarity: "If mathematics describes an objective world just like physics, there is no reason why inductive methods should not be applied in mathematics just the same as in physics".

[23] I hedge my bets here due to subtleties about finite bounds on the number of distinguishable access states for real physical systems. (See Bekenstein 1981b and Chapter 9, page 1.)

[24] As noted in section 1, pinning down precisely a meaning for 'natural law' is difficult, and here I sidestep the tough question of whether a gerrymandered description deserves the name. Likewise, I specifically avoid the question of whether the complexity minimisation strategy can in any way ground a system's *causal organisation* in terms of natural kinds. The definition of functional logical depth already includes information about the behaviour of a system over distinct points in an input domain, and the decomposition strategy requires causal interaction between modules; so the method already presupposes some notion of causal relationship and so cannot by itself ground that relationship.

[25] See, for instance, Fodor and Pylyshyn (1988), Pinker and Prince (1988); also see Fodor and McLaughlin (1990), Smolensky (1988a, 1988b, 1990), and Browne and Pilkington (1994).

CHAPTER SIX

Self Models

All of these models are quite mistaken, and more generally most approaches to phenomenology in the analytical tradition are hopelessly flawed... (McCulloch 1995, p.133)

Recalling the hypothetical zombies of Chapter 2, suppose a zombie could be fashioned not of AND gates and digital memory, like Cosmo, but of exactly the same flesh and blood as conscious Osmo. It *seems* easy enough to imagine a possible world in which a creature physically just like Osmo (or just like me) carries on his everyday business, exhibiting every external symptom of conscious experience, but who nonetheless never experiences anything at all—a creature with so-called 'absent qualia'. The prospect of this apparently logically possible uncoupling of phenomenal experience from the physical details of an organism leads many to believe that a complete description of the latter cannot *explain* the former; it seems logically possible that all the physical details which might figure in an explanation could obtain even without any phenomenal experience at all on the part of the organism.

Of course, in the absence of a very keen understanding of what consciousness *is* and of how it relates to the world, such a 'possibility' doesn't make a very convincing case against a physicalist account of consciousness. Just for fun, suppose that consciousness logically supervenes upon the physical world: suppose that fixing all the physical facts about the world logically fixes all the facts about consciousness, (or, at least, all the positive ones).[1] Further suppose that I just don't happen to know the logical story relating the two, perhaps because I grasp the structure of consciousness only dimly. *Of course* I will in that case find it conceivable that facts about consciousness might vary independently of physical facts. Conceivability in such a case demonstrates only that I don't really understand the full story relating consciousness and physics, not that there is no story. In a similar vein, once upon a time I found it not only conceivable but positively *self-evident* that any statement phrased in the language of a given formal system could be either proven, disproven, or shown to be meaningless within that same system. Until around 1931, nearly every mathematician in the world very probably took a similar view. But holding such a view shows only that neither I nor mathematicians in the early part of this century really grasped all the relevant features of proofs within formal systems. (See section 4 of Chapter 3 on incompleteness.) Yet the ideas of proof and formal system were then and are now *vastly* clearer than ideas about consciousness have been at any time in history.[2]

In general, arguments which begin with the 'conceivability' of uncoupling phenomenal experience from physical details beg the very questions on which some believe they pronounce so clearly. Nonetheless, perhaps a few readers really do understand consciousness itself and the laws of physics[3] sufficiently well—better, for instance, than mathematicians before 1931 understood proofs and formal systems—that for them such arguments do *not* beg the question. Anyone in that position will find little of interest here.

For others, the strategy of the chapter is as follows. Deliberately *refraining* from pinning down precisely a 'definition' of consciousness, the discussion explores the logical supervenience of phenomenal facts on physical facts and outlines one possible cognitive and information theoretic story for making the supervenience relation intelligible. The job is then to evaluate whether the concept of consciousness to which the story links physical facts includes all the most popular features of consciousness or whether it overlooks something. Apart from—I hope!—being better illuminated, the notion of consciousness which will take shape may well *differ* from the initially nebulous pre-theoretic conception. (Recall Aaron Sloman's simultaneity example described in note 2 of Chapter 1.) While purely *a priori* considerations figure in the arguments for logical supervenience, the particular way of understanding the relation constitutes an open and revisable theoretical framework to be checked against constraints suggested by our (hazy pre-theoretic) concept of consciousness, by existing third person empirical data, and by introspectively available phenomenal experience.

1. SUPERVENIENCE AND LEVELS OF EXPLANATION

The notion that consciousness logically supervenes on the physical carries at least one powerful mark in its favour. Namely, it seems that *at some level of description*, we ought to be able to say that something included in the extension of the word 'consciousness' *causes* of some of our behaviour. Consider the statement, "I lost consciousness as soon as the nitrous oxide began to flow" or "When I regained consciousness, I found I had three fewer teeth". At some level of description, it ought to be legitimate to say that a person's *being* conscious, and being able to contrast that with not being conscious, explains in roughly causal terms[4] why that person might make such a verbal report *about* consciousness. The same reasoning ought to apply to statements like, "No physical theory could possibly capture the ineffable redness of my conscious experience of this rose" or "My conscious experience of this particular pain is perversely pleasurable". If so, (ignoring for now the possibility of interactionist dualism), the conclusion that consciousness supervenes logically on the physical looks inescapable. (See, for instance, Kirk 1979 or Seager 1991; Terry Horgan 1987 argues similarly for *metaphysical supervenience*, discussed below.) Assuming both that the cosmos is causally closed and that consciousness is, at some level, causally efficacious delivers immediately the conclusion that consciousness logically supervenes on the physical: after all, if the physical facts could remain the same even while the consciousness facts wandered, then consciousness couldn't be a *cause* of the physical facts.

This needs a little unpacking. First, by 'logical supervenience' I mean it is logically impossible that the facts in the supervenience base obtain without the supervening facts also obtaining. In particular, the relationship is grounded in *a priori* necessity: this requires taking a concept's intension as that which fixes reference for whatever world is actual. In other words, a concept's extension is determined by its intension evaluated *in the actual world*. For instance, there is no possible world physically identical to the actual world in which those things we call 'cats' in the actual world are not also cats, where 'cat' is understood in this-world language. Although there might be other possible worlds in which what we call 'cats' should *in that world* properly be called 'splats', this bears not at all on whether that-world splats are this-world cats. Likewise, if the actual world had turned out to include soft, furry, aloof, cat-like animals with molecular metabolism based on ammonia rather than water, we might still have called *them* 'cats' (to the extent that our concept of *cat* does not preclude ammonia-based metabolism). This intension of 'cat' remains independent of how the actual world turns out and simply specifies how to pick out referents according to what we find in the actual world. On this use of supervenience, facts about cat-hood thus supervene logically on physical facts. Typically (and incorrectly—see section 4.2), this is taken also to mean that anyone who knew the *a priori* intensions of the relevant concepts, those which *determine* reference in the actual world, could in principle derive all supervening facts from exhaustive facts about a particular supervenience base without appealing to any other information.

By contrast, *a posteriori* necessity grounds 'metaphysical supervenience', which returns in more detail in section 4. Metaphysical supervenience features principally in talk about consciousness, with some theorists maintaining that while physical facts do fix facts about consciousness, the *particular* way in which they do so is an *a posteriori* matter. Making sense of metaphysical supervenience depends on understanding intensions as specifying how to pick out referents in counterfactual worlds, *given* actual world reference. (In the language popular in the literature, this is the 'rigid designator', created by 'rigidifying' a concept so that whatever is picked out in the actual world gets picked out in all possible worlds.) It is according to this rigidified twist on intension that Kripke (1980) and Putnam (1975a, b) correctly point out water is H_2O in all possible worlds. On this intension, water is H_2O in all worlds just *because* it is H_2O in the actual world. Typically, metaphysical supervenience of consciousness facts on physical facts is understood to mean that anyone could in principle derive the supervening facts from the supervenience base *provided* they took account of what the relevant concepts pick out in the actual world. The varieties of intension at work in the two sorts of supervenience relation of course need not coincide, and later it turns out to be rather important that candidate *explanations* of supervening facts in terms of the supervenience base acquire a very different status according to whether the supervenience relation is one of *a priori* necessity or *a posteriori* necessity.

Returning to the above idea that consciousness is causally relevant, subtleties tucked into the phrase 'at some level of description' and the word 'cause' also need elaboration. In particular, consciousness appears here as a construction useful for explaining some phenomena at a (supervening) high level, *not* as some separate entity which enjoys causal powers of its own, independent of those of whatever constitutes it. This approach acknowledges the point that several levels of description *and* several levels of explanation may usefully apply to most phenomena. Recall the comment from page 11 on levels of description and the extravagant metaphysics of the emergentists: while consciousness might feature essentially in practically intelligible explanations of high level behaviour, that fact should by no means seduce us into relinquishing ontological grounding in the lowest level physics.

An analogy with a thoroughly uncontroversial third person example draws out some relevant features of explanations couched in terms of higher level descriptions. Consider two alternative explanations of a collection of bits of wood and wire spread in concentric patterns on the road below my window. On one hand, the historical trajectory of every single particle in that collection might in principle be described, at a low level, right back to its original position some time ago, thereby 'explaining' in a certain sense how all that stuff came to be scattered in the road. Such an explanation needn't include any higher level descriptions, although in foregoing them it probably won't be especially intelligible to me if my real target is an account of *bits of wood* rather than of individual particles. Even an explanation including instructions for deriving statements about bits of wood from statements about their component particles might not help very much in that respect. On the other hand, a much simpler explanation relates the stuff on the road to the trajectory, not of myriad subatomic particles, but of a piano dropped from my window by a remarkably strong but clumsy mover. This describes at a higher level what remains physically the very same process.

In an important sense, the piano *as a piano* is really irrelevant (Kim's 1989 'explanatory exclusion'), since I can imagine a very low level account, never even mentioning pianos, of every physical feature of the bits of wood and the dent in the road and so forth. Yet, the alternative explanation in terms of the piano dropped from my window holds its own as perfectly legitimate and much more practically feasible. So, too, for consciousness: while we might imagine a microphysical account—which never invokes consciousness—of any behaviour such as the speech acts described above, consciousness may well feature indispensably in intelligible explanations at higher levels of description.

This difference in levels of explanation easily confuses discussions about the logical supervenience of consciousness on physical facts.[5] I believe it is just *because* we can imagine a microphysical explanation of behaviour without invoking consciousness that so many people think they can imagine nonconscious physical duplicates of conscious organisms. (Ironically, it is of course just such explanations, given in terms of lower level structures which do not by themselves involve consciousness, toward which most contemporary cognitive scientists strive!) Too frequently, the judgement against logical supervenience runs something like this: "I can imagine the activities of all my neurons duplicated in precise microscopic detail in some other body, causing all the same behaviours as I exhibit. Yet nowhere does this imagining seem to require the supposition that the duplicate is actually conscious to exhibit those behaviours, since those behaviours all follow from the precise microscopic details. Therefore, consciousness doesn't supervene logically on those physical activities". Yet, who, if faced with a mountain of data about the motions of trillions upon trillions of molecules, would ever guess it might describe a piano emitting the sounds of its last Rachmaninov concerto before spreading itself in patterns on the pavement? As far as I know, no one has ever denied that piano-hood logically supervenes on the physical (where, as above, the relevant intension specifies how to pick out pianos in the actual world). Once again, the approach begs the question against logical supervenience. In the absence of *independent* reasons for believing consciousness lacks any causal relationships[6] with behaviour—that is, reasons independent of the 'conceivability' of matching myriad microphysical details without matching the consciousness—logical supervenience should appear an attractive theoretical prospect, even if we presently lack a completely comprehensive theory of consciousness and the supervenience relation to clinch the case.

(Chalmers 1996b buys wholeheartedly into epiphenomenal dualism on the basis of conceivability 'arguments' against logical supervenience and in favour of the causal irrelevance of consciousness. The 'arguments', however, are *not* independent in the above sense, and his defence of explanatory irrelevance, pp. 178-179, combined with his Chapter 3 against logical supervenience, veers frightfully close to a vicious circle. Nor does Chalmers anywhere address adequately the limits of human cognizers in attempting to grasp entailment relationships between facts about consciousness and vast arrays of propositions describing microphysical properties of brains and the like; see Mulhauser 1996 on the poor structure of Chalmers's arguments and Cherniak 1984 for one look at relevant limits of rationality.)

While the argument from causal relevance to logical supervenience is straightforward, it still allows that *some* elements of conscious experience might not supervene logically on the physical world, provided we are happy to count *those* elements as causally irrelevant. Moreover, the question of whether *all* features of conscious experience should be counted as causally relevant seems tricky to me; this chapter mostly avoids the issue and advances separate reasons for accepting logical supervenience. Only much later, at the end of section 4, does causal relevance reappear to help settle a near stalemate between metaphysical and logical supervenience.

2. PRELUDE TO A THEORY OF CONSCIOUSNESS

The discussion of Chapter 4 already indicated that assuming there is *some* explanation of how processes in the brain match up with our conscious experience—whether the correlation between the two ultimately comes down to a relationship of logical, metaphysical, or merely natural supervenience—justifies our extrapolating claims about consciousness across relevantly similar cognizers. Limiting the scope of explanations to those which fit within the framework provided by physics, it seems *at the least* that physically identical cognizers are relevantly similar. But we can do better.

The previous chapter described functional similarity in information theoretic terms, laying the foundation for a new kind of functionalism immune to some varieties of trivialisation which infect more familiar versions. Building on that framework, we here ground consciousness in features of the data structures embodied by functional systems, linking similarity of conscious experience to functional similarity. The suggestion has two distinct but closely related components. Making sense of why being conscious is like it is—and why it is like anything at all—rests first on grasping the supervenience relation between consciousness and the underlying cognitive structure. Section 4 argues positively for logical supervenience, but it does not flesh out precisely what that relationship is. Section 5 introduces the 'self model' as one plausible story linking the two; while the self model approach is no doubt imperfect in its fine details, the supervenience suggested by section 4 indicates it probably is at least a close relative residing in the near vicinity of the right story. Both sections expand on the peculiarities of perspective first encountered in Chapter 4; perspectives feature in section 4 to help reconcile the idea of logical supervenience with cognizers' apparent incapacity to derive supervening facts from the supervenience base and then in section 5 in support of a move to slip inside the point of view of the self model.

The gist of the theory of consciousness advanced here is that we may usefully understand the subject of conscious experience as a particular kind of data structure called a *self model*, instantiated physically in a functional system.[7] Conscious experience is effected by *change* in that data structure.[8] (Incidentally, despite the name, the 'self model' plays a rôle in far more than just *self* consciousness, or *self* awareness, which returns briefly in section 4.3 of Chapter 8.) It's helpful to bear in mind some background questions throughout the chapter: Does it make sense to say that *I am* a physically instantiated data structure? If I am not a self model, what else might I be? Does the story told in this chapter adequately explain why it is that the right kind of changing data structure should experience *sensation*? What, if anything, is left to be explained about my own phenomenal experience which being a self model does not explain?

Perhaps the suggestion at this early stage sounds altogether ridiculous. How can a data structure experience anything? But for the moment, I beg a little philosophical indulgence. By the end of this chapter, the suggestion might still sound ridiculous to some, but at least it will be adequately fleshed out so major points of disagreement can be identified easily. The project begins with some conceptual housekeeping.

3. TROUBLES WITH 'MENTAL STATES'

In his classic critique called 'Troubles with Functionalism', Ned Block (1978) exhibits a slew of colourful arguments illustrating how functionalism avoids *liberalism*—the problem of attributing mental states to things which don't have them, such as the Bolivian economy—only at the cost of embracing *chauvinism*—the problem of failing to attribute mental states to things which do have them but which differ markedly from humans, such as creatures with different kinds of brains or psychologies. The closest the article comes to defining 'mental state' (apart from the target identities between functional or physical states and mental states which Block criticises) is "what philosophers have variously called 'qualitative states', 'raw feels', or 'immediate phenomenological qualities'" (p. 281)—or, quoting the jazz great Louis Armstrong pronouncing on what his craft is, "If you got to ask, you ain't never gonna get to know" (p. 281). I believe the bigger problem for philosophy of mind isn't traditional functionalism itself, but the very idea that a 'mental state' as something with phenomenological properties makes any sense. The trouble derives from a conflation of the everyday use of 'state' appearing in statements like "he was in a state of panic waiting to hear the results of the interview" or "I was in *such* a state trying to finish the paper" (where we actually mean phenomenal experience over a span of time) and more technical uses of 'state' meant to pick out instantaneous, well-defined physical or functional states.[9] Two thought experiments below suggest this conflation renders the standard use of 'mental state' in cognitive science and philosophy of mind wholly incoherent.

3.1 Frozen Qualia

Suppose for the sake of an example some correlation, achieved by whatever connection, between facts about phenomenal qualities and facts about the physical structures of subjects experiencing them. This might be almost any brand of logical or natural supervenience—even some kinds of interactionist dualism would do—as long as occurrences of phenomenal experience somehow uniformly correlate with occurrences of particular physical structures.[10]

Given some cognizer, such as Jackson's Mary, for instance, whose phenomenal experiences correlate in some such way with her physical state, suppose we were to *freeze* her physical state right in the midst of some especially vivid phenomenal experience. Of course it isn't *nomologically*, or physically, possible to freeze Mary instantaneously: in real life, she would no doubt shiver and ultimately go unconscious, probably freezing to death, long before cooling down to anything approaching a truly icy state. But surely it's *logically* possible that Mary's physical state just abruptly ceases all motion of any kind, perhaps in the midst of her first vivid visual experience of the colour red. Maybe Mary's greyscale exile proceeds under the supervision of philosophically astute guards, and when her prison keepers suddenly notice her twiddling her visual cortex nerves in forbidden ways, they hastily freeze her lest she discover any more hidden knowledge. (Section 2 of Chapter 4 describes the background for Mary's greyscale existence, introduced by Jackson 1982.)

The question then arises: does frozen Mary have a 'mental state'? Does that 'mental state' have phenomenal qualities? Does she have continual vivid experience of the colour red for all time, until some kind soul unfreezes her? An affirmative answer to all or some of these prompts another curious question: what happens to her mental state (or phenomenal experience, or whatever) if we sneak into the deep freeze and extract, one by one, chunks of whatever makes up her physical state—such as bits of her skin, or eyes, or brain? As an historical side note, it's recorded that in centuries past some 4,000 'witches' were executed in Edinburgh through various hideous means; after death, the bodies of some were actually distributed in pieces to different locations across the country. Flirting with the ghoulish, what happens to frozen Mary's phenomenal experience if we do just that—distribute her frozen bits to the four corners of

Scotland? What if subsequent pangs of guilt lead us to gather up the frozen pieces and reassemble her like a chilled Humpty Dumpty?

To be sure, theorists advocating different varieties of the hypothesised correlation between physical states and mental states will likely tell widely diverging stories of Mary's phenomenal experience through all this. Whatever happens, of course, at least one thing is certain. If we subsequently decide to 'unfreeze' her spontaneously from the initially frozen state, then (other post-thaw factors being equal) Mary will only ever report *one* story about her sequence of experiences beginning before the freezing and ending some time after her 'unfreezing'—regardless of what accounts we might concoct of her frozen phenomenal experience and regardless of different procedures performed on her physical structure during the deep freeze. The reason is that as long as her state at unfreezing matches her state upon freezing, future physical facts about her cannot differ in any way *as a direct result of* alterations to that state while frozen.[11] Indeed, the observation extends far beyond mere reportability: given the hypothesised correlation between physical structure and phenomenal experience, none of Mary's subsequent actual *experience*, reported or otherwise, could differ as an immediate result of different procedures performed on her physical structure while frozen. (Recall that her experience is correlated with her physical structure *ex hypothesi*. Where her physical structure is no different, her experience is no different.) Yet, even though Mary's subsequent experience, such as while reminiscing about the period which included her freezing, must persist unchanged in the face of countless possible occurrences during her frozen moments, the intuition of many remains that there is still some story to tell—that Mary does enjoy phenomenal experience while frozen and that all these different scenarios raise legitimate *and different* questions about her experience for which a good theory of the hypothesised psychophysical correlations must give answers.

I think that's just plain wrong. The right strategy at the point of the very first question is: just say no. There are no 'frozen qualia'. Short of denying the correlation between facts about phenomenal experience and facts about physical structure and allowing that the former may wander with no change in the latter, the right strategy is to accept that instantaneously 'frozen' states correspond to no experience whatsoever. (The following sentence I would print in boldface neon purple if I could.) *There is nothing it is like to be in a particular instantaneous state.* This has nothing to do with the actual far-fetched procedure of freezing poor Mary; the answer is the same if we simply ask hypothetically about Mary's experience at a particular instant of time (without any actual freezing at all). The intuition otherwise is ungainly residue from the dualistic notion that qualia are somehow 'tacked on', parasitic on physical structure or even out there as a linked 'aspect' of that physical structure. The reason early functionalists' machine table analogy seems to miss out something important is that the instantaneous physical states likened to machine states lack 'attached' phenomenal properties in the first place. The right correlation, if one is to be found, should be between *processes* of phenomenal experience and *processes* of physical change, a point which returns shortly.

3.2 Chopped Qualia

Consider first a scenario similar to that outlined above, except that now rather than taking Mary to bits during her frozen moments, we merely freeze and unfreeze her periodically, introducing no changes to her physical structure while frozen. The example is more interesting if we also monitor and control all her environmental stimuli so as to preserve for Mary continuity in all those stimuli which would be continuous, from her perspective, were she not being frozen. For instance, the clock on the wall must stop running during her frozen episodes, and any music or other aural stimulations she might receive would have to stop and start as well. The aim is to chop Mary's experience into segments, separated by frozen interludes. The question then is: how does this treatment feel for Mary? Is her overall experience any different during a period of freezing and unfreezing, as compared to a period during which her physical structure carries on as normal, with no freezing at all? The approach of this chapter suggests that Mary's periods spent frozen feel like *nothing*. More accurately, they fail to feel like *anything*. More importantly, though, it suggests that her experience while being frozen and unfrozen is *no different* from her experience with no freezing at all. Her 'chopped qualia' feel just like normal qualia. To see why, consider the alternatives.

Perhaps the most plausible alternative is that Mary's phenomenal experience somehow 'fades out' or just abruptly 'blinks' into nothingness each time her physical state freezes. Then Mary's experience while undergoing her kinetic ordeal would be that of ordinary experience punctuated with moments of nothingness, perhaps introduced with a quick fade or just blinkingly sudden. How long, I wonder, might she think each of those moments of nothingness lasted? Surely she couldn't be aware of the passage of time while in them, nor could she even form a thought ("The lights have suddenly gone out!") during them. Moreover, following reasoning similar to that above, at no point in her *unfrozen* moments could she ever have an inkling about what it was like to have chopped qualia or about how long the breaks lasted—unless of course she would also spuriously form such bizarre thoughts even when *not* undergoing the freezing treatment. It seems we can make the periods of nothingness arbitrarily protracted, and Mary never notices the difference. Likewise, we may increase their *frequency* as much as we like, and Mary still ascertains no difference: after all, she can scarcely start counting how many stretches of nothingness she feels during the course of a particular experience (such as while watching the second hand mark out exactly one minute on the gerrymandered clock on the wall). Perhaps we *all* experience chopped qualia every day of our lives. This might not be as implausible as it sounds: maybe time (like, say, a particle's charge or spin) is a quantum notion, and our physical structures really do routinely stop dead in their tracks and then jump on to the next state and the next. Phenomenally speaking, *how would we know?*

The point of the original frozen qualia thought experiment was to suggest that there is no experience during a frozen period; the point of the chopped qualia experiment is that ordinary experience punctuated with frozen periods of such nothingness, regardless of the frequency or duration of nothingness, would be phenomenally indistinguishable from ordinary experience not so punctuated (assuming we've done an adequate job of synchronising all the *external* signs so they don't give the game away). The intuitive appeal of strangely frozen or chopped qualia remains, again, in the clutches of Cartesian dualism, deriving from the same pre-theoretic intuitions which envision phenomenal experience hanging off physical structure like some sort of phantom appendage (and which could itself, according to many, be amputated to yield an absent qualia super-zombie).

Recall the background assumption of both thought experiments: we assume *not* one particular theory of mind or another, but merely the existence of *some* correlation (or, preferably, supervenience relation) between physical structure and phenomenal experience. At fault here is not functionalism, dual aspect monism, or any other theory of that correlation—it is the very idea that there is any sense at all in the notion of an instantaneous mental state as something with phenomenal experience 'attached'. Likewise, while Block's (1978) insightful article contains many potent objections to naïve flavours of functionalism, the more insidious culprit is the notion of mental state in terms of which his anti-functionalist arguments are couched.

3.3 Real Qualia Need Cognition in Real Time

In fact, it seems we may dispense with the correlation assumption altogether and argue positively for a supervenience relation of some variety on the grounds of what appears to be a purely conceptual connection between consciousness and underlying cognition. Note that whatever might happen to Mary's phenomenal experience during chopped or frozen qualia experiments, Mary's reported experience does not change, Mary's external behaviour does not change, and Mary's entire cognitive framework and train of thought does not change. And it is, after all, rather peculiar to suggest that different frozen or chopped qualia scenarios could yield radically different phenomenal experiences for poor Mary who, nonetheless, still finds herself verbally reporting the same old story and physically behaving the same old way and carrying on obliviously with all her internal cognitive processes—such as her innermost thoughts and philosophical reflections about the ineffable qualities of her experience—intact. Such a Mary apparently could experience thoroughly bizarre qualia *without ever noticing*. On the view that Mary *does* notice her qualia, outlandish frozen or chopped qualia scenarios lose whatever intuitive appeal they may have held initially.

Some might look kindly on the notion that Mary's always being aware of her qualia is a merely empirically motivated but conceptually inessential supposition: perhaps we know *only* from our own 'direct' experience that we do in fact have a pretty good cognitive grasp on our qualia. In that case, while it might still be bizarre to suggest that Mary's phenomenal experience could vary independently of her cognitive processes, it would *only* be bizarre; such a scenario might still be logically or even naturally possible. But the motivation for believing that differences in qualia require differences in cognition cannot be 'merely empirical'; instead it appears to be a *conceptual* truth that, barring catastrophic failures of rationality and so forth, when someone starts genuinely chopping our qualia, we notice. Suppose it really were logically possible that my phenomenal experience could change significantly without my noticing: in what possible scenario should I imagine myself in order to convince myself of this fact? Obviously, in any such scenario where my experiences changed significantly without my noticing, *I would not notice!* Alternatively, any real life example allegedly empirically illustrating the failure of such a 'merely empirical' maxim that we're intimate with our qualia would not be detectable, even in principle. To put it still another way, it is *constitutive* of qualia that they be noticeable.[12] Or in one last guise: anything which goes unnoticed cannot contribute to the 'ineffable feel' of an experience. Apparently the only escape for any desperate to resist a conceptual connection between consciousness and the underlying cognitive processes involves radically divorcing our *knowledge* of what our qualia are like (knowledge which then degenerates into a mysteriously non-cognitive realm) from our cognitive *beliefs* about what our qualia are like—allowing the logical possibility that we could know our qualia were different (minced, perhaps) while simultaneously believing they remain the same. That route skirts the distant frontiers of philosophical credibility.

Among other problems, such a move opens the floodgates to arbitrarily freakish disharmonies between consciousness and the underlying cognition (which, for present purposes, I mean to include *perception*). For instance, we might all have wound up living our conscious lives with time running backwards, while our cognitive lives went on in the normal direction. Or, everyone might have lived with no phenomenal experience whatsoever except for uniform, agonising, excruciating pain coupled with unceasing, blaringly loud auditory sensations of rave music—yet nuns would still have sat quietly contemplative in convents, and intellectuals would still have argued over whether men or women take greater pleasure in sex. Every word of this book would have been written and cognitively grasped in exactly the same way as in the actual world, even though the phenomenal experience to which it refers would be entirely different. (Indeed, there is every reason to doubt that such a radical divorce of consciousness and cognition could even coexist with the notion of *referring* to phenomenal experience in the first place!) Every *thought* and every argument would proceed as in the actual world; only the subjective feel would run on different tracks entirely. All these 'possibilities', however, are apparently ruled out by the conceptual truth that qualia are noticeable (or, alternatively, that purported qualia which are not noticeable are not qualia at all): *ex hypothesi*, the cognitive lives of the tortured nuns are just the same as those of normal nuns in the actual world. They notice no difference. (The general approach I am advocating might be described as *phenomenological verificationism*, to be distinguished sharply from varieties of verificationism which centre solely on cognition, verbal reporting, or broader aspects of behaviour.)

It seems the only credible alternative to such consequences of rejecting the conceptual connection between consciousness and cognition demands *ad hoc* restrictions on just how wild the disparity between phenomenal experience and cognition may become in the space of logically possible worlds. But the job of justifying such restrictions in any remotely satisfactory way is a tall order. Appeals to *a posteriori* necessary features of consciousness are right out, of course: the semantic ploy of making consciousness a rigid designator has no more bearing on the space of logically possible worlds understood according to intensions fixing reference in the actual world than making 'cat' a rigid designator has on the relation of splats to microphysics. And any *a priori* reasoning about the underlying cognition looks to be a slippery slope leading right back to where things began with a conceptual connection. Finally, any restrictions also press well beyond the realm of natural laws: even an advocate of natural supervenience and extravagant new laws of Nature linking the physical with the phenomenal cannot reject the bizarre (alleged) 'logical possibilities' above on the basis of new *contingent* laws alone.

The two thought experiments and the line of analysis above clearly show, I believe, not only that there can be no instantaneous physical state with associated phenomenal experience, but also that, as a matter of *a priori* reasoning, there can be no alteration to phenomenal experience unsupported by an alteration to cognitive processing. (Note that this does not, however, establish *a priori* logical supervenience; see section 4.1 below.) In other words, difference in cognitive process is a logically necessary condition for difference in phenomenal experience, and phenomenal experience associated with an *instantaneous* physical state straightforwardly does not exist. What remains important to experience is the *change* between instantaneous states. In other words, the phenomenon of experience is inherently *temporal.*

3.4 Nothingness, Nothingness, and More Nothingness?

Someone might object, incidentally, that a nonzero interval of time over which I suggest phenomenal experience occurs looks akin to a host of 'instants' of nothingness placed next to each other, which taken together simply produce a longer span of nothingness, rather than some real phenomenal experience. In this connection, an almost perfect analogy with something like *velocity* helps clarify the point.

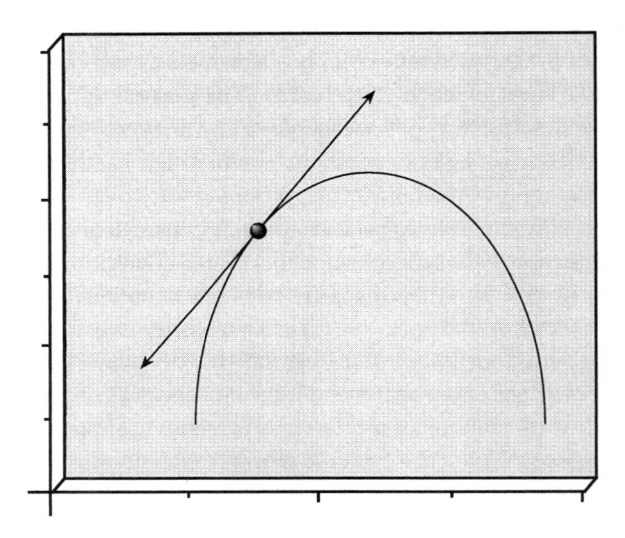

Figure 10. Instantaneous Slope of a Curve

A body can be assigned an abstract measure of 'instantaneous velocity' at a particular instant only in virtue of its motion across some nonzero temporal interval which includes that instant. A 'frozen' particle has instantaneous velocity in this abstract sense, *but it has no velocity.* Attributing instantaneous velocity makes sense only in a context which permits velocity; likewise, attributing 'instantaneous phenomenal experience' makes sense only in a context which permits phenomenal experience. The superficial similarities of the English language constructions run more deeply: in the same sense that instantaneous velocity just *is not* velocity, 'instantaneous phenomenal experience' just *is not* phenomenal experience. Alternatively, an instantaneous physical state just *is not* a mental state, if 'mental state' means a phenomenal state. In each case, the relevant context requires (among other things, such as the existence of an appropriate physical structure in the first place) an interval of time.

Noting that velocity is the derivative of acceleration, another straightforward analogy comes from the slope of a curve at a point—the first derivative of the function describing the curve, evaluated at the desired point. As in Figure 10, a curve may have a slope at a particular point (or more precisely, the line tangent to the curve at a point has a slope), but the fact that there should be any slope at all associated with a point makes sense only in the context of a curve including that point: a particular point on its own, of course, hasn't any slope at all.

The upshot of all this is that to the extent ordinary functionalist theories equate mental states—understood as genuinely instantaneous states in which it is like something to be—with functional states, the approach explored here is not functionalism. The thought experiments of this section suggest that talk about 'instantaneous' mental states outwith the context of the physical change which underlies phenomenal experience breaks down under scrutiny. Moreover, cognition appears to be a necessary condition of conscious experience; in due course, I advance the view that the right kind of cognition is also sufficient.

4. CAPTURING CONSCIOUSNESS

Chapter 4 explored Jackson's 'knowledge argument' against physicalism and concluded that a cognizer's inability to derive an understanding of the phenomenal experience of some object purely on the basis of physical information about that object says nothing about physicalism and little about phenomenal experience. The chapter suggested a more important difference between Mary before and after seeing red or between Margaret before and after smelling a rose: one of *state*, not of information. The different perspectives of the first person and third person emerged as important pieces in the puzzle. Similar considerations now contribute to the view that conscious experience is an immediate feature of change in a physically instantiated dynamic data structure.

4.1 Mileage from a Necessary Connection

First, consider again the line of thought explored in section 3.3, which argued positively for *some* variety of supervenience relation, although without specifying what particular variety. While the relevant modality underpinning the supervenience relation remains so far unsettled, the initial picture taking shape is of 'local' supervenience for a *particular subject*. That is, with respect to a given subject, differences in phenomenology guarantee differences in cognition: cognitive difference is a necessary condition for phenomenal difference, and phenomenal difference is sufficient to guarantee cognitive difference. Alternatively, for a given subject, a *particular* overall change in cognitive substrate can support only one *particular* phenomenology. (Why? If a particular overall cognitive change could support either of two distinct phenomenologies, then the difference in those two phenomenologies would not be sufficient to guarantee a difference in cognitive processes.) While it seems clear that phenomenally equivalent experience may result from different sorts of cognitive change, it is this invariance in the other direction—of phenomenal experience with respect to given cognitive changes—which indicates *some* kind of supervenience relation at work between particular phenomenal experiences and cognition.

At this point, however, the suggested relationship remains very weak. It indicates *only* that an individual subject's phenomenal experience cannot vary, once the set of underlying cognition facts are fixed; it need *not* mean the phenomenal facts can be logically derived from the cognition facts, a prospect about which we may remain for now agnostic. (Readers who find this statement bizarre should skip ahead briefly to 4.2.) It also seems so far to apply only to a particular subject, rather than across all subjects. But this is only the beginning.

Just temporarily, it's useful to adopt one empirical premise: consciousness exists. I know from first person experience that consciousness exists, and I know something of its character. Given that consciousness exists and given that the character of phenomenal experience supervenes, for an individual subject, on cognition, it follows that any other organism 'cognitively equivalent' to me (or functionally equivalent to me, given logical supervenience of cognition on functional organisation) enjoys just the same phenomenal experience as I do. Because the supervenience relation obtains between phenomenal experience and cognitive change, and cognitive or functional equivalence is an equivalence relation, we may relax the restriction that supervenience obtains only 'locally' for single subjects.[13] The qualia of my physical duplicate perfectly match my own, as do those of my functional duplicate built of silicon. The popular 'gradual replacement' experiment (Pylyshyn 1980b, Savitt 1982), which hypothesises swapping a conscious organism's functional units (such as neurons) with structurally alien functional isomorphs (such as silicon chips), describes a scenario in which the victim of surreptitious brain replacement therapy would be none the wiser. No detectable alteration in the subject's phenomenal experience can occur, because change at the cognitive level remains constant throughout. And again, following the phenomenological verificationism of 3.3, changes in qualia which never get detected are not real changes at all. As a corollary, given a world containing a model organism which really is conscious, a functionally identical zombie with absent qualia cannot occur in that world. Osmo's physical twins are conscious, too. (Incidentally, in a succinct and elegant paper, Allin Cottrell 1996 echoes Dennett's 1995b criticism of poor imagination with a direct frontal attack on the 'conceivability' of such a functionally identical zombie which is independent of the considerations here, arguing convincingly that it simply is *not* conceivable to start subtracting off portions of our own consciousness while leaving our cognitive framework unchanged.)

But while the conceptual connection between cognition and consciousness allows handy extrapolation about phenomenal experience on the basis of the equivalence relation operating over the set of functional systems—and, when coupled with the principle of coherence featuring in the brief discussion of scientific explanation (page 68), it even allows extrapolation between functionally *similar* cognizers within a single world—it leaves untouched the question of why any cognizers should experience phenomenal consciousness in the first place. Temporarily borrowing empirical information as in the paragraph above buys only the conditional proposition that *if* consciousness exists at all, *then* functionally identical (or similar) cognizers have identical (or similar) phenomenology. But the question about the antecedent remains: why should consciousness exist at all? Alternatively: why should it have the character it does, rather than some other? The necessity of a difference in cognition underlying every difference of phenomenal experience provides a handy tool for carefully limited *extrapolation*, but it does not, by itself, provide a framework for *explanation*.

The reason is that the approach as developed so far looks very much like the view mentioned briefly in section 1 often called 'metaphysical supervenience', or supervenience grounded in Kripkean (1980) *a posteriori* necessity (also see Evans 1979 and Davies and Humberstone 1980). The central feature—or weakness—of this view is that the necessary connection between consciousness and cognition allows reasoning about phenomenal experience across counterfactual worlds only in the case that we adopt the *a posteriori* twist on intension. In other words, it is only the *a posteriori* knowledge of what consciousness is *in fact* like (and that it exists) which fixes the character of phenomenal experience in those counterfactual worlds.[14] A supervenience relation between consciousness and cognition, the existence of which might be established on the basis of *a priori* considerations, could still be fulfilled even if the character of phenomenal experience differed greatly from what it is; only the *a posteriori* fact of what phenomenal experience is really like fixes the *particulars* of the supervenience relation. Thus, grasping that there is a supervenience relation of *some* kind does not by itself mean that consciousness is 'explained', since it still looks logically possible (but perhaps 'metaphysically impossible'[15]) that physical facts or cognition facts could have been as they are while consciousness facts differed.

Indeed, metaphysical supervenience allows even that there might be no consciousness at all. To my knowledge, no one seriously denies that consciousness exists—even if it has become fashionable in some quarters to buy cheap giggles by misrepresenting hard-headed proponents of cognitive theories of consciousness such as Dan Dennett as if they really don't have conscious experiences like the rest of us do and as if they perhaps think 'real' phenomenal experience doesn't exist—but it is interesting to consider the status of the proposition that it does not. After all, if it were logically possible that a world could be physically just like the actual world except that no consciousness existed (a world of super-zombies), no one in that world would notice! If someone did notice, then her cognitive processes would differ from what they would have been if she had failed to notice, and her world would not be physically just like the actual world. (This assumes that in the actual world she does not spuriously 'notice' that she is not conscious.) This should be uncontroversial, given the considerations above and in particular the *conceptual truth* that we do notice our qualia.

Setting aside the straightforward possibilities for the failure of consciousness to exist—perhaps all nearly conscious organisms spontaneously explode, or the universe contains nothing but double scoops of strawberry ice cream—what should we make of this possibility of a world physically identical to the actual world but lacking any consciousness? (Whatever your answer, it's interesting that your zombified physical duplicate in that hypothetical world would say and think just the same thing.) To put it differently, what should we make of the possibility that all the physical facts of the world remain the same, but consciousness fails to exist (or simply has a different character)? This question requires a return to the problems of perspective of Chapter 4. First, though, a curious logical feature of supervenience relations prompts a brief side trip.

4.2 Implication, Entailment, and the Return of Gödel

It may strike some readers as nonsensical to speak of the supervenience of facts about phenomenal experience on facts about cognition in the absence of some account of *how* cognition facts fix consciousness facts. The reasoning above clearly suggests that phenomenal experience somehow supervenes on cognition, but it leaves open the character of the 'somehow'. It is commonly supposed that there being a supervenience relation between consciousness and the physical world should allow, in principle, an understanding of consciousness to be *logically derived* from facts about the physical world, yielding at least in some sense an explanation of consciousness in physical terms.[16] One clear stumbling block for that view is the case of metaphysical supervenience, where the weakness of the supervenience relation itself compromises the explanation. (Obviously, any candidate explanation or derivation would require the rigidification of the relevant concepts.) Similarly, supervenience might be established by working from the top down, as the discussion above has done—discovering that some facts fix other facts but without exploring precisely *why*, perhaps allowing that the 'why' does exist but just is, for whatever reason, not discoverable. (A specific reason why the 'why' might not be discoverable for the case of the supervenience relation between consciousness and cognition emerges shortly.) While it seems a safe enough assumption that these two cases exhaust the possibilities for supervenience not to guarantee the derivability of supervening facts, the assumption rests on a basic misunderstanding about provability.

Setting aside intensional differences for the moment, the definition of supervenience in terms of fixing one set of (supervening) facts by fixing some other set of facts (the supervenience base) requires only that supervening facts cannot differ, given a fixed supervenience base. It does not require that facts in the supervenience base *imply* the supervening facts, in the sense that the supervening facts should be *derivable* from the supervenience base. That implication (in the sense of there being a derivation) need not equal entailment (in the sense of impossibility) should be clear anyway, but the case of incompleteness provides a helpful example: as explored in section 4 of Chapter 3 (and revisited in section 4.1 of Chapter 5), the set of facts 'fixed' by a suitably strong formal system properly contains the set of facts which can be logically derived within that system. That is, the facts fixed form a proper superset of the logically derivable facts; not all facts fixed by the formal system are theorems of that system.[17] Some of these supervening facts—namely, those not in the intersection of the supervening set and the derivable set—are, it is often said (by Greg Chaitin, for instance), true for no reason at all. A proposition naming those facts which supervene but which are not derivable has greater information content than the formal system itself. In other words, such a proposition carries additional information over and above what is contained in the formal system itself—yet it still supervenes on the facts which describe the formal system. For specificity, consider a proposition naming with n bits of precision the proportion of programs for a particular Turing Machine which halt. The truth of such a proposition is fixed by the basic facts about how the Turing Machine works: there is no extra magic ingredient which determines whether a Turing Machine halts, except the Turing Machine, and it is fully specified by its machine table. Yet the proposition cannot be proven for large n with just that information. (Recall from section 4.1 in Chapter 3 that the random number Ω is violently noncomputable and cannot be compressed; the *only way* to prove its value with greater precision than the number of bits in our chosen axioms is to *assume* what we want to prove.) It is a general feature of such relations that establishing even logical supervenience does not *automatically* guarantee that all supervening facts may be logically derived from the supervenience base.[18] As it happens, however, the more interesting (and probably more relevant) limitation on our capacity to derive supervening facts from the supervenience base in the case of consciousness and cognition results from the perspectival puzzles of Chapter 4.

4.3 The Illusion of Merely Metaphysical Supervenience

Chapter 4 showed that Jackson's 'knowledge argument' cannot be an argument against physicalism, because the story about Mary and the colour red goes just the same even if we *assume* physicalism. Adopting physicalism in no way demands a commitment to the bizarre assertion that possessing all the physical information about the colour red or about roses should enable a normal human cognizer to derive, unaided, a full understanding of 'what it is like' to see red or to smell a rose; thus, the fact that human cognizers cannot perform such feats is no argument against physicalism. Human beings simply lack the capacity to dictate the precise states of their brains, and, as it happens, the states of seeing red or smelling a rose straightforwardly *differ* from states of reading information or of working out logical implications of propositions. On a physicalist account, this merely accidental matter of physiology is exactly analogous to the fact that even a blatantly physical system like a digital computer is not equipped to enter a particular physical state just by reading in information about that state.

Not surprisingly, the line of thought suggesting that a normal organism cannot logically derive an understanding of its own experience of a particular object purely on the basis of physical information about the object applies with similar force to the case of an organism attempting to understand *someone else's* experience. It is the problem of an organism's grasping someone else's experience on the basis of physical information about them but without appealing to its own experience (the 'third person problem' of Chapter 4) which bears most sharply on the project to extend the relationship between consciousness and cognition from one of supervenience based on 'metaphysical' *a posteriori* necessity to full logical supervenience. To be sure, the tools outlined in 4.1 allow one organism to extrapolate from its own experience to propositions about another's experience based on their functional similarity. But our direct observations of what it is like to be any other subject except ourselves are constrained by the fact that we cannot *be* any other subject. Nagel's 'sympathetic imagination', the capacity to place oneself in a state similar to the state of some other being (mentioned on page 67, in the section 'What is it Like to be Nagel?'), marks out an epistemic frontier which we are not equipped to cross.

While the idea wreaks havoc with even the most robust notions of personal identity, it is conceivable on a physicalist view that a bizarre sort of 'shape changing' organism could come to know directly what it is like to be another organism in a fashion analogous to the way Mary comes to see red or Margaret learns how a rose smells: namely, it might gradually enter a physical state identical (or similar) to that of another, thereby transforming itself into the other organism's functional and phenomenal duplicate. But then, arguably, it is not the original cognizer (call him 'Faust'!) who learns what it is like to be a different subject of experience, because the subject with the new understanding is not the same one who started the inquisition. All this can be manipulated extensively, and perfectly plausible examples are not too difficult to concoct. (I regularly learn what it is like to be me five minutes from now, for instance, although that new knowledge always goes out of date.) But the notion does little to raise the supervenience stakes beyond *a posteriori* necessity: just as when Margaret twiddles her olfactory cortex, such a Faustian shape changer still relies upon *a posteriori* knowledge of experience in coming to understand what it is like to be another subject.

What does help present purposes, however, is the recognition that the barrier to observing directly the experience of any other subject is an epistemic matter, one which should not, by itself, seduce anyone into ontological conclusions favouring metaphysical supervenience over stronger logical supervenience based on *a priori* necessity. The trouble with perspectives, or points of view, is quite general. The speculative section 'A Tasty Side Note' in Chapter 4, on 'empirical logic', noted at least one thing not speculative at all: on a physicalist view, deducing one proposition from another—just as coming to understand how a rose smells—is a matter of entering a new physical state (or, more precisely, undergoing a new physical process). It so happens that we are equipped to enter the right physical state when grappling with simple deductions but not when grappling with what it is like to be another subject. (Probably it is rather fortunate that this is so!) This perspectival fact is altogether independent of the sort of supervenience relation which exists between consciousness and cognition. As in section 3, where the assumption of *some* kind of correlation (or, more accurately, some supervenience relation) between phenomenal experience and physical structure prompted the conclusion that there is nothing it is like to be in a particular instantaneous physical state, the phenomenal experience of another subject remains inaccessible to direct observation *whatever* the character of the supervenience relation.

Thus, *even if* the relationship between consciousness and cognition should be exactly the same in all logically possible worlds, the problem of perspectives still means the task of deducing, from physical facts, what most conscious experiences are like lies far beyond what may be achieved by actual, physically instantiated cognizers. In other words, there is nothing at all discordant about suggesting simultaneously that consciousness facts supervene logically on physical facts or cognition facts—i.e., that the relationship between them remains constant across possible worlds—and that, for real cognizers subject to the constraints of an actual physical world, a full understanding of the particular character of conscious experience is *not even in principle* derivable from physical facts. Thus, the appearance of being stuck with merely metaphysical supervenience may be nothing more than an illusion foisted upon us by the perspectival fact that we are just not in a position to derive ourselves into a state of directly experiencing anyone's experiences except our own (or knowing 'what it is like' to be them), regardless of how much information we might collect about the physical world. Whether consciousness facts supervene on physical facts with *a priori* or *a posteriori* necessity, it will *still* only look like they supervene metaphysically. Just as assuming physicalism forces no change in Jackson's story about Mary, even *assuming* logical supervenience, taking intensions as fixing reference for the actual world, looks to have no effect whatsoever on the *appearance* of merely metaphysical supervenience. Given that metaphysical supervenience and full logical supervenience look the same from our perspectives as cognizers, it seems any positive arguments given in terms of how the world or relevant concepts seem to us (as compared to arguments given in terms of the logical structures of the two supervenience relations themselves) count as much in favour of one as the other.

The central questions then concern the structure of the candidate supervenience relations themselves and whether the view of logical supervenience as true but *unuseable* by real cognizers should be more palatable than such a curious sort of metaphysical supervenience. Under either view, the conceptual truth that qualia are noticeable constrains the space of logically possible worlds; neither permits wild disparities within a world between conscious experience and underlying cognitive processes. Metaphysical supervenience, however, allows the logical possibility that facts about the *particular* character of the relationship between consciousness and cognition might differ in distinct possible worlds. It permits, for instance, a possible world in which all the cognition facts are identical to those of the actual world but in which no consciousness exists. On this view, a proposition expressing the relation which consciousness does bear toward cognition in a given possible world carries information over and above the information content of the set of cognition facts in that world. But, unlike the case of a proposition naming facts which logically supervene but are not derivable (as in 4.2), such a proposition does not supervene on the cognition facts.

So, analogies with ordinary incompleteness offer no support for metaphysical supervenience. Indeed, the logical framework underlying metaphysical supervenience for the case of consciousness and cognition seems truly unique. I know of no other instance in which we might be tempted to think some set of facts supervenes with only *a posteriori* necessity *even* after swapping to intensions which fix reference for whatever world is actual ('de-rigidifying' them). The *a posteriori* necessity of the relationship between water and H_2O, for instance, persists only when 'water' is understood as a rigid designator; with intensions taken in their more primitive, actual world incarnation, water-hood supervenes logically on the physical world just as cat-hood or splat-hood does. The reason for the difference is straightforward: the relationship between water and H_2O is one of *a posteriori identity*. (This bears, of course, on the peculiarity of metaphysical supervenience for the case of cognition and consciousness mentioned in note 14 on page 119.) If there is any similar identity for the case of consciousness and cognition, the problem of perspectives occludes it. Suppose consciousness were *defined* as what it is like to be a cognizer with a certain particular cognitive structure: even then, the problem of perspectives tenaciously holds its own, because unless we already have it, we still cannot in the general case transform ourselves into having a particular given cognitive structure.

The uniquely odd status of metaphysical supervenience for the case of consciousness and cognition constitutes a strong mark against it and in favour of logical supervenience, which is straightforward by comparison. Moreover, *whatever* extra propositions we might imagine tacking on to the basic framework of metaphysical supervenience in order to bridge the gap from physical facts or cognition facts to consciousness facts—such as stand-alone logical principles, judgements that consciousness 'just is', or new psychophysical laws of Nature—consciousness remains unexplained by the physical facts. As in 4.1, it would still be logically possible for those facts to obtain without the same positive consciousness facts also obtaining; that is, after all, largely what makes the relationship one of merely metaphysical supervenience. In other words, metaphysical supervenience is an explanatory *cul-de-sac*.

Finally, metaphysical supervenience of the kind explored here, barring modally extravagant 'strong metaphysical necessity' (see note 15 on page 119), cannot accommodate the causal efficacy of consciousness, since it admits the logical possibility of a world physically identical to the actual world but with no consciousness at all. (Again, if adding or subtracting consciousness requires no change in the physical facts, then consciousness cannot be causally relevant to the physical world.) This brand of causal irrelevance is far more worrisome for the case of metaphysical supervenience than the garden variety 'explanatory exclusion' mentioned in section 1, and while it has been the strategy of this chapter *not* to adopt the causal relevance of consciousness specifically as a premise, at this late stage it may nonetheless count for something when the two alternatives otherwise enjoy such similar positive evidence in their favour.

In marked contrast to metaphysical supervenience, logical supervenience brings with it a certain economy, in the sense that it prohibits consciousness facts from wandering about in different garb according to the possible world season; it supplies a firm foundation for exploring possible explanations of consciousness in terms of cognition; and it keeps consciousness at least in the company of pianos when it comes to causal relevance. Moreover, that no candidates for possible explanations could vanquish the problems of perspective is not to say that such explanations mightn't still offer insights into problems of consciousness. Just as the mathematician may search empirically for mathematical truths which lie beyond the reach of reason alone (see section 4.1 of Chapter 5) the cognitive scientist or philosopher of mind may search for insight into a logical supervenience relation between consciousness and cognition even if actually deriving some consciousness facts from cognition facts is beyond the reach of real cognizers. While positive arguments about consciousness and cognition themselves appear to support metaphysical and logical supervenience equally well, the marks against the former constitute a powerful reason to adopt logical supervenience as a working hypothesis: metaphysical supervenience is uniquely odd for the case of consciousness and cognition, it is altogether explanatorily futile, and it renders consciousness causally irrelevant. Logical supervenience, on the other hand, is straightforward and simple, it remains open for explanatory uses, and it preserves the causal efficacy of consciousness. *A priori* considerations lead all the way to the appearance of *either* metaphysical or logical supervenience; the choice is then between packing up and going home with a very peculiar outlook or pressing on with the search for a 'true but unuseable'—but perhaps enlightening!—theory linking consciousness and cognition.

As an aside, notice that if we take the problem of perspectives as establishing conclusively that a proposition describing how consciousness supervenes logically on cognition is impossible to evaluate given the logical framework and empirically accessible facts at our disposal (which I don't think we should *quite* be prepared to do), this amounts to evidence in favour of the conclusion that such a proposition is true. Why? If such a proposition were false, then it ought to have a simple counter-example which *could* in principle be discovered to contradict it using the logical framework and empirical facts at our disposal. For instance, the phenomenal experience of a particular cognizer might differ from that indicated by the proposition, revealing to that subject that the proposition was false. (Of course, disconfirming such a proposition with one negative example is a much easier feat than proving the proposition holds over all possible positive examples!) Establishing conclusively that such a proposition lies entirely beyond the power of our logical framework and empirically accessible facts precludes such a counter-example, which suggests that the proposition is true. This argument mirrors that advanced by Chaitin (1996, p. 28) showing that if the Riemann hypothesis could be proven beyond the power of the usual axioms of mathematics, then it would clearly be true—since, if false, the Riemann hypothesis should have a numerical counter-example which could be easily verified with the usual axioms, were it presented.

The choice I recommend is not metaphysical supervenience; it is logical supervenience grounded in *a priori* necessity, justified by implications of the conceptually necessary noticeability of qualia and perhaps the causal relevance of consciousness. At the same time, the view is tempered by the acknowledgement that the perspectival problems set out in Chapter 4 put a novel spin on the *a priori* part (and, in particular, that they proscribe *a priori* argument all the way to logical supervenience). On this view, absent qualia super-zombies, inverted qualia, temporally twisted qualia, and tortured nuns are all cramped together in the same uncomfortable (but mercifully non-existent) logical boat.

The approach fits with the standard language common in the literature only awkwardly. Typically, logical supervenience with *a priori* necessity as the relevant modality is taken as equivalent to the notion that supervening facts can in principle be logically derived, *a priori*, from the supervenience base. But usage set by philosophical tradition is not always helpful; apart from the fact that the standard position on derivability is just plain wrong, as outlined in 4.2 above, the concerns voiced in the section 'Platonic Hell' in Chapter 3 about elevating logic at the expense of physics return with a vengeance, this time wearing a perspectival hat. The view that *a priori* considerations alone recommend logical supervenience over the metaphysical variety, without establishing any actual derivability, coheres perfectly well with the observation that physics constrains—indeed, *dictates*—what we cognizers can actually derive from the supervenience base. It isn't time to pack up yet; having established a case for logical supervenience with little in the way of actual explanation, the search begins for a way in which the vast chasm between microphysical detail and phenomenal richness may possibly be bridged.

5. INSIDE THE SELF MODEL

This section turns to ways in which the logical relationship between physical facts or cognition facts on one hand and consciousness facts on the other may be made more comprehensible, even in the absence of an explicit derivation of the latter from the former. Allegiance to the 'phantom' view of qualia tacked on to a physical substrate runs deeply into distant nooks and crannies in philosophy of mind, even among many who count themselves staunch allies of strongly reductive physicalist accounts of consciousness: the very fact that so many theorists seem happy with an incoherent notion of 'mental state' attests to this. Section 3 jettisoned this association between instantaneous states of a physical substrate and phenomenal qualities. Where section 4 subsequently explored the link between consciousness and the underlying cognitive structure indirectly and 'from the outside', the task now is to understand 'from the inside' how qualia, liberated from their illegitimate marriage to instantaneous physical states, may be accommodated within an essentially temporal cognitive framework.

5.1 Explanatory Strategy

The arguments above suggest that consciousness supervenes at least metaphysically and probably logically on cognition; in building logical bridges to link consciousness and cognition, it will come as no surprise that I take functional organisation, explored in some detail in Chapter 5, to feature centrally. Functional organisation supervenes logically on the physical world, exactly as outlined in that chapter. As before, I take it that cognition also supervenes logically on the physical world, and it supervenes on functional organisation: there is no sense in which two functionally equivalent organisms could differ cognitively.[19]

(Appealing to supervenience avoids prejudging the question of whether cognition is *identical* to function; it might well be that multiple functionally distinct organisations may instantiate the same cognitive structure.) So, fixing consciousness facts with respect to cognition facts also amounts to fixing consciousness facts with respect to functional facts and physical facts.

The logical landscape within which the self model view finds itself thus looks like this. Assuming the supervenience of consciousness on cognition and on the physical world, one possible account of any given instance of consciousness would be a complete microphysical description of the situation. A good *explanation* of such an instance of consciousness, however, will come at a much higher level, and I suggest the cognitive level of the self model as the relevant one for giving that explanation. (Recall from note 8 on page 109, however, that change in the self model as a cognitive structure need not result solely from processes themselves cognitive in nature.) That the *particular* self model view developed here is the best vehicle of explanation is an open hypothesis, one which is almost certainly wrong in its fine details and which is subject to substantial revision as we achieve a better understanding of relevant logical and empirical constraints. As outlined in 4.2, good explanation does not necessarily come for free even if logical supervenience is established. The self model is just one candidate for filling that explanatory rôle.[20]

As it happens, the self model approach very nearly stands on its own, even in the absence of arguments for logical supervenience—although in appealing to many of the same assumptions, the influence of logical supervenience features implicitly. Even without assuming supervenience on cognition, whatever features of a conscious experience can *make a difference* either to behaviour or to cognition clearly must bear somehow on the functional modules of an organism. Equivalently, it is only those features which affect the functional operation of an organism which can influence externally observable behaviour, such as verbal reports, or alter even the most deeply buried detail of internal cognition. Unless we admit the possibility of uncoupling consciousness and cognition (as in section 3.3) or reject out of hand the causal efficacy of consciousness (as in section 1), thereby reducing it to epiphenomenal cabin luggage, the level of functional organisation seems the most promising place for quantifying not only similarity of cognition but also similarity of conscious experience.

5.2 What Am I?

Conscious experience itself apparently requires a subject: just as no phenomenal qualities float 'out there' independently of changing physical structures, I take it as given that no phenomenal qualities exist which aren't being experienced by a subject. Surely the phenomenal qualities *themselves* cannot take on the rôle of subject. (Consider: what is it like to be *ineffable redness*?) The frozen and chopped qualia examples indicate that whatever the subject might be, it must be changing for it to experience anything. But linking phenomenal experience to change in state rather than to instantaneous state sheds only minimal light on the question of exactly who (or what) it is that is doing the experiencing in the first place. Indeed, even making the case for logical supervenience sheds little light on the question. While one easy answer enjoys a prominent history in the philosophical literature—namely, that *I am* my brain, and my experiences *are* brain states—this identity view is so wracked with difficulties that it plays no rôle here. (For early discussions, see Place 1956 and Smart 1959.) Instead, the considerations above impart a higher level cognitive flavour to the question of where to seat the subject of conscious experience.

However, simply identifying the subject of phenomenal experience with cognitive or functional systems is only marginally more satisfying than identifying it with the specific underlying physical substrate. Vast repertoires of cognitive processes apparently occur all the time without the accompaniment of any phenomenal experience: fortunately for human cognizers driving cars or walking over rough surfaces, for instance, conscious sensation need not be cluttered with the details of every object of perception and every motor action. Likewise, examples abound of 'artificial' processes which are arguably cognitive in nature but remain entirely bereft of conscious experience, such as the processes directed by a computer program normally residing on a CD-ROM in my drawer which emerges occasionally to trounce me in an altogether uninspiring game of chess. Thus, consciousness apparently supervenes on cognition analogously to the way in which functional organisation supervenes on the physical: more than one set of facts about cognition may fix the same set of facts about consciousness, just as one set of facts about function may be fixed by any of several distinct sets of physical facts. This is simply because some facts about cognition can be altered without changing any facts about consciousness. Alternatively, just any old kind of cognition itself, while necessary for consciousness, is not by itself sufficient.

Hence, in searching out a place to seat the subject of conscious experience, what we are really after appears at a slightly higher level of description than that of an organism's basic cognitive framework. At the same time, it should supervene logically on cognition. Rehearsing the supervenience relations of 5.1, functional organisation supervenes logically upon the physical world (and is multiply realisable in terms of it), cognition supervenes logically upon functional organisation (and may be multiply realisable in terms of it), and, finally, what we are after supervenes logically upon cognition (and is multiply realisable in terms of it). This supervening 'something' is the *self model*: a dynamic data structure implemented within a cognitive framework by a functional system. I propose the self model as the seat of conscious experience; *I am a self model*. Phenomenal experience is effected by *change* in the self model.

The gist of the self model view comes to this: it is not the physical substrate *per se* which does the experiencing when the physical substrate changes, and nor is it precisely the functional or cognitive system itself, instantiated by the physical substrate; instead, it is a body of information, physically instantiated by a functional system and changing dynamically with that functional system, which is conscious. On this view, phenomenal experience is an *immediate* feature of that change: it is 'what it is like to be' that body of changing, physically instantiated information.

Very much remains to be explicated about the self model view—including the notion of a data structure and the justification for viewing experience as an *immediate* feature of change in the right sort of one—but another worry takes temporary precedence: excess baggage from too many tours through the history of philosophy waits in the wings, and it must be unloaded now before it begins unfairly burdening the approach on offer.

First, the self model view is not in the business of adding any new properties or aspects of physical structures or of information, nor does it require any new laws of Nature. The self model approach simply offers a way of speaking—like using 'piano' to refer to a collection of wood and wire—intended to make talk of consciousness (and supervenience relations in particular) more intelligible. Thus, it is not dual aspect monism, it is not dualism, and it is not epiphenomenalism; the self model is no more a tacked on construction than 'piano'. (Moreover, to note correctly that the self model logically supervenes on low level physics and thus adds nothing over and above what is implicit at that low level is no more a criticism of a theory of consciousness based on self models than a similarly correct observation about the supervenience of planets is a criticism of Kepler's laws of planetary motion.)

Second, consciousness retains causal efficacy under the self model view by virtue of logical supervenience on cognition and the physical world. Yet, in the company of probably everything else which supervenes non-trivially on microphysical details of the world, consciousness remains 'explanatorily irrelevant' in the limited (and not terribly worrying) sense of section 1.

Finally, in identifying the subject of conscious experience with a changing data structure supervenient on cognition, the self model view is neither a token identity theory nor a type identity theory. It is also rather far removed from the most popular flavours of functionalism, although in common with them it takes the level of functional organisation as an indispensable player in attempts to understand conscious experience.

5.3 Data Structures

Making sense of conscious selves as physically instantiated dynamic data structures requires making sense of data structures in the general case. 'Data structure' needn't be a technical term, by any means, but in using it I do intend a particular quality beyond the straightforward notion of information (the 'data' part), as understood in Chapter 3. With 'structure', I mean to suggest a robust body of information which features *as* a body of information within a functional system. Often, this means the information is specifically maintained by the operation of the functional system. Typically, its rôle in a functional system also means that a data structure can be changed dynamically by the functional system as the system itself changes. The label is of course vague, but it matters not in the least what difficulties might arise in attempting to adjudicate whether a given body of information features *as* information in some functional system. Subsequent discussions about self models always adopt more precise language whenever the vagueness of 'data structure' threatens to become a hindrance.

Understood as a body of information instantiated by a functional system, a data structure need *not* be instantiated continually and directly by some particular functional module or set of modules. In other words, the physical substrate of a data structure need not lie continually within some module at the level of functional decomposition; it might instead be a 'creation' of distinct and far-flung components operating within the system. An example which brings this characteristic into sharper focus is a data structure found in one form or another in almost all modern digital computers: the stack.

In the abstract, a stack is a linear ordered set of numerical values into which new values may be 'pushed' one after another as in a kind of temporary memory storage; later, they may be 'popped' out again—into a register of the central processing unit, or CPU, in the case of a real computer—according to the LIFO (last in, first out) protocol. If values were written on little slips of paper, we might think of a stack as, well...*a stack* of the little slips. 'Pushing' is just placing a new slip of paper on the stack, while 'popping' is taking a slip off again. The contents of the computer's stack may be changed directly by a programmer with instructions to push or pop data; alternatively, they may be changed indirectly in the normal course of the CPU's operation such as when, for instance, the CPU encounters an instruction to visit a subroutine. In this second case, the processor automatically pushes a note of the memory location which comes next after the branching instruction; this is where it will take up reading instructions after returning from the subroutine. The processor then goes on to the address specified for the subroutine, and upon encountering another instruction marking its that routine's endpoint, it automatically pops the appropriate memory location back off the stack and resumes taking up instructions from that point.

For present purposes, the stack's most salient feature is that there is no one place in a computer where the stack must be located, and there is no dedicated 'stack module' to be found. While the particular details depend on the type of computer, in general the stack is implemented by two different sets of numbers ('pointers'), one indicating the start of the stack and one indicating its 'top', together with a chunk of ordinary memory for storing the stack's contents. One or both pointers may be built into the CPU, and actions of the CPU update the contents of the pointers and initiate the copying of information into the section of memory which stores the stack contents. While the operation of the stack *as a data structure* is central to understanding the functioning of virtually any computer program, this 'shared responsibility' for implementing it means the stack almost certainly does not exist *as a functional module*. (Notice that the stack's normally spatially discontiguous physical instantiation is not very amenable to selection by functional decomposition, whereas components like the CPU or memory chips are almost certain to be selected; see section 3.1 of Chapter 5.)

Although the stack is not itself a functional module, it remains entirely legitimate to explain much of the CPU's behaviour (after a branch to a subroutine, for instance) in terms of its contents. In fact, there seems nothing wrong with saying the stack *causes* the CPU to behave in a particular way, even though the 'real' (microphysical) causes of the CPU's behaviour are altogether explicable in terms of properties of electrons roaming in semiconductors and require no mention at all of stacks. This is just another innocent example of explanatory exclusion: we *could* in principle give a microphysical account of CPU behaviour which never mentions the stack, yet its logical supervenience on physical facts earns for the stack a legitimate place in higher level causal accounts. Data structures, such as the stack, which are not themselves functional modules but which play an explanatorily relevant rôle in the system's functioning, we dub *functionally active*. This contrasts with a *functionally passive* data structure which, although it may be maintained by the action of the functional system, does not itself help explain the functioning of that system.[21] Finally, notice that understood as a data structure, the stack stores bit patterns, not clouds of electrons in semiconductors—even though the bit patterns may be implemented by clouds of electrons. In other words, the information stored by the stack understood as a data structure is information at a functional level.

The last quality of the stack relevant for understanding data structures as used here relates closely to this observation: the stack remains 'blind' both to its own implementational details and to the sources of external influences which modify it. The details of what memory chips or current flows store a stack's contents, for instance, are entirely irrelevant to the stack understood *as* a data structure. In addition, relative to the stack, it matters not whether a value is pushed or popped directly by a program using it for temporary storage or by the

CPU as it branches to a subroutine; for the stack, things 'just happen'. Curiously, the stack is, in a sense, just 'along for the ride', even though as above it still may properly be understood as a cause (and an explanation) of CPU behaviour.

What would be interesting would be a data structure with much broader influences both on components of the functional system instantiating it *and* on itself: information within the data structure might interact not only with the rest of the system, as in the case of the stack, but also with other parts of the data structure. An especially interesting data structure would be something like the stack, except it would reside in a system with no CPU (or, alternatively, with very many simpler CPUs) and with primary direction of the instantiating system given over to the data structure itself. It would still be 'blind' to its implementational details and to the sources of external influences which modify it, and while still just 'along for the ride' in the above sense, it could properly be said to be the primary cause of the system's behaviour.

5.4 The Self Model as a Data Structure

On the view advanced here, no simple stack—not even in the largest, fastest, shiniest new multi-processor PowerPC computer—has the tiniest scrap of phenomenal experience as we know it. It does not feel like anything to be pushed or popped, if you are a stack. A sudden bit of 'underflow' (mistakenly trying to pop data from an empty stack) is no more alarming than hours of bug-free storing and regurgitating.

But consider a functional system with all the cognitive capacities normally attributed to sophisticated conscious organisms built in. Suppose the system is functionally similar to an organism I know to be conscious (such as me), thus fixing its conscious experience at least 'metaphysically' and probably logically. What kinds of data structures might such a system instantiate? The self model view is essentially an *empirical* hypothesis about what kinds of data structures such functional systems *do* instantiate, motivated by the desire for a *conceptual* link between their consciousness and their cognition. As a conceptually motivated empirical hypothesis, reasons for rejecting the self model view might appear from either or both of two directions. First, it might turn out that real conscious cognitive systems instantiate nothing like self models. Second, it might turn out that some other hypothesis, while equally empirically plausible, provides a superior conceptual link. If, upon opening the skull of some conscious volunteer, we discovered a brain functionally equivalent to a large, fast, shiny new multi-processor PowerPC computer (complete with stack), the self model view should be rejected because of the difficulties such a setup would encounter implementing a self model. Likewise, if it turns out that self models are of no conceptual help at all in coming to grips with *why* conscious experience should supervene on cognition, then some other more insightful view will surely surpass it.

More specific information theoretic details of self models, together with examples illustrating how some of their simpler capabilities may be implemented neurally, occupy the whole of Chapter 8, but here is a quick sketch. (Those desperate for the details may want to take a detour to that chapter now, especially sections 3 and 4 for low level notes on neural implementation.) From the standpoint of information theory, the central feature of the self model, implemented by a cognitive system much as a stack may be implemented by a digital computer, is its functional representation of large portions of the environmentally situated sensory and motor system of which it is a part. (Recall the definition of *functional* representation from page 44, requiring mutual information content both with the object itself and with information about possible transformations of that object over a given domain.) A sizeable portion of the self model is dedicated to functionally representing the system, 'from the inside' and at a functional level[22], both with respect to information about how the system is affected by its environment and with respect to information about how some components in the system change relative to other components. For systems which include large sensory arrays—the only sort of system, it turns out, in which we should expect self models to develop naturally—this unsurprisingly implies that the self model incorporates considerable quantities of information about its own environment.

The dynamic nature of the system itself suggests that the self model's representation is actively maintained, yet coupling between system and representation is *conditional*, meaning the state of the representation need not mirror that of the system and its environment in perfect lock step. In other words, the state of the representation and that of the represented may temporarily diverge. (By way of comparison, the stack is also an actively maintained data structure, reflecting changes in the system as they occur, but since its rôle understood at the functional level is one of storage rather than representation, the notion of conditional coupling does not apply to it.) The representation is also *lossy*, in that it doesn't preserve the full detail of the object of representation.

The self model amounts to a large store of actively maintained information about the system itself, both as it is at a given moment and as it might be under transformations, and about the system's environment as perceived from the perspective of the system. One way to think of the self model is as a smaller, less-detailed information theoretic copy of the environmentally situated cognitive system of which it is a part, the structure of which normally reflects the functional relationships between some components of the system and the relationships between the system and its environment. The *purpose* of the self model, however, understood as an information theoretic feature of a biological system situated in a real environment, is to harness this information in the direction of the system's behaviour. The self model is not merely an 'informational shadow' following along with the system like a patient scribe; instead, information within the self model plays a functionally active rôle in the temporal evolution both of other parts of the self model and of the system. Much as the stack keeps tabs on information relevant for the functioning of the CPU, gets altered by activities of the CPU, and ultimately directs some of the CPU's behaviour, the self model keeps tabs on information relevant for the functioning of large parts of the system, gets altered by large parts of the system, and ultimately directs large parts of the system's behaviour. One striking difference is that where the stack can be understood as a very simple data structure serving an extremely complex CPU, the self model is an extremely complex data structure serving a large host of functionally simpler components.

For the present purpose of understanding the bearing of self models on consciousness, hand waving about the specific information theoretic properties of such data structures will suffice: self models may simply be thought of as the 'centre of cognitive action' in a functional system which displays all the cognitive capabilities normally attributed to conscious cognitive systems. As a body of information about the system and its environment, however, the self model also constitutes a *context* for the processes of the cognitive framework within which it finds itself. (Conversely, the cognitive framework provides a cognitive context for the information theoretic properties of the self model.) In other words, the centre of cognitive action in a system equipped with a self model is not akin to a CPU, occupying itself with one or two chunks at a time of an instruction stream to which it has access only to local context; instead, the centre of cognitive action is itself a context made up of dynamically updated information about the system and its environment and about interrelationships between and within each.

5.5 What is it Like to be a Self Model?

If, for digital computers, the stack is a low paid assistant and the CPU is the chief executive officer, the self model surely has a fair claim to the title of the executive 'who' or 'what' which does the cognizing in a system featuring such a data structure. A more interesting question is whether it makes sense to think of *being* a self model. Probably it makes no sense to think of *being* a table or a pizza, for instance: although tables and pizzas are exquisitely suited for being tables and pizzas, it doesn't make sense for conscious entities to wonder if they are tables or pizzas. But does it make sense to think that I am a self model? Probably any exploration of this sort of question—the 'what am I?' sort of question—is bound to sound at least a little peculiar, if for no other reason than that the language is not well suited to the job; the discussion below is no different in this respect. Any discussion which doesn't sound a little wild is probably missing something.

Neither the intension nor the extension of 'I' yields straightforwardly to analysis, but this rarely presents difficulties in ordinary use of the word. Here is not the place for extended treatment of what exactly 'I' might mean. But I suggest that thinking of 'I' as referring to a particular self model, the self model which *I am*, misses out nothing, from either the first person or third person perspective, which is captured by the ordinary use of 'I'. Although I admit to finding it bizarre to think of trying to pick it out any other way than with 'I' or its equivalents in other languages, there must be *something* which I am (or, perhaps, some *thing* which I am). There is some thing which any given pizza is, and there is some thing any table is. Since neither tables nor pizzas are *us*, no curiosities of perspective cloud our attempts to identify them; but for the case of 'I', being seems uniquely tricky.

It helps to consider the *prima facie* match between statements applied to you or me and statements applied to a particular self model. Statements like "I have a pain in my tooth" or "I wonder if there will be a conscious machine in my lifetime" or "I am seeing the most gorgeous purple ever" apparently fit the perspective of a self model just fine. The most interesting aspect of the match is of course the having of pain, the condition of wondering, and the seeing of purple: whether for a self model or for the subject picked out by 'I', I take it the statements say something about what it is like to be a particular subject having an experience. If I am a self model, then conscious experience is an immediate feature of change in the self model data structure: it is what it is like for me to *be* the changing data structure. Why? The terse answer, for readers who might have skipped over the preceding couple of dozen pages, is that for the self model, *that's all there is*: changing data. There is no room anywhere else in the data structure to stuff conscious experience, except in the 'what it is like to be' a changing one.

While it seems that must be the source of the conscious experience when the antecedent of the conditional is given—*if I am a self model*—this question differs from that of whether conscious experience should be an immediate feature of change in a self model, considered independently of whether I actually am one. But given the rejection in section 3 of any association between phenomenal qualities and instantaneous states of a physical substrate (and thus between phenomenal qualities and instantaneous states of any supervening data structures), and given the link between consciousness and cognition explored 'from the outside' in section 4, once again there seems nowhere else for conscious experience to go, except into the 'what it is like to be' *perspective* of the changing cognitive structure. Indeed, on the view of logical rather than metaphysical supervenience advocated in section 4, it is here that the phenomenal experience *must* reside. One advantage of the self model slant on that cognitive structure is simply that it provides a way of talking about the particular subject which gets instantiated.

It is also useful to consider features of the self model and their relationship to features of subjects picked out by 'I'. For instance, the self model is (like the stack) 'blind' to the details of its implementation. The self model could no more be aware 'from the inside' of what physical components instantiate it than the stack could; nor could we. Such low level details are invisible at the level of the self model and cannot contribute to its 'ineffable feel'. This blindness bestows on the self model a certain imperviousness to many changes in its underlying physical substrate. As long as the cognitive framework survives unaltered, the self model can detect no difference; likewise for us, for whom the various metabolic processes which turn food into flesh, replacing significant portions of our bodies each day, are usually mercifully uneventful as far as conscious experience goes. Even many—perhaps *most*—details of a system's actual cognitive processes will be lost to the self model, except where the self model explicitly represents those processes at a level where they can play a functionally active rôle. Similarly, we human cognizers can say almost nothing about *how* we perform certain cognitive functions: how exactly do we visualise a pink elephant or count backwards from seven to three in Swedish? From our perspective, such exploits 'just happen'. Along these same lines, the self model is in a way just along for the ride, being automatically updated by the system of which it is a part—yet it remains the cause (and the explanation) of much of the system's behaviour. There is an undeniable sense in which we subjects of conscious experience who get picked out by 'I' are also just along for the ride—and in exactly the same way—as environmental and internal bodily factors impose sensory information upon us; and yet we, too, remain the cause and explanation of much of our behaviour. Finally, just as a tempting distinction seems to exist between what is picked out by 'I' on one hand and my body on the other, there is also a clear distinction between a self model and *its* body (or whatever it finds itself instantiated by). That distinction is as apt to encourage dualistic intuitions in the case of the self model as in the case of 'I'.

Being a self model would account for—would explain, or at least render more comprehensible—many of these features of subjects of conscious experience. But, as in section 4.3, problems of perspective still hinder any straightforward first person logical derivation of an identity. If such a problem of perspective should turn out to be the *only* obstruction to linking the self model and the subject of conscious sensation, then I believe the approach described here deserves serious consideration. Indeed, only half jokingly mimicking Jerry Fodor's attitude toward his language of thought hypothesis (Fodor 1975), as far as I know, the self model view is the only account of the 'I' as the seat of conscious experience which is not *known* to be false.

Of course, almost anything remotely original which could be written about identifying the subject of conscious sensation could also claim that benefit: perhaps the only reason the self model view is not known to be false is simply that it has yet to be very well known at all. But as mentioned above, there must be *something* which is the subject of conscious experience. There ought to be something of which it is true to say that *conscious experience is what it is like to be that thing*. Following the reasoning of section 3, there must be *something* such that conscious experience is a feature of change in that something, from the point of view of that something. I propose the self model as one candidate for filling the blank left by 'something'; objectors are of course invited to field an alternative!

Self models return in some detail in Chapter 8. First, however, while it isn't my purpose to compare and contrast the self model approach with the whole range of existing theories, Chapter 7 takes a brief detour to address (and debunk) one genre of competing theories of consciousness: so-called 'quantum theories of mind'.

NOTES

[1] Getting clear about what *exactly* this means and sorting out the relevant modality is far from trivial, and partial clarification occupies much of the remainder of this chapter. (See Savellos and Yalçin 1995 for several recent attempts.) Fortunately, the intricacies don't bear on the present point. The original applications of 'supervenience' by Hare (1952) and in Davidson's 1970 paper 'Mental Events' (reprinted in his 1980 collection) are distant from those of many contemporary theorists like Kim (1984, 1987). Van Brakel (1994) explores the differences, while Heil and Mele (1993) feature authors working with different meanings and mainly talking past each other. Although many central arguments of Chalmers (1996b) are deeply flawed, that book's second chapter explores supervenience in considerable and useful detail.

[2] Indeed, readers who have absorbed the details of information theory and incompleteness from Chapter 3 will rightly be suspicious of most arguments of the form, 'I cannot think of any argument for *not-X*, therefore *X*', where *X* might be something like 'it is possible that consciousness facts may vary independently of physical facts'.

[3] Including the laws of physics need not slip the discussion into the realm of mere nomic, or natural, supervenience (where the relevant entailment is understood with respect to *natural* possibility): I take it that fixing the facts about the physical world includes fixing physical laws. Natural supervenience requires the addition of some *new laws* (such as, for instance, some psychophysical correlation laws) to fix all the facts about consciousness.

[4] I don't mean here to obscure distinctions between causation and explanation, which Strawson (1985), for instance—and Davidson (1967) before—is at pains to discriminate. (Also see Humphreys 1989.) I mean nothing more controversial than that where consciousness exists as a cause (in the qualified sense of the paragraph following in the text), its causal relationships with behaviour help *explain* that behaviour.

[5] As noted above, even logical supervenience does not *necessarily* mean that the high level facts can be explained in terms of the supervenience base; see section 4.

[6] Here again, by 'causal relationships' I mean to pick out supervening high level relationships: explanatory exclusion should not be terribly worrying as long as the relevant item supervenes logically on the physical world, 'inheriting' causal relevance from it.

[7] The self model as featured here takes distant inspiration from the self models introduced by Sue Blackmore (1993); thanks to Sue for the stimulating conversations acquainting me with the idea. Thomas Metzinger (1993) offers some initial moves to formalise self model descriptions, and a very primitive version of the present view appears in Mulhauser (1995f).

[8] I cannot emphasise too much that while this book explores the self model as a cognitive entity, the state of such a cognitive structure may change not only through cognitive processes but also due to non-cognitive emotional or motivational factors, neurosurgeons poking electrodes in brain tissue, and many other process which directly affects its physical substrate. In other words, a cognitive structure itself may change under other than cognitive influences. Throughout the following sections, 'cognitive change' should be understood in this broad sense.

[9] The loose use of 'state' in section 3 of Chapter 4 falls under the 'everyday' heading.

[10] Following the reasoning of section 1 of Chapter 5, this sort of correlation is in fact trivial, whatever the phenomenal and physical details. I here mean something like what early machine functionalists imagined might work, before Block and others came along to reveal problems; the easiest strategy for the moment is to think solely in terms of supervenience rather than appealing to mathematically naïve varieties of correlation. The discussion in 3.3 discharges the assumption anyway, arguing for it positively.

[11] That is, provided her physical state at unfreezing matches that at freezing, her physical state at unfreezing is 'blind' to all that may have happened in the chilly interlude. Her future physical state might differ for many different reasons, but *not* as an immediate result of anything which happened while her physical structure was frozen.

[12] I mean this line of thought specifically to apply without rigidifying any of the relevant concepts such as consciousness, cognition, or whatever. That is, I take it that *however* we imagine our qualia might have been in the actual world, it remains constitutive of any qualia worthy of the name that they be noticeable.

[13] Because, as it emerges below, the weak view outlined up to now remains compatible with a merely metaphysical supervenience relation, such extrapolations are for the moment confined to a set of possible worlds (such as the single member set including the actual world) across which the same particular relation holds. At this point, given strong *a priori* intension, it would be modally fallacious to argue from the observation that 'across all possible worlds, if my phenomenal experience is such and such, then my physical duplicate's phenomenal experience is such and such' to 'if my phenomenal experience is such and such, then across all possible worlds my physical duplicate's phenomenal experience is such and such'.

[14] The case of the *connection* between cognition and consciousness differs from the case of cats and splats or water and H_2O. While cat-hood is a matter for our linguistic choosing—i.e., *we* decide what makes a cat a cat and thereby determine the extension of the concept for the actual world—'connection-hood' is not. The disanalogy between discovering cats in the actual world and discovering a connection between cognition and consciousness in the actual world puts a funny twist on the difference between *a posteriori* metaphysical supervenience and logical supervenience for the particular case of consciousness and cognition. This curiosity deserves more attention than it receives here. (Also see Horgan 1993.)

[15] Distinctions between 'metaphysical possibility' and logical possibility are dubious: for present purposes, it is better to say that logically possible but 'metaphysically impossible' worlds are simply *misdescribed* with intensions which fix reference in whatever world is actual rather than with the more fitting rigid designators. So-called 'strong metaphysical necessity' and possibility, according to which some meaningful distinction can be maintained at the cost of rationalising additional modalities, does not feature here.

[16] See Block and Stalnaker (1996) for a sparkling argument against the assumption of theorists like Jackson and Chalmers that some *conceptual analysis* of consciousness in physical or functional terms is required to close the so-called 'explanatory gap' (as it was dubbed by Levine 1983). They go on to argue that (contrary to popular belief) the facts about water are *not* entailed *a priori* by the facts about microphysics and that even if the notion of an absent qualia zombie (or 'super-zombie') is not internally contradictory, it is irrelevant to the issue of whether consciousness facts bear the same relationship to physics facts as water facts do.

[17] Hofstadter's (1979) figure 18 attractively illustrates the relationship between provably true or false and unprovably true or false propositions within formal systems.

[18] Just to be explicit, I take it that the set of propositions stating physical facts, together with rules of inference allowing one to check whether a given proof is valid, constitute a formal system just as much as other sets of propositions and inference rules constitute formal systems. The point of the noted paragraph is of course independent of the banal observation that an effective decision procedure exists for tautological soundness in systems like the propositional calculus. (For first order logic, a decision procedure remains impossible.)

[19] This supervenience relation has nothing to do with the distinction between broad and narrow content (Putnam 1975a): as understood here, cognitive structure and cognitive processes are all 'in the head' and do not change in any way as a result of a cognizer's surreptitious teleportation to Twin Earth. For now, concern focuses on the 'syntax' of cognition, not the semantics of meaning or reference (or of rigid designators in particular).

[20] The relationship between the self model view and the existence of a logical supervenience relation between consciousness and cognition bears some similarities to the relationship described in section 4.1 of Chapter 5 between particular functional decompositions searched out by empirical means and the existence of functional decompositions with certain precise mathematical properties (established solely by reasoning).

[21] Of course, *any* data structure which affects the system's operation conditionally (such that differences in the data structure are reflected in differences in the system's behaviour) should properly be understood as functionally active; digital computer examples of entirely functionally passive data structures are thus difficult to find. However, some clearly are more passive or more active than others. For instance, while the stack is very active, the data structure indicating the total number of free megabytes on a disk drive (*not* the structure indicating what particular blocks are free) is comparatively passive.

[22] Similarly to the case of the stack, the self model represents primarily functional rather than microphysical features of the system.

Chapter Seven

Schrödinger's Cat is Dead

I can safely say that nobody understands quantum mechanics... Do not keep saying to yourself... 'But how can it be like that?'... Nobody knows how it can be like that. (Feynman 1967, p. 128)

In our description of nature the purpose is not to disclose the real essence of the phenomena but only to track down, so far as it is possible, relations between the manifold aspects of our experience. (Bohr 1934, p. 18)

The previous chapter explored how consciousness might supervene on cognition and on the physical world, and the next chapter outlines how cognitive structures capable of supporting consciousness might be implemented neurally—but neither treatment has anything at all to say about quantum mechanics. I think that speaks well for both chapters, but many theorists in contemporary cognitive science and philosophy of mind would disagree.

Two principle concerns prompt some theorists to link quantum mechanics and theories of consciousness. On the one hand, it is commonly thought that consciousness plays some important rôle in processes of quantum measurement and that any good account of consciousness ought to explain how it can play such a rôle. And on the other hand, a few people think that the framework of classical physics is not, by itself, up to the task of explaining consciousness and that a good explanation must rely on some special features of quantum mechanics. Usually these features include either state vector reduction—the 'collapse of the wavefunction'—or purported special capabilities (often computational or 'super-computational' ones) of physical substrates while existing in hypothesised coherent states of linear superposition. Talk of Schrödinger's cat, an example many believe shows quantum mechanics on its own predicts bizarre superposed states for ordinary objects like cats, is often mixed in for good measure. The trouble is, comparatively large and high temperature items like cats (or neurons) do not exist in persisting states of linear superposition capable of exhibiting interference effects, and quantum mechanics—contrary to received wisdom—offers no reason to think they should. Cats, bats, lumps of wax, and even physicists' friends spend their time in well defined classical states, and their behaviour, even after interaction with thoroughly quantum systems like decaying atoms, can be described perfectly well with ordinary probability calculus. Indeed, effective classicality extends, under almost all conditions, far below the neural level to that of medium-sized molecules. Yet all of this is consistent with the notion that modern quantum mechanics is a universal theory applicable to everything in the cosmos and even to the cosmos itself. Making sense of that consistency as it bears on philosophy of mind is a primary goal of this chapter.

At a glance, reasons abound for rejecting outright any ostensible bridge between quantum and classical. After all, the classical physics of Newton seems all wrong when applied to very small systems: straight line trajectories and statistical distributions, for instance, clearly don't mesh with the actual interference patterns of the two slit experiment. Likewise, the interference effects so prevalent in quantum theory are noticeably absent from the behaviour of most everyday objects: crowds of last-minute shoppers squeezing through doors on Christmas Eve do not distribute themselves across shopkeepers' walls in messy wavelike bands of high and low concentrations of person matter.

Yet, as this chapter shows, we needn't follow the Copenhagen school in simply dividing the world into classical items and quantum ones, each with different and incommensurable sets of natural laws. Schrödinger meant his cat example to reveal an incompleteness in the quantum theory by deliberately coaxing quantum properties across such a dividing line and into a macroscopic world made to behave counterintuitively. While it doesn't show incompleteness, the example does suggest it is wrong to rest our understanding of the world on such a demarcation at all. It reveals a shortcoming not in the theory but in our understanding, a shortcoming against which theorists have begun only recently to make headway.

This chapter shows why Schrödinger's cat is, despite the title, *either* dead *or* alive—but certainly not both at once. From a modern approach to quantum mechanics based on standard logic, the classical approximation of everyday objects surfaces automatically, and courtesy of an effect called *interactive decoherence*, interference effects between classical states disappear; within a rigorous mathematical context, the opportunity to drop state vector reduction as a real physical process appears.[1] Consciousness plays no rôle, and no uniquely quantum effects persist at the neural levels where some believe they exist.

The following section outlines the relationship between philosophy of mind and quantum mechanics—what little there is. Since much philosophical writing on quantum mechanics takes place in a mathematical near vacuum, I include in section 2 a whirlwind summary of the quantum formalism; this sets out the terminology and background for the rest of the chapter so that each later technical statement may be traced directly back to here. Section 3 outlines a specific approach to interpreting quantum mechanics which, while not essential for present purposes, makes the mechanisms to be described somewhat easier to understand. The next two sections explain the relationship between classical and quantum processes and, in particular, show how the two are smoothly linked at a low level. After section 1, readers familiar with the mathematical foundations of quantum mechanics may skip ahead to section 3 (starting on page 152) without missing a beat, and those who prefer to avoid the technical details altogether might skip all the way to section 6 (starting on page 164), perhaps with a brief stopover for section 5.3, just to get a general feel for the mechanisms underlying decoherence in the general case. The last section summarises the impact of the chapter on the kinds of connections so frequently entertained in the literature between philosophy of mind or cognitive science and quantum mechanics and finds them entirely lacking in rigour.

1. Two Problems of Interpretation

Two questions summarise the current concerns. First, does the process of quantum measurement require a conscious observer? Second, does quantum mechanics allow for special effects in, say, neural microtubules, which could play a rôle in instantiating consciousness in biological systems? It turns out that the particular character of the negative answer suggested here for the first question also prompts a negative answer for the second. Below, I briefly describe the background of each.

1.1 Conscious Observers

In the early days of quantum theory, perhaps it seemed less peculiar than it does today that consciousness itself should be introduced directly into the process of measurement. More important was the fact that quantum mechanics offered an active rôle for an *observer* at all; unlike its isolated and, in principle, non-interfering rôle in classical mechanics, the observer suddenly took centre stage in the theory of quantum measurement. In the 'new physics', the properties a quantum system displays often depend upon what properties an observer chooses to measure. The emphasis on a conscious observer originated with von Neumann's (1932) 'projection postulate', according to which interaction with a conscious observer precipitates, somewhere along the line, a discontinuous and probabilistic jump in the state of a quantum system from what might have been a 'virtual' state of linearly superposed possibilities into a single determinate actual state. The peculiarities of the formalism at the time of this probabilistic jump from 'pure' quantum state to determinate actual state (the 'collapse of the wavefunction') were worked out in some detail by London and Bauer (1939), who referred to the reluctance of textbooks to engage the matter by noting that "physicists are to some extent sleepwalkers, who try to avoid such issues and are accustomed to concentrate on concrete problems" (pp. 218-219).

A range of quantum theories of mind has grown up around the idea that consciousness itself plays some part in collapsing the wavefunction. Perhaps the best known work is that of neuroscientist Sir John Eccles (1986, 1990; also see Popper and Eccles 1977). In one variation, units of causally prior mind stuff called 'psychons' directly collapse the wavefunctions of grids of presynaptic vesicles in neural bundles. A self-avowed advocate of interactionist dualism, Eccles believes the nonphysical conscious mind thus exercises direct control over the activity of cortical structures which, in turn, control observable behaviour.

The view that a specifically conscious observer must step in to 'finish off' a measurement and that a quantum system continues to evolve in a pure quantum state until such a measurement takes place also leads naturally to questions about just how big a system displaying uniquely quantum properties can be; in particular, if quantum mechanics is a universal theory, does it apply to *everything* until it is measured? Schrödinger's (1935) cat, caught in a superposed state by a decaying atom until someone opens a box and observes it, became a launching point for dozens of popular works on quantum theory. Largely under the influence of Wigner (1961, 1963, 1967), even more outlandish scenarios of macroscopic objects in pure quantum states captured the imagination of physicists, philosophers, and lay readers alike. Wild stories entered the popular folklore, including the notion that quantum mechanics suggests the cosmos evolved for billions of years in a quantum mechanical superposition until, in some far-flung corner of the superposition, the first conscious organism appeared, looked around, and collapsed the wavefunction of the whole thing, bringing into actuality the very path of history which made its own existence possible.[2]

This chapter dismisses such stories and accounts for the processes of quantum measurement and the appearance of a process indistinguishable from state vector reduction without recourse to consciousness or even specifically to an observer *per se*. If consciousness plays no rôle in quantum mechanics, then a theory of consciousness has no questions to answer about such a rôle.

1.2 Quantum Consciousness

The alternative, of course, is that perhaps quantum mechanics has a rôle to play in consciousness. Sir Roger Penrose (1989, 1994) and Stuart Hameroff (Hameroff and Watt 1982; Hameroff 1994; Hameroff and Penrose 1995, 1996) offer what seem to me the most intriguing of current theories on such a prospect. On their view, large groups of proteins within neural microtubules sustain coherent superposed evolution in a pure state, recruiting other superposed proteins to the group until the group reaches a certain size, at which point they all undergo spontaneous state vector reduction as a side effect of Penrose's speculative rendition of 'quantum gravity'. The evolving protein lattice is, for them, nothing short of a molecular-level cellular automaton. The phase of coherent superposed evolution they equate with 'pre-conscious quantum computing', while the moment of wavefunction collapse either corresponds to or *is*—it can be hard to tell—a discrete conscious event. *Somehow*, all these cytoskeletal events are meant to add up to conscious experience.

Powerful arguments on both philosophical and scientific grounds (Grush and Churchland 1995) render the idea of simply equating events of conscious experience with state vector reduction, gravitationally induced or otherwise, somewhat dubious. (See Penrose and Hameroff 1995 for a reply.) Grush and Churchland (pp. 24-26) also offer several reasons for discounting uniquely quantum effects in real microtubules, and this chapter suggests other reasons why objects the size of microtubule components could not sustain coherent evolution except under the most extraordinary of circumstances—like, say, being frozen near absolute zero. Crucially, Hameroff-Penrose style coherent evolution can take place only in the absence of an object's interaction with its environment. While the two offer some speculative mechanisms for isolating microtubule components from an *external* environment, despite repeated urgings in personal communication with Stuart Hameroff, I have never managed to persuade them to address the problem of the *internal* environment, which turns out to be similarly important in the view of decoherence outlined here. My impression (Hameroff, personal communication) is that both happily live with the working assumption that nothing causes either 'real' state vector reduction or any effect experimentally indistinguishable from it except observation or quantum gravity—in other words, the assumption that the mechanisms summarised here do not exist.

2. QUANTUM FORMALISM IN A NUTSHELL

This section ignores huge tracts of mathematical territory for the sake of simplicity and concentrates only on those aspects of the quantum formalism required to preserve some kind of rigorous base for the chapter's explorations. A search for more thorough treatments of the quantum foundations might start with Sakurai (1994) or d'Espagnat (1989), while Kostrikin and Manin (1989) provide an excellent and occasionally entertaining introduction to the requisite linear algebra. (In the interest of brevity but at the risk of losing audience, this section takes the linear algebra as given.)

2.1 States and Geometry

Quantum mechanics describes the state of a system with the famous Schrödinger wave equation, a partial differential equation depending, in part, on the system's total energy given by its Hamiltonian function. We represent the equation as a vector in a Hilbert space $\mathcal{H}$. Hilbert space is a complex vector space with an inner product and a norm. Geometrically, the inner product is related to the angle between two vectors—more on that in a moment—while we may think of the norm as a vector's 'length' or magnitude. Hilbert space itself is a *unitary* space, which for present purposes means that it preserves the normal character of addition and of distribution of multiplication over addition. A function can be represented as a vector according to the values it returns for each value in its domain, so that in a simple finite discrete case, we might associate $f{:}S \rightarrow K$, where $S = \{1,\ldots,n\}$, with a vector formed by its values $(f\ (1),\ldots,\ f\ (n))$.

Vectors in $\mathcal{H}$ appear in the *bra* and *ket* notation of Dirac, where a vector α represented in matrix notation as a column becomes $|\alpha\rangle$. Such a vector is called a 'ket', while the corresponding $\langle\beta|$, represented as a row matrix, is called a 'bra'. Multiplication by a scalar a is straightforward:

$$a(|\alpha\rangle) = a|\alpha\rangle = |\alpha\rangle a$$

The inner product of a bra and a ket, akin to the Euclidean inner product except that it may return complex values, is written $\langle\beta|\alpha\rangle$, revealing the roots of Dirac's terminology in $\langle\ \rangle$, called a 'bra(c)ket'. The inner product is a Hermitian operation, meaning transposed complex conjugates are equivalent:

$$\langle\beta|\alpha\rangle = (\langle\alpha|\beta\rangle)^*$$

where the star indicates complex conjugation.

The square root of the inner product of a vector with itself is its *norm*, analogous to the Euclidean magnitude, and a *normalised* vector $|\tilde{\alpha}\rangle$ is proportioned so its norm is unity:

$$|\tilde{\alpha}\rangle = \left(\frac{1}{\sqrt{\langle\alpha|\alpha\rangle}}\right)|\alpha\rangle, \qquad \text{so that} \qquad \sqrt{\langle\tilde{\alpha}|\tilde{\alpha}\rangle} = \||\tilde{\alpha}\rangle\| = 1$$

By convention, since scalar multiples of vectors represent the same state (i.e., properly speaking, states correspond to *rays*), all vectors are taken to be normalised. And thus we may treat the inner product both as the cosine of the angle between two vectors

$$\frac{\langle\beta|\alpha\rangle}{|\langle\beta|\||\alpha\rangle|} = \cos\theta \qquad \text{(where } \theta \text{ is the angle between } \alpha \text{ and } \beta\text{)}$$

and as the length of the *orthogonal projection* of the ket along the bra (alternatively, the 'representation' of the bra in the ket).

2.2 Observables and Matrix Operations

While the quantum formalism represents physical systems with state vectors, the observable properties of those systems appear as *operators*. Observable operators A are self-adjoint, or Hermitian (equivalent to their own complex conjugates transposed), and they operate upon kets from the left and bras from the right. Operators associate but do not generally commute, while the product of an operator and a ket or bra is just another ket or bra, respectively.

Representing vectors and operators as matrices requires first the idea of an *orthonormal basis*, a set of normalised vectors $\{e_1,\ldots e_n\}$, each orthogonal to every other:

$$\langle e_k | e_i \rangle = \delta_{e_k e_i}$$

where δ, the *Kronecker delta function*, is equal to one when $i = k$ and zero otherwise. The basis vectors are said to span Hilbert space in the sense that any vector is a linear combination, or *superposition*, of basis vectors:

$$|\alpha\rangle = \sum_{i=1}^{n} c_{e_i} |e_i\rangle$$

where the complex coefficients

$$c_{e_i} = \langle e_i | \alpha \rangle$$

corresponding to the length of the orthogonal projection of $|\alpha\rangle$ along $\langle e_i|$, indicate the representation of the given basis vector in the overall superposition. Thus, shifting around scalar multiplication,

$$|\alpha\rangle = \sum_{i=1}^{n} |e_i\rangle\langle e_i|\alpha\rangle$$

The first part of the summation

$$|e_i\rangle\langle e_i| = \Lambda_{e_i}$$

is formed by the *outer product* of identical basis vectors and is called a *projection operator*. It selects the portion of a vector parallel to the given basis vector. Conveniently, inner and outer products in Dirac's notation associate:

$$|e_i\rangle\langle e_i|\alpha\rangle = (\langle e_i|e_i\rangle)|\alpha\rangle = \langle e_i|(\langle e_i|\alpha\rangle)$$

so we may view the above summation either as a series of scalars multiplying $|e_i\rangle$ kets or as a series of projection operators $\{\Lambda_{e_i}\}$ working on $|\alpha\rangle$ kets. (More generally, with non-identical vectors, the outer product yields an operator which *rotates* a ket in the direction of the bra vector.) Association works analogously for operators, so that

$$\langle\beta|X|\alpha\rangle = (\langle\beta|X)|\alpha\rangle = \langle\beta|(X|\alpha\rangle) = \langle\alpha|X^{\dagger}|\beta\rangle^{*} = \langle\alpha|X|\beta\rangle^{*}$$

where the last equality holds just for the case of a Hermitian operator X which is its own adjoint such that

$$X = X^{\dagger}$$

Given a particular set of basis vectors $\{|e_i\rangle\}$, we can now represent bras, kets, and operators in matrix form. Kets are straightforward, with

$$|\alpha\rangle \doteq \begin{bmatrix} \langle e_1|\alpha\rangle \\ \vdots \\ \langle e_n|\alpha\rangle \end{bmatrix}$$

where the dotted equals sign indicates 'is represented by', since the matrix representation may differ according to our choice of basis. A bra vector comes as a row matrix

$$\langle\beta| \doteq \begin{bmatrix} \langle\beta|e_1\rangle & \cdots & \langle\beta|e_n\rangle \end{bmatrix} = \begin{bmatrix} \langle e_1|\beta\rangle^{*} & \cdots & \langle e_n|\beta\rangle^{*} \end{bmatrix}$$

so that

$$\langle\beta|\alpha\rangle = \sum_{i=1}^{n} \langle\beta|e_i\rangle\langle e_i|\alpha\rangle = \begin{bmatrix} \langle e_1|\beta\rangle^{*} & \cdots & \langle e_n|\beta\rangle^{*} \end{bmatrix} \begin{bmatrix} \langle e_1|\alpha\rangle \\ \vdots \\ \langle e_n|\alpha\rangle \end{bmatrix}$$

(The inner product is independent of basis.) Notice that the outer product in the summation forms a series of projection operators, which sum to the unity operator:

$$\sum_{i=1}^{n} |e_i\rangle\langle e_i| = \sum_{i=1}^{n} \Lambda_{e_1} = I$$

Obviously, the inner product returns by ordinary matrix multiplication simply a scalar, while the outer product returns a square matrix:

$$|\beta\rangle\langle\alpha| \doteq \begin{bmatrix} \langle e_1|\beta\rangle\langle e_1|\alpha\rangle^{*} & \cdots & \langle e_1|\beta\rangle\langle e_n|\alpha\rangle^{*} \\ \vdots & \ddots & \\ \langle e_n|\beta\rangle\langle e_1|\alpha\rangle^{*} & \cdots & \langle e_n|\beta\rangle\langle e_n|\alpha\rangle^{*} \end{bmatrix}$$

Finally, we represent an operator X in the general case with a square matrix as

$$X \doteq \begin{bmatrix} \langle e_1|X|e_1\rangle & \cdots & \langle e_1|X|e_n\rangle \\ \vdots & \ddots & \vdots \\ \langle e_n|X|e_1\rangle & \cdots & \langle e_n|X|e_n\rangle \end{bmatrix}$$

2.3 Spectra and Probabilities

The *eigenvectors* of an observable A are a special set of orthonormal basis vectors $\{|a_i\rangle\}$ with the unique property that

$$A|a_i\rangle = a_i|a_i\rangle$$

where the scalars $\{a_i\}$ are dubbed the *eigenvalues* of the operator. (By convention, the same letter denotes both eigenvectors and corresponding eigenvalues.) In other words, the operator projects into the subspace spanned by the eigenvectors, and its effect over the space of each eigenvector is equivalent to multiplication by a scalar—the eigenvalue. The set of eigenvalues $\{a_i\}$ of an observable A, called its *spectrum*, is the set of *possible values* which a quantum system may display with respect to the property associated with A. The matrix representation of an observable operator in terms of its own eigenvectors takes on a special *diagonal* form:

$$A = \sum_{i=1}^{n} a_i \Lambda_{a_i} \doteq \begin{bmatrix} \langle a_1|A|a_1\rangle & 0 & 0 \\ 0 & \ddots & 0 \\ 0 & 0 & \langle a_n|A|a_n\rangle \end{bmatrix}$$

A fundamental postulate of quantum theory[3] is that the *probability* of finding a quantum system $|\psi\rangle$ in a state corresponding to an eigenvector for an observable A (i.e., that it has the value of the associated eigenvalue with respect to that property) is:

$$p(a_i) = |\langle a_i|\psi\rangle|^2$$

In other words, the probability for finding value a_i is the square of the 'length' of the projection of the system's state vector along the associated eigenvector. The *expectation value* $\langle A\rangle$ of an observable A with respect to a state $|\psi\rangle$ is then the sum of the possible eigenvalues times the probability of finding each:

$$\langle A\rangle = \langle\psi|A|\psi\rangle = \sum_{i=1}^{n} a_i|\langle a_i|\psi\rangle|^2$$

All this so far applies to a quantum system in a *pure state*, where the system can be described with exactly one wave function. Often, we are concerned with statistical mixtures, in which a whole system consists of various non-interacting quantum systems with different individual state vectors. In that case, we use the *ensemble average* $[A]$ in preference to the expectation value $\langle A\rangle$ for a pure state (although the term 'expectation value' is often applied to both cases):

$$[A] = \sum_{j=1}^{m} w_j \langle \psi^j | A | \psi^j \rangle = \sum_{j=1}^{m}\sum_{i=1}^{n} w_j \left| \langle a_i | \psi^j \rangle \right|^2 a_i$$

where $\{w_j\}$ are the fractional populations of state kets $\{|\psi^j\rangle\}$ in a mixed ensemble. Notice probability here enters twice:

$$\left| \langle a_i | \psi^j \rangle \right|^2$$

is the quantum mechanical probability for a system in $|\psi^j\rangle$ to be found in an A eigenstate $|a_i\rangle$, while w_j is the ordinary probability for finding a system characterised by $|\psi^j\rangle$ in the ensemble.

Probabilities can be calculated conveniently with the last two mathematical tools: the density operator and the trace. The first is defined in the general case as

$$\rho = \sum_{j=1}^{m} w_j |\psi^j\rangle\langle\psi^j| = \sum_{j=1}^{n} w_j \Lambda_{\psi^j}$$

reducing in the pure case to

$$\rho = |\psi\rangle\langle\psi|,$$

which in matrix form is sometimes called an *einzelmatrix*. The density matrix is a projection operator if and only if the system is in a pure state. The einzelmatrix has the special property that, in an appropriate basis, it has one diagonal element equal to one and all other elements as zero; thus it equals its own square. (Indeed, all projectors, having eigenvalues of only zero or one, share this last feature. This connects to the earlier discussion of projection operators $\{\Lambda_{a_i}\}$ where the operator corresponds to a yes/no question as to whether a system displays the associated property.) The density operator allows calculation of an ensemble average conveniently as

$$[A] = \mathrm{Tr}(A\rho)$$

where the *trace* of an operator X is the sum of its diagonal elements (returning a value which is *independent* of the choice of basis):

$$\mathrm{Tr}(X) = \sum_{i=1}^{n} \langle e_i | X | e_i \rangle$$

The trace has several other useful properties:

$$\mathrm{Tr}(XY) = \mathrm{Tr}(YX)$$

$$\mathrm{Tr}(|a_i\rangle\langle a_i|) = \delta_{a_i a_j}$$

$$\mathrm{Tr}(|\beta\rangle\langle\alpha|) = \langle\alpha|\beta\rangle$$

$$\mathrm{Tr}(U^\dagger XU) = \mathrm{Tr}(X)$$

where the *unitary operator* U appearing in the last has the special properties that

$$U^\dagger U = I \quad \text{and} \quad UU^\dagger = I$$

Unitary operators, which conserve inner products and norms (and thus probabilities) are often used to translate from one basis into another, and the Schrödinger equation of a system evolves in time under the action of a unitary operator.

2.4 Wavepacket Reduction

With these tools to hand, the standard account of wavepacket reduction, which will later turn out superfluous on a modern approach to quantum theory, comes easily. The probability given above for finding a system $|\psi\rangle$ in a state corresponding to an eigenvalue a_i for an observable A can also be written as

$$p(a_i) = \langle\psi|a_i\rangle\langle a_i|\psi\rangle = \langle\psi|\Lambda_{a_i}|\psi\rangle$$

To cover the case of continuous rather than discrete spectra while ignoring the subtleties, we rewrite the probability for finding $|\psi\rangle$ in some possibly continuous domain J with respect to an observable A as the expectation value for the projector for that domain:

$$p(J) = \langle\psi|E(J)|\psi\rangle$$

(Again, the projector now corresponds to a yes/no question as to whether the value lies in a particular domain.) Then wavepacket reduction means that immediately *subsequent* measurement of a different observable B gives the probability of finding $|\psi\rangle$ with a value in domain K as

$$p(K) = \langle\phi|F(K)|\phi\rangle$$

where $F(K)$ is the projector for domain K of B and

$$|\phi\rangle = CE(J)|\psi\rangle$$

Here C just ensures normalisation of $|\phi\rangle$ such that

$$\frac{1}{|C|^2} = \langle\psi|E(J)|\psi\rangle = p(J)$$

In other words, probabilities for the second measurement depend only upon the projection of $|\psi\rangle$ into the space spanned by the eigenvectors of J rather than on the whole thing. The method is similar with an initial state characterised by a density matrix; given

$$p(J) = \mathrm{Tr}\big(E(J)\rho\big)$$

the outgoing reduced density operator is given by

$$\rho' = CE(J)\rho E(J)$$

where, again for normalisation,

$$\frac{1}{C} = \mathrm{Tr}\big(E(J)\rho E(J)\big)$$

In textbooks, state vector reduction often appears without comment as to its physical significance. After all, it serves marvellously for predicting experimental outcomes, and so long as it does perhaps interpretation amounts to unwelcome philosophising. (And recall London and Bauer's comment from page 143!) But without adding extravagant philosophical baggage, the simplest path from mathematics to reality seems to be that *before* the measurement process, the quantum system is in some state $|\psi\rangle$, while *after* the measurement process, it is in some *different* state $|\phi\rangle$, which in the simplest discrete case might just be one of the eigenstates $|a_i\rangle$ of an observable A. Before measurement, a system evolves (entirely deterministically) under the action of a unitary operator, but the very *act of measurement* apparently precipitates a discontinuous (and probabilistic) jump into (in the discrete case) a single determinate eigenstate. Eventually it turns out that state vector reduction is not a real physical process at all, although it suffices as a handy calculational tool. First, however, we turn to the logical framework within which the further discussions of quantum mechanics will be fixed and with respect to which we describe the classical approximation for the behaviour of macroscopic objects.

3. CONSISTENT HISTORIES

While my own previous papers on interactive decoherence managed without the language of the consistent histories framework, I now think the method makes understanding decoherence far easier. Robert Griffiths introduced the original version of the consistent histories approach to quantum mechanics in 1984, and while it has multiplied into slightly different flavours under the influence of subsequent authors, the basic ideas remain the same. This section introduces the formulation of Roland Omnès (1994), whose technique is both more accessible on account of its recent presentation in the first book-length treatment of the topic and, I believe, preferable for technical reasons.

The basics are simple. Hilbert space provides the framework for everything which can sensibly be said about a system—i.e., there are no hidden variables. Groups of *histories*—sequences of properties occurring at certain times—to which we can assign well-defined probabilities form internally *consistent families*; within a consistent family, histories are mutually exclusive and jointly exhaustive real possibilities. Histories which don't appear in any such group are *inconsistent*, or meaningless.

The consistent histories approach offers explicit methods for checking the reasoning behind statements about quantum systems. In quantum mechanics, it is all too easy to reason to wrong conclusions when words outweigh mathematics. But once the consistency of a family of histories is ensured, by computing a few traces, the efficacy of 'ordinary logic' is guaranteed, and neither speculation nor special quantum logics have any place.

3.1 Consistency Conditions

With the choice of consistency conditions comes the greatest potential for dispute between different proponents of the consistent histories framework. I follow the method of Omnès (1994), who requires that the probability of a history is always positive or zero, the probability of a trivial history is unity, and the probabilities of two disjoint histories are additive. (Some other stipulations, such as suppressing certain redundancies in history descriptions and automatically returning zero probability for a history where two consecutive properties lead to nonsense, are for now safely left to the side.) The probability p of a particular history h is

$$p(h) = \mathrm{Tr}\left(\Omega^{\dagger} \rho \Omega\right)$$

where ρ is the density matrix of the system in question and Ω is defined by the series of projection operators for different properties at different times:

$$\Omega = E_1(t_1) \cdots E_k(t_k) \cdots E_n(t_n), \qquad t_1 < t_k < t_n$$

For a system in a pure state $|\psi\rangle$, with $\rho = |\psi\rangle\langle\psi|$, the probability reduces to

$$p = \left|\Omega^{\dagger}|\psi\rangle\right|^2 = \left|E_n(t_n) \cdots E_k(t_k) \cdots E_1(t_1)|\psi\rangle\right|^2$$

Thus, each new projector acts on the previous state; the final probability comes from the projection of whatever vector emerges from the previous projectors. Omnès neglects the blatant connection with state vector reduction, but this construction clearly foreshadows his later proof that something like state vector reduction is, from a calculational standpoint if not from a metaphysical one, essentially correct. Although the formulation does not explicitly *presuppose* state vector reduction, it's worth noticing that what might otherwise appear later as an utterly astounding proof is plainly unremarkable in view of the probability's basic definition. As an aside, some may also like to know that such probabilities derive very nicely in the context of Feynman path integrals (Omnès 1994, pp. 130-132), although as before this section avoids such complications for the sake of discrete case simplicity.

With these probabilities to hand, the consistency condition for additivity for a simple family of two histories (disjoint with contiguous union) is very concise in terms of double commutators:

$$\mathrm{Tr}\{E_2(t_2)[E_1'(t_1),[\rho,E_1''(t_1)]]\}=0$$

where the histories are

$$\{E_1'(t_1),E_2(t_2)\} \quad \text{and} \quad \{E_1''(t_1),E_2(t_2)\}$$

with the union

$$\{E_1(t_1),E_2(t_2)\}, \quad \text{where } E_1=E_1'+E_1''$$

The commutator relationship is the usual:

$$[A,B]=AB-BA$$

When the trace above is zero, this guarantees additivity of disjoint histories, represented in the general case by:

$$p(h')+p(h'')=p(h), \qquad h=h'\cup h''$$

Such definitions of the probability for a distinct history and the consistency conditions are of course *time asymmetric*. This differs markedly from the time symmetric formulation of Griffiths (1984). Anticipating the coming foray into reasoning built around the consistent histories approach, one advantage of the time asymmetric formulation is that it preserves the normal logical equivalence between an implication and the *modus tollens* of the same—i.e., the implication from the negation of the consequent to the negation of the antecedent—while the alternative does not. Although Griffiths (1995) claims no advantage comes from Omnès's time asymmetry (especially given the underlying *dynamical* equivalence of quantum mechanics under time reversal), it is attractive at least for preserving the normal understanding that non-commuting observables (those for whom the commutator relationship above does not return zero), such as position and momentum, annihilate information with respect to each other. That is, we might contrive histories wherein relative arrangements of non-commuting observables suggest time-reversed probabilities *ought* to differ. (Perhaps the nuances of time reversal have not yet become an issue because the reliability of the general consistency conditions has only yet been proven rigorously for the $n = 2$ case.)

It is also worth noticing that consistency here emphatically *does not* presuppose decoherence. Some authors, notably Gell-Mann and Hartle (1990) rest consistency explicitly on decoherence. But on the present view, consistency is a *logical* constraint on discourse about histories, while decoherence is simply a *dynamical effect*—exactly as it should be.

3.2 Complementary Logics

Having seen how to select consistent families of histories—of which there might be several for a given system—it's easy to construct logics with them. Propositions are elementary histories (such as one property occurring at one time), and they are manipulated with standard logical connectives.

For instance, AND, OR, and NOT are the usual set theoretic operations, now applied to subspaces of $\mathcal{H}$ picked out by projectors. Likewise, we take entailment between two propositions a and b associated with domains α and β either as the probability measure of the domains' intersection or as the conditional probability

$$p(b|a) = \frac{p(a \,\&\, b)}{p(a)}$$

When the probability of a is non-zero and the conditional probability above is unity, we say $a \rightarrow b$. Not surprisingly, fuzzy logic inspires a refinement; in particular, we allow entailment when

$$p(b|a) \geq 1 - \varepsilon$$

where ε is a small probability of error, the rôle of which becomes more important in section 4 with the discussion of the classical approximation. Logical equivalence, similarly, requires that two propositions each entail the other.

Crucially, a logic built upon a consistent family is itself consistent and all the usual rules, including DeMorgan's laws, are satisfied. Rather than creating new logics with new rules of inference—hallmark of the so-called 'quantum logics'—the method defines different logics simply by reference to the field of propositions over which normal logic operates. Resting different logics upon different fields of propositions naturally gives rise to the well known quantum feature of *complementarity*.

Two logics are said to be *mutually consistent* when some larger logic contains their union, while they are called *complementary* when there is no such larger logic based on a consistent family. Significantly, we are *guaranteed* (Omnès 1994, p. 162) that if two complementary logics each contain propositions a and b, a valid implication between them in one logic is likewise valid in the other.

This convenient arrangement allows us to reason, and to reason *correctly* and without contradiction, about properties of quantum systems even when those systems can be described in different and mutually incompatible ways. For instance, given a two slit or interferometer experiment, no bizarreness clouds the question of which path a particle followed, because there is no consistent logic which contains *both* interferences and information about individual paths. The two are complementary, and we may reason within one logic about interferences or within another about paths, *but never both*. Omnès makes more precise the comment at the beginning of this section that within the consistent histories framework, everything there is to say about a quantum system is phrased in terms of subspaces of $\mathcal{H}$:

> Any description of the properties of an isolated physical system must consist of propositions belonging together to a common consistent logic. Any reasoning to be drawn from the consideration of these properties should be the result of a valid implication or of a chain of implications in this common logic. (Omnès 1994, p. 163)

Although here isn't the place to explore the issue further, notice that in spite of the no-contradiction feature, it is possible to reason to different conclusions in complementary logics. Omnès prefers to reconcile this with what is the case in the real world by distinguishing between *true* and *reliable* properties. Dubbing 'sensible' a logic which includes propositions describing results of actual measurements, *true* describes properties which are those results or are logically equivalent to them *and* which don't generate contradictions if *added to* any other sensible logic. A *reliable* property, by contrast, is contained in one or more sensible logics or can be added to them (and in these logics it is logically equivalent to an actual measurement), but there exists at least one sensible logic where adding it would generate a contradiction. This separates side effects of mathematical complementarity from properties of the real world, and it accommodates sticky features of scenarios with correlated but space-like separated subsystems, such as the famous 'paradox' of Einstein, Podolsky, and Rosen (1935)—which, incidentally, inspired Schrödinger's own *gedanken* experiment with the cat. It also underwrites the categorical assertion that *quantum mechanics is separable with respect to true properties*.

While I find such an arrangement appealing, Griffiths (1995) cites unpublished work indicating problems as well as more unpublished work by Omnès arguing the problems can be overcome. The case certainly looks far from closed; but for present purposes, we pursue the matter no further and recount Omnès's treatment of the emergence of a classical approximation from quantum mechanics.

4. THE CLASSICAL APPROXIMATION

In 1927, Paul Ehrenfest proved a quantum mechanical analogue of Newton's second law of motion, showing that the centre of a wavepacket moves like a classical particle and confirming—at least in terms of *dynamics*—Bohr's *correspondence principle* that quantum mechanics should reproduce the correct classical equations in the classical limit (i.e., when the value of Planck's constant can be considered negligibly small). Yet, Ehrenfest's theorem doesn't explain classical determinism, and by itself, it doesn't allow us to establish a 'logical equivalence' (in terms of the two-way entailment of the previous section) between a present measured datum and the past fact of which it is supposed to be a true indication. This requires some 'translation' between classical properties expressed in terms of ordinary position and momentum (q, p) in phase space—which Omnès calls *collective observables* because they are observable properties of a collective quantum system—and quantum properties expressed as projectors in $\mathcal{H}$. While many different approaches address the problem within the consistent histories framework, the solution outlined here again follows Omnès (1994).

The method is straightforward. With each 'cell' C in phase space (closed, bounded, and simply connected), associate a family of practically equivalent 'quasi-projectors' F, where 'associated' means that each wavefunction $|\psi_c\rangle$ well inside the given cell is almost exactly an eigenfunction of F with eigenvalue one. Quasi-projectors are self-adjoint with discrete eigenvalues, all of which are very near the edges of the range $[0\ldots1]$; then $\mathrm{Tr}(F) = N$, where N (or $N(C)$) corresponds to the number of quantum states localised in the cell. Quasi-projectors have the property that

$$\mathrm{Tr}\left(F - F^2\right) = NO(\eta)$$

where η is a (hopefully small) *effective classicality parameter*. We say F is of *rank* N and *order* η. Finally, define the equivalence class for quasi-projectors with the requirements that

$$N - N' = NO(\eta)$$

$$\mathrm{Tr}|F - F'| = NO(\eta)$$

where Tr$| \; |$ is the so-called *trace norm*, the sum of the absolute values of the eigenvalues of the argument. Using the 'real' initial density operator for the system, we say the system displays the classical property corresponding to the quasi-projector F when

$$\mathrm{Tr}|F\rho F - \rho| < \eta$$

It turns out there is a very good correspondence between quantum evolution and the quasi-projectors on the one hand and classical evolution and classical properties on the other, to within an error quantified by $\zeta > \eta$, the *dynamic classicality parameter*, which depends on the regularity of the initial cell, the system's Hamiltonian, and the elapsed time. Omnès notes that it is provable—using microlocal analysis, also known as 'pseudo-differential calculus' (well beyond the present scope because I have no grasp of it!)—that we can *always* in principle find an appropriate family of quasi-projectors of rank $N(C)$ and order η which yields the classical approximation.

The approximation only works for significant time scales, however, for *regularly shaped* cells (where $N(C)$ is large and η is small). Classical systems in which phase space cells become very distorted, with chaotic systems the paradigm example, can be approximated over only a very short span of time. This is to be expected, of course. Also, the actual definition of the effective classicality parameter η as given by Omnès reveals a feature on which he offers no comment. Specifically, the parameter depends partly on the ratio of the boundary of the cell to the cell's volume (recall the use of 'associate' above and the requirement that a vector be 'well inside'), and the volume concentrated near the boundary of a space grows nonlinearly with the dimensionality of the space. For example, Kostrikin and Manin note (1989, p. 124) that a 20-dimensional watermelon with a radius of 20 cm and a skin thickness of 1 cm is nearly two-thirds skin. This suggests that as we approximate the temporal evolution of more degrees of freedom in higher effective dimensionalities, the effective classicality parameter grows, and the useful life of the approximation likewise decays nonlinearly.

These *caveats* notwithstanding, the established correspondence amounts to a refined and more general form of Ehrenfest's theorem which recovers the essential validity of 'classical logic'—the ordinary logic of the previous section applied to the propositions of classical physics. Classical determinism derives, to within small quantum corrections, directly from the probabilistic laws of quantum mechanics.

Due to the minute corrections, reasoning built upon classical physics suffers always at least a small risk of error when applied to the real world. Under regular dynamics, this is of no practical consequence. There is no practical danger that the roses on my table will suddenly leap into orbit about Alpha Centauri courtesy of quantum tunnelling or that they will suddenly become indistinguishable from a very small version of Mary the neurophysiologist due to some spontaneous reorganisation of nuclear matter. Despite the effective diffusion of a quantum mechanical state so that non-vanishing probabilities might exist for finding objects in all sorts of bizarre states, the correspondence between classical physics and the evolution of quasi-classical collective observables remains *extremely* reliable. This is an important step in bridging the gap between quantum mechanics and reality, but it is not yet the whole story.

5. INTERACTIVE DECOHERENCE

The reason a classical approximation only partly bridges the gap between quantum mechanics and the reality we observe around us is that the discussion above began by assuming a classically sensible state for the collective observables and then finding families of quasi-projectors. But suppose we *don't* begin with a classically sensible state. Although the previous section described how quantum mechanics may reliably follow the evolution of collective observables, why should we expect well defined collective observables to exist in the first place? This is a far more interesting question than the classical approximation itself.

5.1 Macroscopic Interference

The problem shows up clearly with a measurement example which, incidentally, reveals an explicit inconsistency within the standard Copenhagen interpretation. Consider an experiment to measure an observable A of a quantum system $|\psi\rangle$ with macroscopic apparatus M. Suppose the quantum system begins in a linear superposition of eigenstates of the observable, where for simplicity we assume the system may display only one of two values with respect to A:

$$|\psi\rangle = a_1|a_1\rangle + a_2|a_2\rangle$$

Assuming the apparatus works correctly, immediately after a measurement, the macroscopic M is in a state

$$|\theta\rangle = a_1|\theta_1\rangle + a_2|\theta_2\rangle$$

where $|\theta_1\rangle$ and $|\theta_2\rangle$ indicate states of a pointer reflecting corresponding values for the eigenvalue of A. (This ignores the uninteresting states of the rest of M; of course M isn't *really* in state $|\theta\rangle$.) The probabilities for M indicating either state correspond to $|a_1|^2$ and $|a_2|^2$.

Yet the reduced density operator for pointer states is

$$\rho = |\theta\rangle\langle\theta| = |a_1|^2|\theta_1\rangle\langle\theta_1| + |a_2|^2|\theta_2\rangle\langle\theta_2| + a_1^* a_2|\theta_1\rangle\langle\theta_2| + a_1 a_2^*|\theta_2\rangle\langle\theta_1|$$

where the extra terms, the off-diagonal terms of the density matrix, indicate the potential for possible *interference effects between macroscopically distinct states of the measuring apparatus*. It is exactly the presence of analogous non-vanishing off-diagonal terms which accounts for the pretty patterns in interferometer experiments. If M were governed by ordinary probability calculus, we would expect the off-diagonal terms of the density operator for M (rather, the *reduced density operator*, the *partial trace*—see below—of the full density operator taken over the irrelevant degrees of freedom within the apparatus) to vanish. But quantum mechanics offers no immediately obvious reason why they should.

The Copenhagen interpretation of quantum mechanics is *explicitly inconsistent* because it treats the (apparently nonvanishing!) off-diagonal terms as if they did not exist. Recall the 'fundamental postulate' from section 2.3 that

$$\langle A \rangle = \langle \psi | A | \psi \rangle = \sum_{i=1}^{n} a_i |\langle a_i | \psi \rangle|^2$$

from which, decomposing $|\psi\rangle$ into a superposition of eigenstates with coefficients $\{a_i\}$, it turns out the off-diagonal terms are ignored entirely. The 'fundamental postulate' relies on a strict division between classical and quantum; ignoring the off-diagonal terms amounts to stipulating by decree that classical systems *shall* obey ordinary probability calculus. Such a stipulation, however, is both inelegant and superfluous; it can be derived directly from the dynamical evolution of quantum systems.

5.2 A Simplified Example

One reason it has taken more than six decades for decoherence as a dynamical effect to be understood and numerical estimates of its efficacy obtained is that increasingly realistic models of physical systems liable to decoherence become hugely complex. Outlining the mathematical tools to understand the simplest example of proper decoherence could easily double the size of this chapter before even getting to the example. To get a general flavour of the effect, this section outlines an example offered by Omnès in which decoherence does occur but in which the effect depends on only a subset of those dynamical features responsible for it in the general case. This in hand, 5.3 proceeds with a broader description, in words, of decoherence more generally.

Omnès's example first requires the idea of a *partial trace*. When considering together two systems, each with their own Hilbert spaces, the overall system is described in a space which is the *tensor product* $\mathcal{H} \otimes \mathcal{H}'$ of the two Hilbert spaces. The details of the tensor product don't concern the present project, but given two systems $|\psi\rangle$ and $|\psi'\rangle$ in a combined Hilbert space, with $\{|u_k\rangle\}$ and $\{|v_j\rangle\}$ orthonormal bases of the two individual spaces, respectively, consider the partial trace of an operator A defined in $\mathcal{H} \otimes \mathcal{H}'$. The partial trace with respect to the basis $\{|u_k\rangle\}$ is:

$$\mathrm{Tr}_u A = \sum_{i=1}^{k} \langle u_i | A | u_i \rangle$$

If ρ_1 is the density operator for the whole system, then

$$\rho = \mathrm{Tr}_u \rho_1$$

$$\rho' = \mathrm{Tr}_v \rho_1$$

and, in the pure case, the density operator for the whole system is the tensor product of the two individual density operators:

$$\rho_1 = \rho' \otimes \rho$$

(This reflects the absence of correlations between the two systems; joint probabilities are just *products*.)

Consider a simple idealised pendulum, coupled to its *internal* environment by classical friction. (I.e., some energy of the pendulum is siphoned off in the form of heat dissipated into its mass.) For simplicity, the pendulum begins with zero initial momentum at angle θ_0. Since the pendulum can be treated as a harmonic oscillator analogous to the quantum case, the initial time dependent state of the system appears in terms of two systems, the pendulum itself (described with collective variables) and its internal environment

$$|\psi(0)\rangle = |\phi_{\theta_0}\rangle_c \otimes |0\rangle_E$$

Here the subscripts denote 'collective' and 'environment', and the environment begins at absolute zero, or the ground state. Suppose the pendulum's initial state is a superposition of two macroscopically distinct angles, yielding wavefunctions

$$|\psi_1(0)\rangle = |\phi_{\theta_1}\rangle_c \otimes |0\rangle_E$$

$$|\psi_2(0)\rangle = |\phi_{\theta_2}\rangle_c \otimes |0\rangle_E$$

so that the whole system is described by

$$|\psi(0)\rangle = a|\psi_1(0)\rangle + b|\psi_2(0)\rangle$$

At a later time t, the whole system is described by

$$|\psi(t)\rangle = a\left(\sum_j c_j^{(1)}(t)|\phi_j^{(1)}(t)\rangle_c \otimes |w_j\rangle_E\right) + b\left(\sum_j c_j^{(2)}(t)|\phi_j^{(2)}(t)\rangle_c \otimes |w_j\rangle_E\right)$$

where terms inside the brackets indicate the relevant vectors expanded in terms of energy eigenstates for the environment. The vectors $\{|\phi_j^{(1)}\rangle_c\}$ give average values for the angle and change in angle which are practically equal to those for classical motion starting at θ_1, $\{|w_j\rangle_E\}$ are the permitted energy states for the mass of the pendulum, and the $\{c_j\}$ are probability amplitudes which, crucially, differ significantly from zero only for values of $\{|w_j\rangle_E\}$ near the classically dissipated energy.

It gets interesting when we consider the time dependent density matrix

$$\rho(t) = |\psi(t)\rangle\langle\psi(t)|$$

Initially, the reduced density matrix for the *collective* observables (i.e., the partial trace of the complete density matrix over the environment) is

$$\rho_c(0) = |a|^2|\phi_{\theta_1}\rangle\langle\phi_{\theta_1}| + |b|^2|\phi_{\theta_2}\rangle\langle\phi_{\theta_2}| + ab^*|\phi_{\theta_1}\rangle\langle\phi_{\theta_2}| + a^*b|\phi_{\theta_2}\rangle\langle\phi_{\theta_1}|$$

indicating a pure state, the last two quantities once more representing non-zero off-diagonal terms. But after some time, the collective density operator becomes

$$\rho_c(t) = |a|^2\sum_j |c_j^{(1)}|^2 |\phi_j^{(1)}\rangle\langle\phi_j^{(1)}| + |b|^2\sum_j |c_j^{(2)}|^2 |\phi_j^{(2)}\rangle\langle\phi_j^{(2)}|$$

$$+ab^*\sum_j c_j^{(1)}c_j^{*(2)}|\phi_j^{(1)}\rangle\langle\phi_j^{(2)}| + a^*b\sum_j c_j^{*(1)}c_j^{(2)}|\phi_j^{(2)}\rangle\langle\phi_j^{(1)}|$$

And here the coefficients $\{c_j^{(1)}\}$ are non-vanishing only when the w_j is near the right classically dissipated energy $W_1(t)$, in which case the corresponding $\{c_j^{(2)}\}$ (which depend on $W_2(t)$) vanish. In other words, the off-diagonal terms in the density matrix for the collective observables must vanish once the difference $W_1(t) - W_2(t)$ becomes large enough that the corresponding coefficients don't overlap. Only the first two terms in the collective density operator survive this process. This points to the characteristic properties of decoherence in general.

5.3 Decoherence in the General Case

The most prominent aspect of the example above is of course the relationship between classically dissipated energies and the density of the energy spectrum for the environment. Specifically, the difference $W_1(t) - W_2(t)$ is highly sensitive to the difference between the corresponding initial angles and large compared to the quantum fluctuations in the pendulum's internal energy. This reflects basic features of quantum perturbation theory, where two slightly different perturbations give rise to extremely different wavefunctions (to an extent inversely proportional to the separation between adjacent energy levels): wavefunctions describing internal environments coupled to two different macroscopic states are liable to be *vastly* different.

It is this huge difference, together with the enormous dimensionality of the Hilbert space for a macroscopic object—a dimensionality equal to n^N for a collection of N subsystems, each with n degrees of freedom—which implies that wavefunctions for macroscopically distinct states *very rapidly* become orthogonal. The effect is analogous to the fact in ordinary geometry that as the dimensionality of a space approaches infinity, the proportion of its subspaces intersecting transversally, or aligned to each other in the *general position,* approaches unity. ('General position' indicates that two spaces intersect in the smallest possible dimension with a union of the greatest possible dimension, such as two planes in three dimensions intersecting at a line or two planes in four dimensions intersecting at a point). Simplifying greatly, the probability of randomly selecting two wavefunctions in a high dimensional Hilbert space which are *not* orthogonal is vanishingly small. So long as there is even a very small coupling between an object and its environment—*internal or external*—this tendency of wavefunctions for the environment rapidly to become orthogonal means the phase relationship between macroscopically distinct states is destroyed, off-diagonal terms of the reduced density matrix vanish, and interference effects become impossible. This amounts to an extremely rapid and efficient decrease in the number of its possible states which can be distinguished through their effects on the environment (i.e., which are *experimentally meaningful*). To give an idea of the efficiency of the process, Joos and Zeh (1985) calculate a decoherence time of just 10^{-36} seconds for a spherical grain of dust one micron in radius at standard temperature and pressure.

Once the off-diagonal terms of the reduced density matrix disappear, of course, we can treat different eigenstates of the collective observables with ordinary probability calculus. The philosophical implications of this return in a moment, but first the discussion of decoherence ends with a note of its importance for measurement theory.

5.4 Decoherence and Measurement Theory

Most importantly, it is decoherence *in the measuring apparatus* which transfers the quantum property of a microscopic system into something real and distinguishable—and observationally meaningful—in the macroscopic world. Ignoring fine details of the internal arrangement of the measuring apparatus, ordinary probability calculus applies straightforwardly with respect to the data, and truth 'flows' backwards from the data shown by the measuring apparatus to the measured properties of the microscopic system.

After impressive proofs of a string of theorems establishing the equivalence of measurement data and microscopic properties, the equivalence of the respective probabilities, and the outcome of repeated measurements of the same system with respect to the same observable, Omnès (1994, p. 338) *derives* from the consistent histories approach the general form for state vector reduction. Ironically, in light of the comment sometimes made that *decoherence* offers merely a "calculational tool" (Kiefer 1991, p. 379), this rule for state vector reduction emphatically *does not describe a real physical process*; it is merely a *computational convenience* for predicting the outcomes of measurements. The framework as outlined here permits state vector reduction to be dispensed with altogether; instead, the wavefunctions of measuring apparatuses and the like could, in principle, be followed in minute detail, nonetheless turning up—on account of decoherence in the measuring apparatus—*the very same results*. With its explicitly inconsistent *stipulation* that macroscopic objects behave classically, the Copenhagen interpretation guarantees the same calculational result, but on the present view both the quasi-classical macroscopic behaviour and the mathematical rule of state vector reduction can instead be *derived*. We are rescued from all the trickiness one might worry could arise as a result of taking measurements without physically reducing state vectors by the logical structure of the consistent histories approach. Veterans of the measurement problem will, however, notice this can't *quite* be all there is to it.

6. IS THIS A FAPP?

> *I am quite prepared to talk of the spiritual life of an electronic computer... What really matters is the unambiguous description of its behaviour, which is what we really observe. The question as to whether the machine* really *feels, or whether it merely looks as though it did, is absolutely as meaningless as to ask whether light is "in reality" waves or particles. We must never forget that "reality" too is a human word just like "wave" or "consciousness". Our task is to learn to use these words correctly—that is, unambiguously and consistently. (Neils Bohr as quoted by Kalckar 1967, p. 234)*

John Bell (1990) coined the moniker 'FAPP'—For All Practical Purposes—for approaches to quantum theory which start and finish by pointing out that it gives the right answers. The old state vector reduction might be justified, for instance, by pointing out that its predictions correspond extraordinarily well with experimental results. Never mind that it treats off-diagonal elements of the density matrix for the measurement apparatus as if they didn't exist and offers a puzzling and rather unsatisfying picture of the way the world really is. Some maintain that the approach to decoherence summarised above smacks of FAPP as well. Bell himself (1990, p. 36) argues with respect to decoherence that if we weren't on the lookout for probabilities in the first place, we'd have little reason to suppose that diagonal elements of a reduced matrix correspond to a disjunctive 'this possibility *or* that possibility *or*...' instead of the conjunctive 'this possibility *and* that possibility *and*...' normally attributed to pure states of linear superposition. In other words, mightn't a decohered state *still* be a superposed state? After all, no unitary process can turn a pure state into a mixture.

For my part, I find it remarkable that virtually everyone happily accepts the notion that an electron, say, may 'really' be in infinitely many places at once, yet some then balk at the proposed mechanism for putting it in one 'real' and determinate place. Consider: what is the evidence for electrons *really* existing in infinitely many places at once? Has anyone ever *seen* an electron in infinitely many places (or even just two) at once? Of course not. But literally thousands of experiments, by way of macroscopic measuring instruments, have verified *signs* of their existence in such a form (i.e., a pure state capable of exhibiting interference effects); so, we infer that they really do exist that way. And so it goes for many other entities in physics, including all the sub-microscopic ones: we see only the macroscopic signs or effects of their existence.

But why then should the available empirical data constitute evidence for the notion that decohered states for *collective observables* still indicate an object in more than one place, rather than one *actually* located somewhere? (Alternatively: how could empirical evidence adjudicate on the two alternatives?) If a decohered object cannot exhibit interference effects, what could possibly constitute evidence that it really was in more than one place? The standard answer is that interpreting decohered states as *real* disjunctive alternatives (as opposed to 'virtual' conjunctive ones) demands an additional quantum mechanical axiom; namely, it requires some principle which directly connects the underlying 'reality' of von Neumann's type I process (unitary evolution) and type II process (state vector reduction).[4] That is, it apparently requires an axiom which connects structures *after* decoherence to reality in a new and different way than before, stipulating that decoherence *is* state actualisation.

Such a judgement makes sense, of course, provided we *begin* with the assumption that the mathematical structures of the quantum formalism *before* decoherence already correspond to reality in some particular preferred way. But suppose we begin instead with the assumption that macroscopic objects, the ones we observe around us, behave in accordance with the normal probability calculus. (It's a sneaky trick, appealing to everyday reality when the topic is heady quantum mechanics, but suppose we try it on for size.) Then the quantum formalism after decoherence automatically corresponds nicely with reality. A collective density matrix which has become diagonalised through interactive decoherence offers numbers which behave like real probabilities, and there is every reason to believe they *are* real probabilities. (Or, trying to judge competing assumptions even-handedly, is there any reason to believe they are *not* real probabilities?) Now it is instead the alternative view which requires a new axiom: namely, that the mathematical structures *before* decoherence correspond to reality in any particular way.

And nothing more clearly illuminates the absurdity of trying to *justify* an axiom (as opposed to simply assuming it) stating this correspondence—i.e., the correspondence between pre-decoherence formalism and reality—than Neils Bohr's above question of whether an electron *really* is a particle or a wave. Just because an electron can be *described as* a wave (or particle), does that mean it *is* a wave (or particle)? Does it change being what it is when we describe it in a different way? I don't think so. Moreover, it is helpful always to keep in mind when considering the question that superposition and the other features of the quantum formalism are just mathematical devices for describing the results of observations: particular details of the formalism are in no way essential, because in the space of all mathematical statements, there is an infinite host of other mathematical descriptions which match any finite set of empirical data just as well. (This is the general problem of underdetermination first mentioned on page 26 in Chapter 2.) The existing quantum formalism just happens to be an especially concise description. Suppose we could compress the quantum formalism into a more compact form—and it scarcely needs pointing out that *no one knows* if this is possible, because as far as I am aware no one has proven the algorithmic information content of the formalism. (See Chapter 3.) Would we then have to ask a whole *different* set of questions about the correspondence between parts of the new compressed description and elements of reality? Indeed, the very assumption that each part of the quantum formalism must correspond to some element of reality was the famous 'classical' mistake of Einstein and his two graduate students, Boris Podolsky and Nathan Rosen (Einstein *et al.* 1935).

Finally, it might be useful to phrase things this way: the processes of decoherence can turn a 'pure state' *for quasi-classical collective observables* into a mixture. Of course, no unitary process can turn a 'real' pure state into a mixture, but states of collective observables defined in terms of quasi-projectors aren't 'real' pure states. This provides what is required to make sense of an 'actual' macroscopic world while simultaneously permitting an agnostic view of the 'actuality' at the very lowest levels. (Perhaps at the very lowest levels *there is no actuality* in the macroscopic sense of the word.) And if there is anything about which one ought to remain agnostic, I would much sooner give up a particular view of the 'hidden' microscopic 'actuality' than of the real macroscopic world I observe around me every day behaving in accordance with the ordinary probability calculus.

I believe the view which emerges from all this complements nicely Bohr's sentiment at the start of this chapter as well as a particular view of scientific explanation with a long tradition. E.W. Hobson echoes Bohr:

> The function of Natural Science is to describe conceptually the sequence of events which are to be observed in Nature... Natural Science describes, so far as it can, *how*, or in accordance with what rules, phenomena happen, but it is wholly incompetent to answer the question *why* they happen. (Hobson 1923, pp. 81-82, emphasis original)

Likewise for present purposes: quantum mechanics describes how electrons *behave* but neither why they behave that way (as opposed to some other way) nor what they 'really' are apart from their behaviour. Quantum mechanics does not explain *why* the behaviour of an electron can be predicted with a Schrödinger wave equation or *why* Planck's constant is what it is. It shows merely that such a description enables correct prediction of experimental results.[5] (In this carefully limited sense, *all* scientific theories are FAPPs.) While a determination to move away from descriptive views of the scientific enterprise toward more 'substantive' explanation has been a hallmark of nearly all analytic philosophy of science in the second half of this century, not even the most sophisticated contemporary approach overcomes the basic insight of Hobson, Bohr, and others like them. Modern scientific explanation may 'explain' in the sense of unification or in the sense of logical implications from sets of propositions describing natural laws (together with empirical data), but still it does not begin to bear on the fundamental question of the *overall* 'why?'. Demanding something more from the quantum formalism strikes me not as a sign of some sort of philosophical maturity; instead, it seems precisely analogous to Jonathan Swift's eager scientist ambitiously set on the ludicrous enterprise of extracting sunbeams from cucumbers!

7. MIND—WHAT'S QM GOT TO DO WITH IT?

The chapter began by noting that quantum mechanics and classical physics each look terribly unprepared for any but the most uncomfortable existence in the other's domain. Newton's physics goes all wrong, for instance, giving sub-microscopic particles classical trajectories, while the waves of quantum theory seem all out of place in a macroscopic world of distinct and well defined objects. But it turns out that the time evolution of macroscopic classical observables is mirrored by the quantum evolution of families of quasi-projectors, and quasi-classical determinism (with only very tiny corrections) emerges automatically from a probabilistic foundation. Through decoherence, we've followed the process of measurement *without* state vector reduction and discovered that the interaction between a macroscopic system and its environment is sufficient to provide macroscopic reflections of quantum properties which may be treated with ordinary probability calculus. Interactive decoherence enables the automatic emergence from the quantum substrate of a macroscopic reality described by macroscopic collective observables. Quantum theory applies universally to large and small objects alike, but the dynamical properties which emerge for macroscopic quasi-classical bodies are simply different from those of elementary particles like electrons.

With respect to philosophy of mind, the chapter has shown, from the standpoint of physics itself, why neither of the two reasons canvassed in section 1 for linking quantum mechanics and consciousness ought to be convincing. For quantum theories of mind which require some special quantum mechanical effects in protein lattices or elsewhere in brain-temperature or microtubule-size objects, the case is very nearly closed. Current work on quantum effects in microtubules simply does not take account of decoherence due to dynamical coupling between an object and its *internal* environment (Hameroff, personal communication). To be sure, it is far from impossible to sustain coherent quantum effects in macroscopic objects. For instance, lasers—originally an acronym for 'light amplification by stimulated emission of radiation'—work by keeping transitions in electron energy states in phase throughout a whole macroscopic substrate. SQUIDs—for 'superconducting quantum interference devices'—likewise exhibit quantum properties both in virtue of being superconductors and as a result of quantum effects over their Josephson junctions. But each requires *very* special conditions to perform its feats—including continuous energy input for the first, supercooling for the second, and special geometries for both. Showing that similarly special conditions could plausibly exist in real biological systems requires doing a little more work, particularly since the mechanisms described to date do not even acknowledge interactive decoherence as a possibility.

For the case of approaches to quantum mechanics which involve conscious minds in some central rôle, things are only slightly less clear cut. The most important point is that the quantum formalism can do all the work itself without recourse to a conscious observer or even to an *observer* at all. If, for some reason, someone were really desperate to link quantum mechanics and consciousness, perhaps it could be done; somewhere within the probabilities describing what becomes actual after decoherence, perhaps some (causally irrelevant, epiphenomenal) rôle for consciousness could be contrived. But such a rôle would differ markedly to that which consciousness has played in quantum mechanics to date, where so far it has simply appeared in tandem with the notion of an observer. In the absence of some independent reason that consciousness should suddenly acquire such a new rôle, even though quantum mechanics works fine without it, trying to make the connection seems a desperate and altogether unjustified strategy.

The contemporary literature on 'quantum theories of mind', as well as on what might be called 'mind-based theories of quantum mechanics'—including what is written by physicists—is almost universally devoid of real details about the physics underlying the alleged connections. More often than not, authors seem to expect the philosophical and cognitive science communities simply to take it on faith that someone must really have worked out all those details, especially when physicists are involved—after all, they're physicists! But to the extent they allow their faithfully believing colleagues outwith physics to go on thinking those details are there when they are not, the physicists, like the proverbial emperor, have no clothes. Echoing the sentiment of Dennett with respect to zombies (see Chapter 2), if quantum mechanics is important to consciousness or *vice versa*, some physicist ought to be able to say why in non-question begging terms. Doing so does not mean citing the authority of other physicists, it does not mean speculating about Bose-Einstein condensates, and it does not mean simply deferring to one or another aspect of the Copenhagen interpretation; it means doing some real physics, showing how the mechanisms summarised in this chapter can be circumvented, and displaying the results for all to see. To be sure, I am no expert on physics, and in the end it may turn out that my understanding of what is principally Roland Omnès's work is all wrong or irrelevant. But it will take effort to show that, effort which so far has not been made. Until it is, the kinds of links between quantum mechanics and consciousness so frequently entertained in the literature remain, from the vantage point of this chapter, unsubstantiated fancy.

NOTES

[1] Some relevant work includes Albrecht (1992), Mulhauser (1995a, c, and e; 1997b), Omnès (1990, 1994), Paz *et al.* (1993), Paz and Sinha (1992), and Zurek (1991, 1993). Of these, Omnès (1994) is the most significant and complete.

[2] Philosophy and popular folklore are full of perversions. In fact, the notion on which this story rests, that the full wavefunction of the cosmos, representing all its possible states, evolved in a pure state until the first conscious organism appeared in some component of that wavefunction, reveals clearly that we do not live in a universe actualised in this way. The story is simply wrong: a conscious organism ought to have appeared almost *immediately* after the creation of the universe on this view, since, while of very low probability, such a freak occurrence would still have been one possibility latent in the wavefunction. If this were the case, the history of the cosmos ought to be altogether different than it looks to be, and even now almost any well isolated system with enough mass ought spontaneously to reorganise itself into a conscious entity and become actual. Curiously, I have never encountered this straightforward reply in the literature.

[3] See section 5.1 for an explanation of why the 'standard interpretation' of probabilities is explicitly inconsistent.

[4] Views vary on how to account for the actualisation of one determinate reality. In personal communication, Brian Josephson suggests that what is really needed is a mechanism (such as consciousness) whereby some *on-diagonal* elements of the density matrix go to zero, effectively rendering quantum mechanics more deterministic. Saunders (1993) points to evolutionary constraints on the development of complex systems as a superselection rule, while Omnès (1994) and Griffiths (1995) both seem to adopt a straightforward view that reality is unique and

simply unfolds probabilistically in accordance with the predictions of quantum theory. It is this last view to which I am most sympathetic.

[5] It will come as no surprise that I disagree with Bohr's behaviourism with respect to conscious computers; but as for physics, it seems spot on. The intuition otherwise calls to mind the Aristotelian metaphysics of 'substance' and the urge to find a thing's 'essence' apart from its interactions with other things, an urge the original Aristotelian form of which—despite tension with the text's next sentence on philosophy of science—has been *almost* entirely expunged from contemporary science.

CHAPTER EIGHT

Building Conscious Data Structures

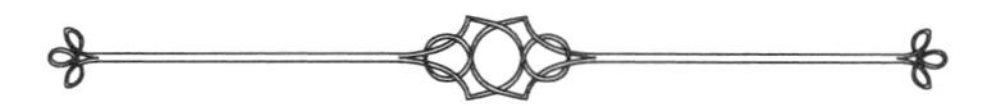

While Chapter 6 introduced self models as a candidate link between cognition and consciousness, helping to make more intelligible a supervenience relation clouded by problems of perspective, the particular relationship between the information theoretic properties of self models and their underlying physical substrate featured in that discussion only marginally. Where that chapter mainly explored the conceptual space from the self model 'up', toward supervening consciousness, the present chapter works in the opposite direction, 'down' toward a supervenience base, a biologically plausible framework for implementing self models in neural tissue.

The journey begins with some expansions on the information theoretic aspects of self models described in section 5.4 of Chapter 6 and continues with brief introductory remarks on neural modelling. Developments in contemporary neuroscience, and some advances due to Stephen Grossberg and his colleagues in particular, provide a significant portion of the ideas required to make sense of self models implemented in real brains. Along the way, both representation and the notion of hypothesis testing acquire a neural basis, and the symbol grounding problem dissolves. Readers who have skipped over the details of the information theoretic approach to representation outlined in Chapter 3 may prefer also to skip the information theoretic properties of the self model described here in the following section, instead going straight on to the neural basics in section 2, which begins on page 174.

1. INFORMATION AND THE SELF MODEL

As outlined near the end of Chapter 6, one way to think of the self model is as a functionally active, less detailed information theoretic image of the environmentally situated sensory and motor system of which it is itself a part. Its structure normally reflects the functional relationships between many components of the system and the relationships between the system and its environment. Perhaps the label 'self-in-a-world-as-it-looks-from-here model' would have been more descriptive, since the dynamic data structure essentially reflects a centred *perspective* on the world and the system within it. Indeed, the 'self' part of the self model might be understood as an implicit creation of the global consistencies within, or underlying invariant properties of, three sets of information: about the environment as filtered by the system's sensory tissue, about the system's motor actions, and about the system itself, both with respect to proprioception and in terms of internal functional relationships between the organism's cognitive subsystems.

Metaphorically, the self model amounts to an information theoretic 'window' on the world as seen through the organism's sensory systems and a window on the organism itself as seen through proprioception, motor output, and the relationships of each to the other and to the perceived outside world. At the same time, the self model is itself the structure which 'looks through' these windows; the self model both lodges and, partly with the help of the rest of the system, answers 'queries' about the system and about its relationships to the environment. These queries and answers underwrite what is perhaps the most significant characteristic of the self model understood within an evolutionary context: its ability to test hypotheses about the organism itself, about its environment, and about the relationships between and within each—and to direct behaviour accordingly in a functionally active way. This hypothesis testing capability takes centre stage in section 4, which addresses its neuroanatomical preconditions, but for now this section develops more carefully some information theoretic features of the self model.

1.1 Lossy Compression and Conditional Coupling

The self model's functional representation of its own environmentally situated perceptual and motor system amounts to *lossy* compression of information about the system, its environment, and the relationships between them. (See section 3 of Chapter 3 for functional representation.) 'Lossy' denotes a compressed representation from which it is not possible to recover the full details of what is represented. Alternatively, the transformation from object to representation is many-to-one and thus non-invertible. (Lossy compression features in, for instance, the popular MPEG—for 'Moving Picture Experts Group'—format for encoding digital video data. By recording only the changes between individual frames in a video stream and by ignoring substantial amounts of relatively unimportant detail in each frame, MPEG achieves compression ratios for typical video streams of around 30:1.) For the moment, this indicates simply that the mutual information content of the self model's representation with respect to that which it represents is, while significant, not maximal.[1]

The temporal evolution of the representation itself is also partially (and conditionally) *coupled* to that of the system. By 'coupled', I mean the activity of the self model's representation is linked to the activity of the overall system and the environment represented. A reflection in a mirror offers a degenerate case of a coupled representation: when the object reflected changes, the reflection itself changes. (In the general case, however, such an image is neither functional nor compressed.) The conditional information relation between the representation and that which is represented usefully quantifies the notion of coupling: when the information content of a program for deriving a description of the object of representation from a minimal description of the representation itself remains consistently low over time, this suggests the representation persistently bears useful similarity to what is represented. Conversely, when the conditional information relation deviates above the minimum, this indicates the need for extra information to 'fill in the gaps'. Given a functional representation, this extra information may, in the most straightforward case, amount to specifying appropriate application of the transformation axioms with which the functional representation has high mutual information content. In such a case, specifying application of the transformation axioms may come to little more than detailing the particular environmental conditions within which the system finds itself. Alternatively, a deviation of the conditional information relation may reflect a change in the system in response to environmental conditions outwith the domain of application for the transformation axioms. (As in note 11 on page 44, recall that representing the information content of transformation axioms may be achieved entirely *implicitly*; nothing even remotely resembling an explicit transformation axiom appears in the implementation strategy outlined later in this chapter.)

Finally, the information theoretic 'match' between the self model and the system or environment may vary for either of two reasons. First, the coupling between representation and represented is only *partial,* meaning that the self model is not normally a perfect 'mirror image' of the system and environment. Second, coupling is *conditional*: what matching there is may be temporarily suspended to a large extent, inducing a correspondingly large increase in the conditional information relation described above.

1.2 Coupling and Hypothesis Testing

Such suspension occurs in the case of the hypothesis testing mentioned previously (and which returns in greater detail below). From an information theoretic perspective, hypothesis testing is characterised by the kind of deviation in the conditional information relation described above and a subsequent 'extraction', or utilisation, of information from a functional representation about the effects of non-actual environmental or internal conditions. In other words, hypothesis testing is a process of suspending the link between the representation and the represented in order to use the compressed representation as a 'model' for testing the hypothesis. If this could be done with the image in a mirror (which of course it cannot), hypothesis testing would amount to temporarily suspending the link between the reflection and that which is reflected and applying some transformation to the reflection itself so as to glean information about how the object reflected might behave under a similar perturbation.

Significantly, the self model takes responsibility both for hypothesis testing *and* for generating scenarios to test. This becomes much clearer once a neural architecture which may support it is laid out below, but for now it will do simply to note that while the activity of the self model answers 'queries', it is also the activity of the self model which poses them. Given its functionally active rôle in directing behaviour, there are good reasons for thinking such a capacity to generate and test hypotheses about itself and its environment garners for the organism equipped with a self model robust selective advantages, *ceteris paribus*, relative to those lacking such a facility.

Finally, recall the observation from the section 'Grounding Representation' in Chapter 3 that, understood as objects with high mutual information content, representations are comparatively rare. Just as the overwhelming majority of individual strings approach maximal randomness—fewer than one in a million can be compressed by 20 bits or more—the overwhelming majority of pairs of strings are also maximally random relative to one another. Most pairs, being algorithmically independent, do not represent each other.[2] For simple reasons of cardinality, instantiating anything with the information theoretic properties here attributed to the self model is *extraordinarily* nontrivial.

2. REPRESENTATION AND FUNCTION IN NEURAL SYSTEMS

One of the primary puzzle pieces which any advocate of a roughly representational approach to consciousness ought to provide is a biologically plausible account both of how any structure within an organism can *be* a representation and, moreover, of how such a thing can be *used as* a representation. Present purposes suggest two *caveats*, however. First, by 'representation', I mean strictly the precise and objective information theoretic notion of the word set out in section 3 of Chapter 3; I do not mean anything like the formal symbols of abstract computational algorithms, chosen by the programmer to 'represent' quantities. Moreover, I explicitly reject as a cognitive science research strategy the much maligned formal computational approach of old symbolic artificial intelligence. Such rejection is actually shared by most contemporary AI researchers; as Inman Harvey (forthcoming) suggests, "many whose knowledge of AI is secondhand do not appreciate that for a decade or so the new thrust in AI has been towards situated, embodied cognition, and a recognition that formal computation theory relates solely to the constraints and possibilities of machines (or people) carrying out algorithmic procedures".

Along the same lines, nothing here presupposes that in order to use a representation as a representation, an organism must apply to it some inherently 'computational' transformation. On the contrary, in the spirit of Beer and Gallagher (1992), the approach I outline here is thoroughly dynamical in nature: representations, instantiated by physical structures in real organisms, are created and transformed by the dynamical evolution of neural systems; a representation is used as such when the physical features which instantiate it within an organism play a causal rôle in shaping the organism's internal (i.e., cognitive) or external behaviour with respect to the object of representation. That is, using a representation as a representation requires first that the information *itself* is functionally relevant or functionally active in the sense of the stack example from section 5.3 of Chapter 6. However, a functional rôle alone is necessary but not sufficient, because a given representation instantiates significant mutual information content with many other objects which it represents but does not represent *for the organism*. Here we understand it to represent an object for the organism when the representation specifically bears on the temporal evolution of the organism's cognition or behaviour with respect to the object represented. This receives some more careful cleaning up in 2.3, starting on page 185, but for the moment the more immediate job is to explore how specifically representational structures may be implemented neurally in the first place and to outline some basic ideas for making sense of the neural architectures which appear in section 3.

2.1 Neural Basics

Many excellent textbooks and collective volumes cover human knowledge of our own brains from the nuances of excitable cell membranes up through gross anatomy.[3] This body of knowledge appears, on one hand, remarkably robust, mature, and detailed and, on the other hand, startlingly incomplete. This section cannot begin to recount even in summary all that is available on the topic elsewhere, and as mainly an outsider to the field and certainly no authority on cognitive neuroscience, I am not the best to offer such a summary anyway. (My own published papers on artificial neural networks are so primitive as not to be worth citing!) My modest aim is only to give enough of a flavour of neural models to lend some credence to the notion of neurally instantiated representation and to illustrate that the self model is not unlike what we could reasonably expect to find in real biological systems; it is *not* to argue that self models definitely exist in biological systems because of some particular set of ironclad neuroscientific facts. The nomenclature and style of presentation adopted here is a quirky mix of standards in both neuroscience and artificial neural network research and thus suits neither perfectly, but for present purposes I believe it captures the most immediately salient features of each.

We start in the middle, by appeal to an analogy with digital circuitry which is nearly as far removed from real neural behaviour as possible. Figure 11 shows a simple set of AND gates, each of which takes two inputs and produces a '1' output if and only if both inputs are '1'. So, the rightmost AND gate produces a '1' if and only if all the inputs $\{x_1 \ldots x_4\}$ are '1'.

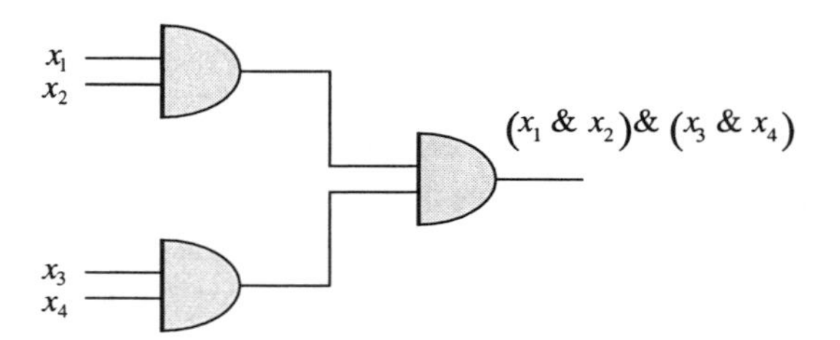

Figure 11. Basic AND Gates

Real neurons[4] resemble this picture only very faintly, but the example offers somewhere to begin. In particular, ignoring the biological details of cellular function, it can be useful to think of a neuron as a 'black box' which does something *remotely* like a logic gate.

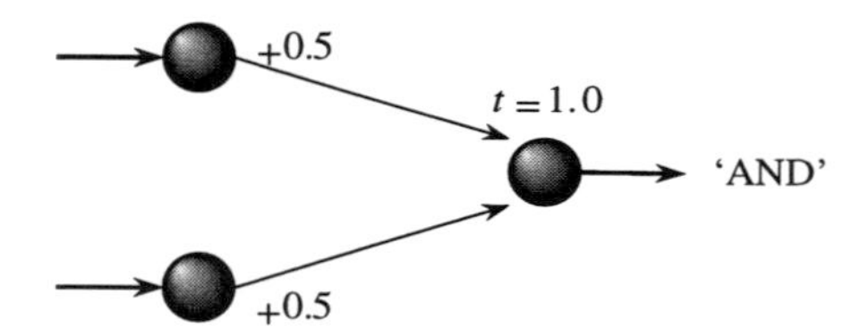

Figure 12. Logical Connective in Naïve Neural Form

In reality, a neuron receives 'inputs' in the form of neurotransmitter chemicals delivered to its dendrites and soma, or cell body, across junctions with other cells called *synapses*. These induce either excitatory post-synaptic potentials (EPSPs) or inhibitory post-synaptic potentials (IPSPs), which on the present simplified picture can be understood as electrical signals, conducted by the exchange of charged particles across the cell membrane, which either incline the cell to fire outgoing action potentials ('spikes') along its axon or discourage such firing. The overall balance of incoming potentials affects whether the neuron fires (in a process of 'depolarisation', during which the cell membrane's electrical negativity decreases) at an increased spiking frequency or remains at its base frequency. Finally, it is normally assumed that neurons conduct signals in only one direction: inputs, or afferents, impinge on the cell body or the dendrites, which passively funnel signals to the cell body; and outgoing signals, or efferents, traverse the axon from its proximal to distal end. (Contrary to popular tale, this is apparently wrong—see Stuart and Sakmann 1994—but the evidence against it is quite new, and for this context no harm comes from retaining the normal assumption.) Thus, abstracting away from the details suggests a unit which takes inputs in one area and provides in another area spiking behaviour related to some overall function of the inputs. Based on this abstraction, Figure 12 depicts a very simplified neural analogue of the AND gate.

Here, afferent signals enter two 'neurons' from the left, and each passes on efferent excitatory signals to a third neuron. The numbers '+0.5' indicate the strength of the excitatory connection between cells; this synaptic efficacy, or 'weight' in the language of artificial neural nets, works something like a multiplier applied to the output behaviour of the neurons. The label 't = 1.0' indicates the firing *threshold* of the third cell; if afferent signals sum to a total below this level, it fires at its base frequency, while total signals at or above this level induce it to fire at a higher frequency. For this example, the threshold could be set anywhere in the interval (.5,1.0]—i.e., above .5 and at or below 1.0. It's easy to see that if afferent signals excite the initial cells to fire at a normalised suprathreshold rate of '1', total excitatory signals they provide to the 'AND' cell will excite it to a suprathreshold level as well. Normalising suprathreshold cell outputs at the unit level allows this set of three cells to be understood as a sanitised neural version of the familiar AND gate. (Of course, a single cell with multiple inputs and an appropriately tuned threshold may function as an analogue of the AND gate as well.) Lowering the threshold of the 'AND' cell to .5 or below or, more generally, to below the level of the product of a single synaptic weight and the strength of the 'ON' signal it might carry, effectively transforms it into an 'OR' cell. A single cell with a zero threshold and producing inhibitory transmitter chemicals, corresponding to *negative* efferent connection weights, acts as a 'NOT' for a single input, and the whole repertoire of standard logical connectives follows straightforwardly. (As an aside, real nerve cells are thought to produce either inhibitory or excitatory neurotransmitter chemicals but not both; thus, properly speaking, positive and negative connection weights should not emanate from the same cell. Instead, an end result like that of mixing positive and negative efferent signals may be achieved by adding an inhibitory 'interneuron', like a NOT gate, which converts excitatory signals to inhibitory ones.)

Such a simple digitised rendition very roughly matches early pictures of neuronal functioning (McCulloch and Pitts 1943) and the interpretation which features in the first modern approach to artificial neural networks from a computer science perspective (Rosenblatt 1962). But apart from blatant biological implausibility—see below—the primitive 'perceptron' model of neurons working like binary ON/OFF threshold units suffers the well known limitation of producing networks computationally equivalent to finite automata, a limitation which, due to Minsky and Papert (1969), effectively killed research by computer scientists into artificial neural networks until the revival spawned by more capable models in Rumelhart and McClelland (1986). (Fortunately, Minsky and Papert's work went largely unnoticed by neuroscientists, who had long since advanced beyond such grossly oversimplified models anyway.)

Injecting a little more biological plausibility into the picture requires a host of modifications, of which I mention here only two. First, the relationship between total input strength and spiking behaviour in real cells is neither perfectly deterministic nor well modelled with a simple discontinuous 'all or nothing' threshold. Figure 13 compares the simple threshold (a) with a 'sigmoid', or 'S'-shaped (b), response function, which, while still too naïve for biological reality, at least does it a little less violence. Here, total inputs appear along the horizontal axis, while (normalised) output appears on the vertical.

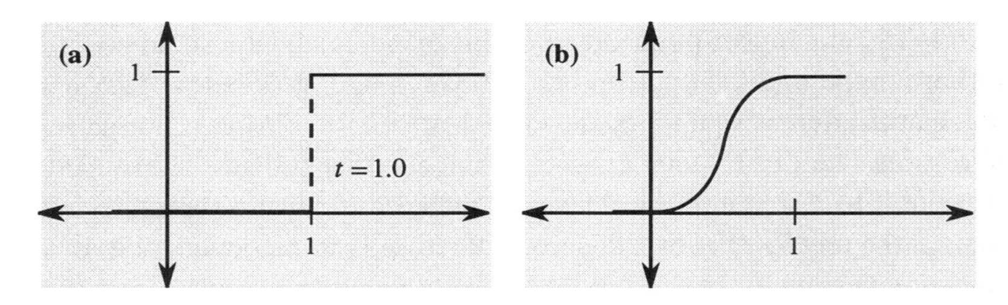

Figure 13. Threshold (a) vs. Sigmoid (b) Response Functions

A graded analogue response for individual nodes is much more physically reasonable than that of the classic binary threshold unit. Not only are real physical systems incapable of distinguishing two arbitrarily similar values as required by a discontinuous threshold function, but the spiking frequency of real neurons is a continuous value (Shepherd 1990; Kosko 1992; also see Gustafsson *et al.* 1992 and references therein). That is, spiking frequency varies in real time over a continuous range, rather than simply assuming one of two possible values. Perhaps under the influence of the early 'perceptron' model, it is sometimes asserted incorrectly that real neurons are, in the end, merely discrete ON/OFF switches because they are either firing or not firing. But this ignores an expanse of neural dynamical features at least as vast as what is ignored by saying that rain is a discrete phenomenon because either it is raining or it isn't. In any case, in comparison to simple discontinuous thresholds, the minor improvement in biological plausibility achieved by the sigmoid response function makes possible much richer dynamics within networks of neurons and increases theoretical[5] network power markedly, to at least the level of Turing Machines (Siegelmann and Sontag 1992, 1995). (Chapter 9 addresses possibilities for extending the power of neural networks even further.)

The second major missing feature is *plasticity*: individual connection weights, or synaptic efficacies, in real neural networks change over time. In other words, neural networks do not merely 'compute' constant (or stochastic) transformations of their input signals; they adapt. Rules governing change in synaptic efficacy, both permanently and transiently and over both the short and longer term, remain the subject of enormous research attention. The early association rule of Hebb (1949; also see Hayek 1952), suggesting that synaptic strength increases straightforwardly according to the degree of correlation between activity in the pre- and postsynaptic cells, still maintains its place as one of the favourite 'biologically plausible' rules applied in artificial neural net research—despite the fact that its biological *im*plausibility as anything but a very crude approximation is almost universally acknowledged in the neuroscience community.[6] Likewise, the formal equivalence between the Rescorla-Wagner (1972) description of behavioural learning during classical conditioning and a special case of the 'delta' network learning rule of Widrow and Hoff (1960) has earned for rules of the latter type some biological credibility and a central rôle in artificial neural net research, usually under the guise of 'backpropagation' (as the method was dubbed by Rumelhart and McClelland 1986).[7] But while the Rescorla-Wagner rule is useful, if not entirely correct, as a description of *higher level* conditioning (see Grossberg 1982c for an historical review), empirical evidence of a neuroanatomical basis for the corresponding Widrow-Hoff rule is nonexistent. (Lynch *et al.* 1989, p. 183 indicate simply that the rule "may more fruitfully be thought of as a behavior-level rule".)

In fact, synaptic efficacy itself is properly divided into at least two components: presynaptic efficacy, or the amount of neurotransmitter chemical released by the presynaptic cell per depolarisation, and postsynaptic efficacy, reflecting the voltage generated in the postsynaptic cell for a given quantity of transmitter delivered. In all likelihood, separate sets of mechanisms at both molecular and structural levels drive changes in the two components independently. For specificity, one model (Finkel and Edelman 1985, Finkel *et al.* 1989) understands presynaptic efficacy as a combination of a baseline efficacy and the short term variations in the availability of neurotransmitter due to the relationship between its production and depletion; longer term modification in presynaptic efficacy amounts to a shift in the baseline efficacy taking place across all the synaptic terminals of a cell's axon (as might be effected, for instance, by a change in gene expression). The model links change in postsynaptic efficacy to the relationship between signals arriving at nearby synapses on the same cell (heterosynaptic inputs) and signals arriving at the particular synapse in question (homosynaptic inputs). Similarly to mechanisms suggested by Lynch and Baudry (1984), heterosynaptic signals interact with modifying substances produced locally in response to homosynaptic inputs so as to yield a complex interaction between the spatial and temporal properties of both sets of signals. Unlike blanket changes in baseline presynaptic efficacy, occurring nonspecifically across a cell's entire axonal arborisation, modifications in postsynaptic efficacy affect individual synapses differentially and reflect both spatial and temporal properties of the afferent signals. (For summaries of other approaches to long term change in synaptic efficacy, see note 19 on page 193.)

The upshot of this and other more biologically plausible models of learning in real networks is that *somehow*—a brazen gloss which suffices for present purposes—networks of nerve cells modify themselves over time in a way which relates significantly to the signals they process and to the balance of a whole host of intra- and intercellular chemical substances. In particular, by way of many different mechanisms, some of which appear in greater detail in section 3, specific populations of neurons may come to respond preferentially and consistently to *particular* spatial or temporal features of activity in the receptive fields of their dendritic trees.

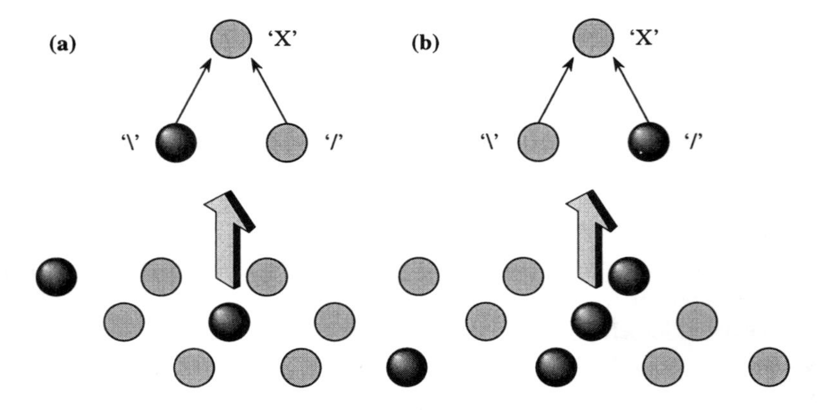

Figure 14. Simplified Neural Feature Extraction

For example, Figure 14 depicts grossly simplified feature extraction by cells which respond differentially to particular spatial patterns in a receptive field. Consider a simple 'layer' of neurons, such as an early stage of some sensory receptor system, with efferent projections to the receptive fields of distinct cells or cell populations by way of connections which excite the two cells or populations differently. (In this illustration, individual 'cells' should be understood as cell populations.) For instance, the '\' cell in Figure 14(a) might respond only to particular corresponding activity in the presynaptic layer, while the '/' cell might respond to different activity, shown in (b). One basic way to achieve such response is simply to provide the relevant cells excitatory connections with all and only those cells active in the appropriate pattern and null connections with the rest, but it's safe to assume that the organisation of biological networks will not be so tidy. (It is also safe to assume that, in keeping with the discussion above on response functions, real cells may code information in various ways, perhaps by means of modulation of their action potential frequencies or perhaps through the temporal relationships between action potentials of distinct cells—for example, see Aertsen and Preissl 1991, Judd and Aihara 1993 or Fujii *et al.* 1996—rather than solely through the rudimentary ON/OFF business of either firing or not firing.) Just as activity in the '\' and '/' cells thus depends on the presence of a particular pattern of activity in the initial neural array, further feed-forward connections can yield an 'X' cell or population which responds preferentially to the 'pattern' made when both '\' and '/' populations are active. Of course, the full story requires filling in several details—such as inhibitory connections to filter the response of the '\' and '/' cells to all *and only* relevant patterns, rather than any superset of relevant patterns—but the specific details don't bear on the main point.

That main point is this: within the context of the above network—i.e., given the connection patterns of the network, the particular response properties of the individual cells, and the physical laws governing their interaction—the activity of the 'X' cell (or the '\' cell or the '/' cell) *represents* the pattern of activity in the initial neuronal layer. Why? Information about both the architecture and, for instance, the fact that the 'X' cell is excited to suprathreshold levels (or is firing at a particular frequency), implies that the pattern of activity in the initial neuronal layer is a member of whatever class of such patterns can elicit such activity in 'X'. (Why 'class of patterns'? The transformation describing the relation between activity in the initial layer and that in subsequent populations need not be invertible: it may be that any of several distinct input patterns elicit roughly the same response in the 'X' cell, for instance.) Implication between the two bodies of information indicates mutual information content and thus representation. In other words, given a minimal description of one body of information, whatever portion of information about the other body corresponds to the implication from the first effectively comes for free.[8] So, within the context of a neural system (its interconnections and so forth), activity in a particular neural population can represent the activity of whatever other neural populations caused the activity of the given one.

Treating the initial layer of the example network in Figure 14 as an actual sensory receptor sheet, analogous reasoning suggests that activity of the 'X' cell represents the *object* which caused the activity in the receptor sheet. Indeed, on the assumption that some (nonmaximal) subset of the features of the particular object brings about the activity in the receptor sheet, we can even say the 'X' cell or population activity represents that particular subset of features. Alternatively, the collective activity of an ensemble of such feature extractors, each responsive to different subsets of an object's features, may *together* represent that object even more precisely.[9] For a quick and tidy discussion of representation specifically in the visual system, see Marrocco (1986). (For more detail, see the classic Hubel and Wiesel 1962 and Lund *et al.* 1975.) Marrocco (pp. 48-57) includes a useful evaluation of an issue left aside by the present discussion: whether the match between a stimulus in a given visual cell's receptive field and that cell's activity is best understood in terms of stimulus shape or spatial frequency. On the first alternative, neural populations 'analyse' visual stimuli in terms of object features, and on the second, in terms of underlying spatial frequencies. While frequency theory is undoubtedly more biologically plausible than feature theory, the latter is simpler for present purposes; in any case, remaining agnostic on the question will do for this context, because the approach here can be rephrased easily to fit with the language of either. (For instance, rather than speaking of representation of an object in terms of its topological features extracted by neural populations, we swap to speaking of representation in terms of sine wave components extracted through what can be described as Fourier analysis performed by neural populations.)

Leaving aside the feature vs. frequency subtleties, two characteristics of this simplified view of representation bear especially on present purposes. First, the sort of architecture described above amounts to a system for categorising stimuli; representing a pattern as an 'X' comes to *categorising* it as an 'X'. (After all, activity in the 'X' cell represents not only the particular pattern which did in fact elicit it, but the whole class of patterns which might counterfactually have elicited it.) Depending on the invertibility of the transformation which yields activity in 'X' and on the projection of those cell's efferent signals, a system containing a rudimentary network like that from Figure 14 may lack any ability whatsoever to reconstruct any details of activity in the initial neuronal layer or any details of the object which caused that activity. It may simply display the capacity to categorise future presentations of similar stimuli in a similar way, and nothing more. Giving the observation a slightly different twist, such a system might categorise properties of stimuli and behave toward stimuli as if it possessed an internal 'picture' to which it could refer, yet such behaviour in no way implies the availability of such an internal 'picture'. Likewise, the presence of a *representation*, as understood in the objective framework offered in this book, also in no way implies the presence of such an internal 'picture'.

Second, from the vantage point of the system, the activity of a single neural population which represents efferents of several others effectively *compresses* information about those other cells. That is, just one or a few smaller neuronal populations may essentially compress information about activity across an entire sensory receptor sheet. Where activity in smaller populations depends differentially on particular features of afferent signal patterns as in the example above, we can understand this information compression as smaller populations representing an activity pattern in terms of its different features. As in the previous paragraph, however, it may well be that a system incorporating such compression in terms of features lacks the capacity to 'decompress' the information directly, recovering the fine details of the compressed signals. Both because a system may lack the architecture to perform subsequent processing on the representation and because the representation itself may respond similarly to any of a disjunctive class of distinct patterns, such compression is *lossy*.

(As an aside, an end result related to but distinct from actual decompression may be achieved as a side effect of the more detailed architectures of section 3—see note 22 on page 196. Similar effects may also follow selective disinhibition of direct projections from a neuronal array x to some later array y; for whatever reason, signalling across such a pathway might normally be kept in check by inhibitory signals from active compressed representations which would usually provide their own signals elsewhere in the receptive field of y. In this way, y could *either* receive afferents from a compressed representation *or* more directly from the source x.)

2.2 Neural Darwinism

Two programmes of neuroscience research deserve particular attention from those interested in pursuing further the neural basis for the kinds of ideas outlined in this book. The first is due to Gerald M. Edelman and the second to Stephen Grossberg, two theorists whose approaches contrast in every way. Edelman emphasises the rôle of what amount to positive and negative selection pressures on large populations of neurons in adapting an organism's nerve tissue during its lifetime. (Thus it may come as no surprise that Edelman migrated to neuroscience from research in the selectionist mechanisms of the immune system, for which he was awarded the Nobel prize in 1972.) Grossberg, whose path to neuroscience began with mathematics and psychology, aims instead to model directly the simplest possible network configurations which display properties analogous to those of higher level cognitive and motivational processes (such as primacy and recency effects, etc.). Grossberg's style is uncluttered and accessible, while Edelman writes almost as if his theories are and should be some of the most complex and difficult material in history. Books and articles by the two display remarkable consistency in ignoring each other. But while their methods clash rather significantly, I believe the actual mechanisms each describes complement nicely. This section explores Edelman's work very briefly, while Grossberg's occupies the whole of section 3.

Edelman's aspiration with the theory of neuronal group selection,[10] or 'neural Darwinism', is to explain how an organism can categorise its perceptions in a world which obviously doesn't come with labels pre-fitted. Many of an organism's basic actions demand the ability to differentiate object from background, friend from foe, desirable from harmful. Many capacities here attributed to self models presuppose the very same abilities, both for categorising exogenous stimuli and for categorising the results of internal associations. Edelman's concern focuses particularly on the kinds of complex polymorphous categorisations (i.e., those lacking singly necessary and jointly sufficient conditions for category membership) characteristic of human psychology (see Smith and Medin 1981 and the contributions in Rosch and Lloyd 1978; on the slipperiness of category boundaries in an attribute space, also see Bongard 1970).

The selectionist aspects of Edelman's approach bear on groups of hundreds to thousands of functionally related neurons, with similar receptive fields, maintaining strong connections with other cells in the group and weaker connections to those outside the group. In contrast to the work of Changeux and colleagues (Changeux and Danchin 1976, Changeux *et al.* 1984), whose selectionist model centres on the eliminative selection of *individual* neurons, the theory describes mechanisms implementing both positive and negative selection of entire populations. Two central concepts underlie the group selectionist approach: *reentrance* and *degeneracy.*

The first refers broadly to axonal arborisation, or projection, from one group into the input areas of another (or, alternatively, dendritic arborisation into the appropriate efferent areas). Projection back into the input areas of the originating group is a special case of anatomical reentrance referred to variously as *recursivity*, *recurrence*, or simply *feedback.*

'Degeneracy' refers to the availability within each repertoire of functionally related neuronal groups of an abundance of approximately isofunctional but not isomorphic structural variants. That is, within each large 'grouping of groupings' of neurons, there exist many similarly responding groups with differing architectures. (Degeneracy relates closely to *redundancy*, which describes groups of structures both isofunctional and, to some close approximation, isomorphic.) In this context, degeneracy acts as the source of variability required within any population which owes its constitution to selective pressures, and the same variability helps ensure that the overall repertoire retains the ability to respond to new stimuli in novel ways.

Edelman argues that interaction between different synaptic modification rules at work within degenerate repertoires (such as the pre- and postsynaptic rules described above on page 179) results in the *effective* selection—through the strengthening or weakening of their connections—of some groups over others for the job of representing properties of their afferents. That is, while the particular mechanisms *implementing* selection operate at cellular and molecular levels, the theory suggests that the resulting change in neuronal repertoires can be interpreted usefully in terms of selection acting at the group level.[11] Edelman has demonstrated basic artificial feature mapping networks operating on this principle which, even when restricted to postsynaptic rules alone, mimic notably well the results of several experiments illustrating the emergent dynamic organisational properties of areas of somatosensory cortex mapping the hands and fingers in monkeys (Merzenich *et al.* 1983a, b, and 1984).

Reciprocal reentrant signalling between distinct repertoires, according to the theory, coordinates representations in separate areas of the brain, ultimately creating coherent relationships between representations of stimuli in different modalities and between those representations and behaviour. Anatomical reentrance also refines the properties of receptive fields in different repertoires and sharpens their topography. Edelman (1989, p. 289) is at pains to point out that the rudimentary 'classification couple'—a pair of basic feature maps with reentrant signalling—in his Darwin II network overcomes artificial neural network limitations mentioned above due to Minsky and Papert (1969). In fairness, however, there now exists in the literature a vast selection of other artificial neural networks which also overcome the early limitations—a vast selection which, much to the consternation of his critics, Edelman routinely ignores in his own writing.

Finally, along the lines of the note above that neural representation can be understood as classification, or categorisation, Edelman and his colleagues view long term memory strictly in terms of such classification:

> Long-term changes do not "store" activity patterns. They are necessarily crude because of the biochemical and molecular constraints on long-term storage. Instead, long-term changes have a population property by which they can differentially bias the system toward recreating short-term changes similar to those that precipitated the long-term change. The constant introduction of variability makes perfect recall unlikely, but it allows for continued adaptability and paves the way for generalization. Our view of memory, in fact, is...one of enhanced bases for *recategorization* of objects or situations. (Finkel, *et al.* p. 161)

It is clear, however, that such an approach, which Finkel *et al.* liken to the old 'reconstructionist' school[12] (Oldfield and Zangwill 1942a, b; Bartlett 1964), remains altogether compatible with understanding representation in terms of mutual information content. In particular, whatever mechanisms are responsible for the "differential bias" also bear responsibility for retaining mutual information content in just the way described in 2.1.

2.3 Representation—Being One vs. Serving as One

Before proceeding to some more detailed examples of neural circuitry, the interpretation of how a representation comes to be used as such, outlined at the start of this section, requires some clarification. In particular, the approach begs no questions about symbol grounding, a problem which returns in section 4.4. At first it might be thought that the view suggests a representation is used as a representation of something precisely when it is grounded in that something; taking the symbol grounding problem as one of understanding how a syntactic representation can bear semantic content about something, this yields the altogether unhelpful 'a representation is used as a representation of something precisely when it bears semantic content about that something'. On the contrary, representations here become *automatically* grounded in many 'somethings' in virtue of the definition of representation in terms of mutual information content. But that same definition also means any particular representation bears content about many things. The view described at the start of the section is that the rôle of the representation in an organism's cognitive structure picks out the *relevant* 'something'. Neural activity in my visual cortex representing a tree I am about to climb also represents many other tree-shaped things, but not *for me*; my behaviour and internal cognitive manipulations of the information in the pattern of activity reveal that for me, the activity pattern represents *this* tree and not a tree in Morocco (I am not in Morocco) or a tree-shaped fissure in a piece of quartz.

However—and this is crucial—descriptions of such behaviour and cognitive manipulations *already appear implicitly in the self model*, because the self model functionally represents much of my cognitive apparatus and large portions of my environment. Thus, the context of the organism as a whole answers the question of whether a particular representation is *used as* a representation and, moreover, reveals what in particular it represents for the organism. Of course, just as section 3 of Chapter 3 noted originally that a representation may *be* quantitatively good or bad, so too may a representation as understood here be *used* well or poorly to represent a particular object. (Accounting for the philosophical bugbear of *mis*representation offers no spectacular challenges.) The account applies equally well for existent things, imaginary things, and abstractions (where, for the last, it makes sense to analyse their information content at least to the extent that they can be *described*). For the case of existent things, the story goes just as for the tree example above, according to which the self model functionally represents my cognitive and behavioural disposition toward a particular tree, thus pinning down what the representation represents *for me* out of a large class of tree-like things. But in my brain are also found representations of wizards, unicorns, and talking butterflies, probably none of which are to be found 'out there'. On the view advanced here, however, my representations nonetheless instantiate substantial mutual information content with *descriptions* of all these things, and likewise for abstractions.[13]

(This may clarify the sense in which the approach to representation and content advanced here differs radically from the popularly denigrated Aristotelian 'correspondence' theories. The approach adopted in this book is so far removed from the mainstream philosophical literature on representation that even a modest discussion situating the present view within that context might well double the size of the chapter! For a broad look at the field, see Stich and Warfield 1994, for instance; Godfrey-Smith 1996, pp. 166-203—although in the end almost certainly opposed to the methods described here—also offers a very tidy discussion which complements this section nicely. In Godfrey-Smith's taxonomy, the present view is an *immodest, success-independent* theory.)

Finally, in some cases, it is clear that fixing the 'extension' of a representation—i.e., fixing a particular item to which my representation is appropriately linked—rather unsurprisingly requires appeal to environmental information. Consider a representation which for me indicates something roughly like 'that thing three feet in front of me': to establish what particular thing this refers to obviously requires information about what it is that is three feet in front of me. Yet this in no way compromises the points above, because for me the representation is *not* of the *particular* thing, but of that thing three feet in front of me, whatever it might be. So, on the present view, whether a representation is used as such in the cognitive framework of an organism is fixed objectively and unmysteriously by the structure of the representation understood specifically within the context of the organism; the representation, together with its context, always retains mutual information content with the item represented. (Suppose, on the contrary, that it did not. Then the conjunction of the representation and the rest of the organism would be algorithmically independent of the object of representation, and neither the organism nor its autonomously initiated temporal evolution—recalling the self model's *functional* representation—could instantiate any information about the object. The organism could neither report nor mentally construct any description whatsoever of the object, unless by appeal to indeterminacies, functional aberrations, or environmental sources of information not themselves represented by the self model. This is clearly inconsistent with possessing a representation of the object, used as such, in the first place.)

All this can be made a little more rigorous by recalling the terms of the approach to functional decomposition described in Chapter 5. We may understand an object to be represented by a self model or other cognitive system specifically when the physical structures instantiating I/O sets for some functional module or—more likely—some *set* of modules identified by the complexity minimisation strategy maintain high mutual information content with the object. (The emphasis on *sets* of modules reflects the trickiness of pinning down the physical location of a representation within a functional system, as exemplified by the stack example of Chapter 6.[14]) Such a representation becomes a functionally active representation of a *particular* object when the same I/O sets are identified by the complexity minimisation strategy applied for overall input domains and output ranges which specifically include interaction with the object.

It is worth noticing that this approach expressly accommodates the case of a neural population learning an activity pattern initially in response to sensory signals, akin to the primitive feature extraction arrangement of Figure 14, but the subject later suffering a lesioning of the relevant afferent pathways. In that case, activity in the population may continue to represent the same sort of object it always did, even though that activity can no longer be interpreted in the context of the afferent pathways which originally created it. For instance, a subject who learns to recognise objects visually but subsequently loses visual acuity might nonetheless elicit activity endogenously in neural populations responsible for representing objects perceived visually, even though those populations lose the capacity to respond to genuine exogenous visual stimuli. Even without the sensory receptors and associated pathways, which together with the representing neural population would, by themselves, maintain substantial mutual information content with the object of representation, the connections which *still* give the representing population an appropriate rôle in the activity of the rest of the system nonetheless retain the mutual information content. Fortunately, the present way of understanding representation use does not break down simply because a subject loses eyesight or just shuts her eyes!

Back on page 42, I suggested it might sometimes be useful to step out of the Turing Machine framework and consider the effectiveness of representations directly with respect to other systems such as brains. Some might wonder at the motivation for keeping the formal definition of representation in terms of mutual information content even now, when a representation's being used as such is so clearly bound up with an organism's behaviour, environment, and individual cognitive processes. Mightn't it be easier and simpler to say that a representation or a symbol stands for some object just when the organism itself thinks, says, or otherwise behaviourally indicates it does? Wouldn't it do simply to say that a pattern of activity serves as a representation just when an organism's nervous system can use it to 'calculate' features of the object and behave accordingly? As a non-rigorous shorthand, no doubt it is easier. But the formal definition carries many advantages, the most significant of which for the moment is that the present approach preserves the status of a particular representation as an actual matter of objective fact. The facts about a representation depend in no way on the subjective preferences of an experimenter or on the interpretation of testimony. Having said that, actually *finding* representations in an organism remains of course, following section 4.1 of Chapter 5, a clearly empirical task.

3. GROSSBERG'S ADAPTIVE RESONANCE THEORY

In contrast to Edelman's scarcely penetrable style, Stephen Grossberg's articles on cognitive neuroscience are concise, transparent, and understated. My first encounter with Grossberg came with his invited lecture at the 1995 International Workshop on Artificial Neural Networks, during which he quietly and modestly—almost blandly—summarised the high points of the most recent two decades of his work. Excited exchanges with other participants later revealed that I was not alone amongst younger members of the audience in shrinking into my seat with the sudden realisation that a man many of us had never even heard of[15] had already modelled what we were trying to model, had done it better, and had done it while we were still in primary school. The best way to avoid reinventing wheels in the field is simple: read Grossberg first!

Grossberg's adaptive resonance theory[16], or ART, offers a beautiful example of neural mechanisms at once simple, powerful, and versatile enough that they may feature in some way in the implementation of nearly every significant cognitive capacity of the human species. More than any other neuroscientific theory of which I am aware, ART in its various incarnations describes what deserve to be called 'universal principles of brain computation' (Grossberg 1995). This section explores some simple parts of Grossberg's work at a rather non-technical level, hopefully including just enough detail to clarify the rôle this type of theory can play in the approach to consciousness developed in this book.

A major motivation behind adaptive resonance theory is the need for a neural structure permitting novel sensory patterns to be stored quickly and effectively without compromising the integrity of patterns already coded in the neural tissue. Grossberg phrases the 'stability-plasticity dilemma' this way:

> How can a system's adaptive mechanisms be stable enough to resist environmental fluctuations which do not alter its behavioral success, but plastic enough to change rapidly in response to environmental demands that do alter its behavioral success...If adaptive memory elements can change quickly in response to behaviorally important environmental events, then what prevents them from changing quickly in response to *all* signals that they process, whether meaningful or not...? (Grossberg 1984, section 2, emphasis original)

(Interestingly, later in the same paragraph, Grossberg goes on to say that "To define behavioral relevance, we need to choose a level of discourse that focuses on the interactions within the system as a whole, rather than on local computations at each cell", making for nice echoes with Chapter 5.)

A neural solution to the stability-plasticity dilemma must also address the question of how errors in the coding of representations can be detected (or how existing codings can be adapted) when individual cells lack any information about whether an error has occurred or whether an adaptation is appropriate. It turns out that Grossberg's solution to the dilemma also yields a system capable of rapid testing of hypotheses about matches between representations.

3.1 Competitive Feedback Networks

The first small building block addresses the more fundamental question of how a neural network can distinguish a relevant signal from background noise, amplifying the former while suppressing the latter. In the context of real neural wetware, the danger is that low intensity signals will be washed out by cellular noise, while higher intensity signals will saturate the tissue, exciting cells to a 'uniform hum' within which the real signals are likewise lost. Hebb's (1949) early cell assemblies, units of mutually reinforcing neurons with overlapping receptive fields, are subject to the problem in a particularly nasty way: although excitatory feedback between adjacent cells effectively renders the cell assembly more sensitive to afferent signals, it does so at the risk of near certain saturation. Lacking a mechanism for offsetting the excitatory feedback, activity of the cell assembly persists long after offset of the original afferent signal and fades only in the face of comparatively slow neurotransmitter depletion.

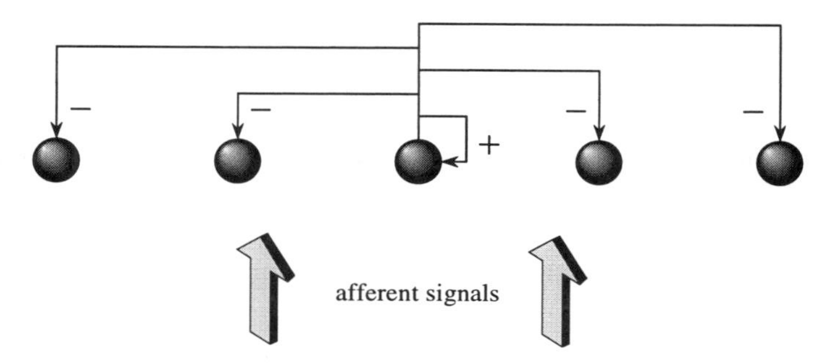

Figure 15. On-Centre Off-Surround Architecture

One solution to what Grossberg (1980) has called the 'noise-saturation dilemma' combines excitatory feedback with inhibitory feedback. As in Figure 15, consider an 'on-centre, off-surround' network in which nodes deliver excitatory feedback to themselves and inhibitory feedback (via interneurons, as in 2.1, not shown here) to their neighbours.

In simulating such networks, the signal provided to neighbouring cells is often calculated according to a function which is the difference of two Gaussian functions, a larger and a smaller. Frequently called DOGs (for difference of Gaussian), or Mexican hat functions, members of this class of functions appear roughly like Figure 16 in general shape. The idea is that the centre cell or population of cells receives strong excitatory feedback, while nearby neighbouring cells receive strong inhibitory feedback, tapering off as a function of distance from the active centre. Literally hundreds of experiments have verified a rôle for DOG functions in the receptive fields of neurons in widespread areas of the brain. (See Kuffler 1953 or Hubel 1988 for the visual system, for instance.)

Given feedback scaled with a sigmoidal response function, which turns out to be essential for their noise suppression properties, on-centre off-surround competitive networks of this kind exhibit a host of biologically useful properties, including contrast enhancement, pattern matching, and automatic sensitivity retuning. (For rigorous *proofs*, see Grossberg 1973, 1981b.)

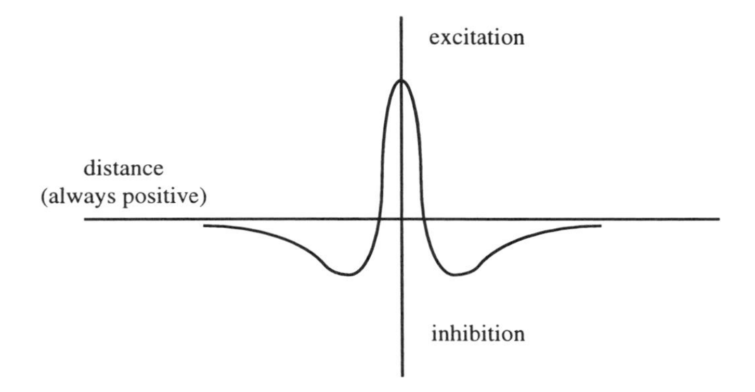

Figure 16. Mexican Hat (DOG) Function

Mathematically, one central feature which enables these properties is the tendency for total excitatory response in the network to be conserved. Loosely speaking, with only a certain amount of excitatory output to go around, the end result is to suppress responses to minor afferent anomalies and enhance the activity of 'winning' populations of neurons, isolating their response from that of their neighbours. Once a particular cell population responds to an input signal, it becomes much more difficult for others to do so, and the network 'locks on' to that response pattern. Uniform input, whether of high or low intensity, is uniformly suppressed. In terms of pattern matching, this behaviour means that when sets of afferent signals presented simultaneously are similar (and thus additively more robust), the network enhances the signal, and when signals are different (interfering destructively), they are treated as noise and suppressed. Variation in nonspecific input to the inhibitory interneurons shifts such a network's quenching threshold (QT), which determines the minimum afferent signal contrast to which the network will respond, thus also fixing the degree of similarity which two patterns must exhibit to count as a match. Finally, excitatory feedback within the network amounts to a limited capacity to sustain firing, the direct physiological analogue of short term memory, or STM.

3.2 The Gated Dipole

Replacing the nodes of an on-centre off-surround feedback network with *gated dipoles* expands the capabilities and biological plausibility of the basic architecture enormously. Depicted in Figure 17, the gated dipole includes distinct ON and OFF cells driven by an arrangement of competing nodes which are in turn driven by signals gated by comparatively slowly varying neurotransmitter chemicals (shown here with a broken line).[17] The gated dipole displays, among other useful properties, antagonistic rebound; i.e., if previously on, the gated dipole initiates a transient 'OFF' output following *either* the sudden offset of the signal at x or the delivery of nonspecific arousal to both x and y. (No doubt the appealing superficial similarities between the gated dipole structure and the classic bistable, or 'flip-flop', at the heart of modern digital circuitry will not be lost on readers with an interest in the latter.)

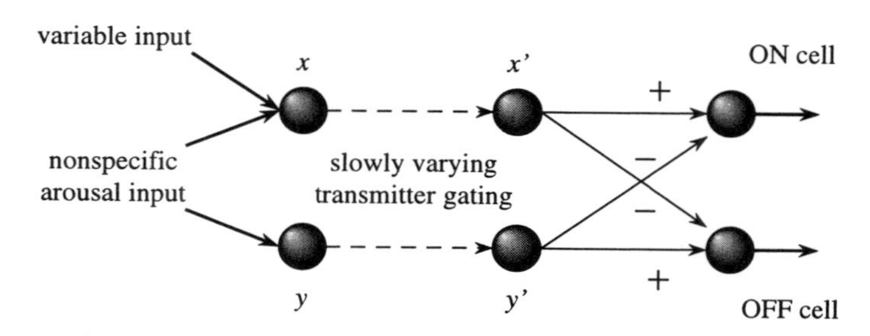

Figure 17. Grossberg's Gated Dipole

To see why this happens, consider the action of the dipole when x receives an excitatory signal. The slow transmitter by which x and y each communicate their signals *gates* their output. In other words, higher signals mean the delivery of more neurotransmitter; the excitatory signal ultimately passed on can be thought of as the product of the potential and the quantity of neurotransmitter. Prior to onset of the excitatory signal, the transmitter will have accumulated at x and y roughly equally, according to a constant production rate inhibited by negative feedback proportional to the transmitter concentration. Thus, when x provides an excitatory signal, x' receives more excitatory neurotransmitter than y', and the ON cell wins the competition. The slow accumulation of transmitter for x and y, however, means that although x' continues to receive greater input, the amount of transmitter available in the synaptic knobs of x decreases; this effectively calibrates the dipole for subsequent antagonistic rebound. Why? If the variable, or *phasic*,[18] input to x is suddenly removed, while the same level of nonspecific arousal to x and y continues, the potentials of x and y equalise but the depleted transmitter at x means that y' receives more excitatory input from y than x' receives from x; the OFF channel wins the competition. Since the potentials at x and y are now equal, the neurotransmitter balance returns to equilibrium after a short interval, and the rebound signal disappears.

The more interesting rebound effect results from a sudden increase in the nonspecific arousal to both x and y. In particular, if the change in nonspecific input exceeds a certain constant (which depends on the particular network equations but *not* on the level of the signal to x), the dipole rebounds with an intensity described by an increasing function of the potential at x. Thus, while an increase in nonspecific arousal sufficiently large to rebound any dipole in a field can rebound them all, the rebound affects only those providing ON signals at the time of the nonspecific arousal burst. Given sigmoidal response functions, however, this rebound behaviour acquires a subtle but remarkable twist: namely, it turns out that an arousal burst which rebounds strongly active members of a dipole field actually *enhances* the ON response of those which are very weakly active. It is this property which, as we see below, allows weakly active representations to emerge out of the background when a mismatch occurs between expectations and input signals. That these rebound features fall out of the dipole's basic structure and a sigmoidal response function Grossberg (1982c, section 23) calls "a minor mathematical miracle".

3.3 ART Dynamics

Considering two different on-centre off-surround dipole fields linked by conditionable feedforward and feedback pathways leads to one simple rendition of the basic architecture of adaptive resonance theory, shown in Figure 18. This arrangement solves the problems outlined at the start of this section and dissolves the mystery of the stability-plasticity dilemma.

The ART architecture comes in several flavours, but the central feature of the basic layout sketched here is that feedforward signal patterns from one competitive feedback network combine with feedback signal patterns from a second to establish 'resonance' between the active suprathreshold STM traces in each. Sustained excitatory exchange between the two allows the STM activity to be encoded in long term memory (LTM). The particular mechanisms of adaptive filtering (Malsburg 1973, Pérez *et al.* 1975) which takes place along the LTM pathways[19] aren't important to this general picture; what is significant is the basic idea of specific patterns active across conditionable pathways resonating through what Edelman (section 2.2 above) calls anatomical reentrance. For specificity, the earlier network (F_1 in the figure) is sometimes considered as a particular thalamic nucleus connected via ascending cortical projections to the later (F_2), which in turn connects back to the former via descending limbic projections; alternatively, the two may be thought of simply as successive cortical stages. On such a reading, feedback from F_2 is understood as top-down expectancy, which combines with the initial bottom-up coding of F_1 for a match or a mismatch. But, of course, the general arrangement considered here could just as easily be implemented between networks in distinct modalities—such as mapping sheets serving separate sensory modalities—connecting again with Edelman's notion of phasic reentrant signalling between distinct maps. This is jumping ahead, however; much remains to be explored with the architecture of Figure 18. The system operates as follows.

The afferent pathway to the system bifurcates into a specific pathway, carrying signals to what is here labelled the 'attentional subsystem', and a nonspecific pathway, activating the 'orienting subsystem', which can be understood for the moment simply as an arousal source. The attentional subsystem refines the network's processing of expected events and codes correlations between the patterns elicited in the two separate layers in LTM. The orienting subsystem, on the other hand, exists to help the system reorganise in response to novel events.

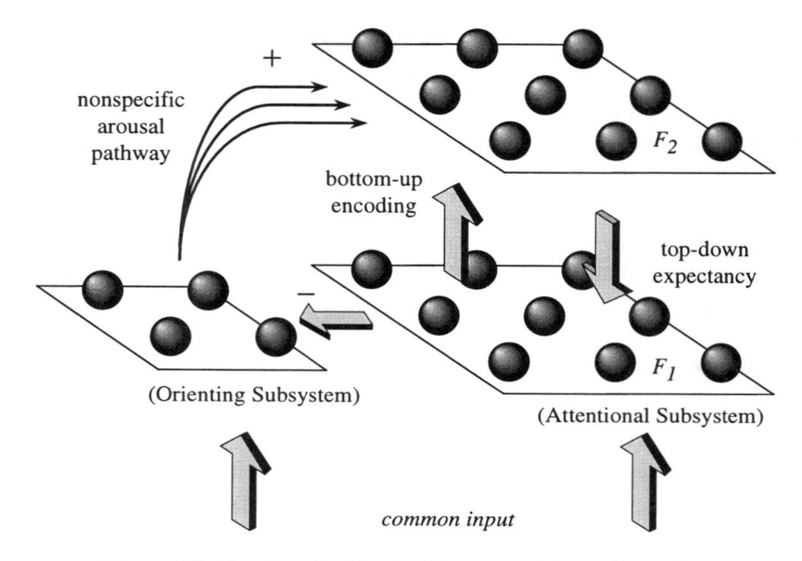

Figure 18. Grossberg's Adaptive Resonance Theory in Action

Just to have somewhere to begin, suppose that some input elicits a particular response pattern across F_1, which passes that pattern on to F_2. Suppose this signal elicits another pattern across F_2, which likewise returns feedback to the first. In the case of a stable and appropriate learned expectancy, these two sets of signals reverberate, or resonate, between the two networks, allowing LTM adaptive filter sampling and effectively fixing the 'attention' of the system on the particular input. At the same time, the output of F_1, delivered nonspecifically across the orienting subsystem, inhibits that subsystem, preventing it from delivering any strong signal to F_2. (Note that the match between the signal provided by F_2 and the existing STM trace across F_1 presupposes only appropriate filtering along the LTM pathway and nothing about the *particular* STM traces active across either network; in other words, F_1 and F_2 each may—and no doubt *will*—respond with entirely disparate patterns, but the right filtering through the LTM pathway still ensures resonance.)

The case in which the top-down expectancy of F_2 doesn't match the coding of F_1 is more interesting. When such a mismatch occurs, the otherwise stable STM trace across F_1 is rapidly diminished by the basic contrast enhancement and noise suppression properties of the competitive feedback network; this suppression of the signal across F_1 due to the mismatch results in a corresponding elimination of the inhibitory signal keeping the orienting subsystem in check. Consequently, the orienting subsystem delivers a sudden burst of nonspecific arousal through a nonconditionable pathway to F_2.[20] (Recall that it, too, receives excitatory afferents from the branching pathway which feeds F_1.) As described above, this burst resets the strongly active dipoles in F_1 and, as a net result, effectively passes their suprathreshold activity on to any other dipoles which had been very weakly active. The mismatching STM trace across F_2 is thus replaced with some other pattern which, as before, is delivered back to F_1 as a new top-down expectancy. Meanwhile, removal of the previous offending expectancy allows the original STM trace to be reestablished across F_1, priming the system for a new test between top-down expectancy and the bottom-up encoding. Eventually, either a correct match emerges between the STM traces active over F_1 and F_2, or a new set of cells in F_2 is recruited for a fresh pattern with which the F_1 STM trace can build a new LTM association and top-down expectancy.

As Grossberg points out, incidentally, this last process—whereby compatible STM resonances in both F_1 and F_2 are finally established and LTM conditioning can begin—matches the psychological observation that a subject's ability to code data in LTM is correlated with their paying attention to it (Craik and Lockhart 1972, Craik and Tulving 1975). Finally, in terms of linking the formal properties of the ART architecture described here to observable features of brain tissue, robust evidence from human ERP (event-related potential) studies suggests the nonspecific arousal burst from the orienting subsystem is the formal analogue of the ERP complex known as N200, while the STM reset it triggers in the competitive dipole field shows up in real brains as a P300.

All this amounts to a system which solves the stability-plasticity dilemma, protecting the longer term integrity of learned representations while enabling novel inputs to be encoded quickly and efficiently. The system requires no special properties of individual neurons or any unexplained 'supervisor' to detect or correct errors in representation; the overall interactions within the system handle error correction and control the coding of novel patterns.[21] The fundamental ideas at work, and especially the basic gated dipole, help explain a remarkably varied range of experimental data and suggest both modifications to and rigorous neuroscientific foundations for diverse cognitive and motivational theories. (See Grossberg's publications cited above in note 16 on page 189 for just some of the applications.) For present purposes, it is also interesting that the activity of such an ART structure amounts to a system for rapid hypothesis testing: each successive pattern elicited at F_2 and described in the present context as a top-down expectancy constitutes a hypothesis about information coded by the activity pattern at F_1. ART offers an attractive account in very mathematically and neurophysiologically specific terms of how hypotheses may *seem* to be processed in serial when in fact the underlying neural substrate remains entirely parallel in nature. (Grossberg 1984, section 19, appeals to this property in arguing for ERP testing of information processing theories of cognition more generally, expressing the hope that the approach will "dampen the depressing dogma that any instantiation of an information processing concept is as good as any other".)

4. CIRCUITS OF THE SELF MODEL

Although working out specific connection diagrams is by no means trivial, a general picture of broader applications of and extensions to the basic ART architectures is appealingly straightforward. Significantly, nothing restricts the sort of hypothesis testing described above to simple perceptual categorisations in a single modality; nothing (except insufficient connectivity in given neural tissue, of course) precludes ART style reentrance between dipole fields serving different modalities or between fields which themselves represent higher level associations between distinct modalities.[22] For instance, bottom-up processing of visual cues might elicit conditioned top-down expectancy of particular auditory input or might prime motor control networks for an appropriate response.[23] Related applications have attracted a great deal of research, and adaptive resonance theory continues to enjoy widespread and robust predictive success.

4.1 Representational Efficacy and Selective Advantage

Where the activity of nerve tissue processing sensory information bears in some way on an organism's behaviour—which of course it normally does—the existence of even simple top-down expectancy no doubt carries, *ceteris paribus*, powerful selective advantages. A nervous system exhibiting such expectancies may effectively endow the organism with the capacity to learn about regularities in its environment and to anticipate future environmental events on the basis of past regularities and present sensory cues. The very process of evolution by natural selection presupposes some degree of environmental regularity,[24] and there can be no doubt that behaviourally effective information about that regularity, instantiated in functionally relevant nerve tissue (whether by ontogenetic learning or phylogenetic 'hard wiring'), garners evolutionary advantage.[25]

For the case of phylogenetic branches along which the advantages of nervous systems offset their immediate physiological costs (such as demand for oxygen and nutrients, etc.), both *a priori* reasoning and the known fossil record suggest that history will, with a few exceptions, have favoured progressively expansive arrays of nerve tissue. We might imagine an historical progression during which organisms develop to incorporate increasing amounts of information about environmental regularities relevant to their survival and reproduction. It seems a safe bet that nervous systems will have continued to develop in this respect, achieving higher resolution and flexibility across the most selectively relevant sensory modalities, until such time as further improvements could no longer be 'cashed out' for a net advantage within an individual organism's lifetime. Barring any tough physiological impracticalities with the architectures described by adaptive resonance theory, impracticalities which have yet to be discovered, we should expect to see either exactly those architectures or something with similar functional capacities emerge for processing both selectively relevant single modality sensory information and cross modal information for senses where useful correlations between stimuli exist. (See the section 'A Tasty Side Note' beginning on page 62 for examples of anomalous 'irrelevant' cross modal associations.) There is every reason to think ART style architecture may plausibly have developed through the course of natural selection, implemented by nervous systems increasingly capable of gathering environmental information, anticipating change in that environment, and testing hypotheses about that environment. (Indeed, a very large body of empirical evidence suggests that ART style architectures *have* developed through the course of natural selection and are now to be found in, among other creatures, *Homo sapiens*!)

So, on a very simplistic reading of evolutionary progression, unfettered by the subtleties and complexities of biological reality, the natural path of history can plausibly (and unmysteriously) be expected to have created organisms prepared to face their environments with a rather sophisticated cognitive framework for hypothesis testing. (This is, obviously, merely a 'just so story', intended only to establish *prima facie* plausibility without rigorously filling in the details!) But consider the selective advantages which would accrue to an organism with a nervous system equipped not only to test hypotheses in response to afferent signals provided exogenously by the environment but also to test hypotheses about signals induced *endogenously*. By selectively inhibiting signals at some stage along the normal afferent sensory pathways and substituting endogenously initiated signals (perhaps after the fashion described in the parenthetical paragraph on page 183), such an organism acquires the capacity literally to test counterfactual hypotheses about what *would* be the case *if* its sensory inputs (and/or, alternatively, its motor outputs) were such and such. The capability becomes especially interesting in the case of cross modal processing, where an endogenously produced 'mock' input in one modality may excite activity patterns which amount to hypotheses about correlated inputs in another modality or other patterns which prime motor outputs (for instance, by altering a quenching threshold) for a response appropriate to the mock input. Coupled with a comprehensive system of sensory and proprioceptive maps, terminal motor maps, and cross modal associations between them, such a capability becomes very significant indeed. As it is frequently observed when considering such hypothesis testing, an organism so equipped may enjoy a safety net which competitors lack: by 'trying out' prospective courses of action in advance, it may (borrowing Karl Popper's image) allow its hypotheses to die in its stead.

4.2 Natural Emergence of the Self Model

The capacity for endogenously initiated hypothesis testing over a wide range of sensory maps and association areas suggests a coherent functional representation not only of the organism's environment but also of the organism itself. This is not *quite* right, of course: to an extent, the broad coverage of the maps and association areas already functionally represents the organism itself, since this information already reflects how parts of the organism change in response to changes in its environment. This is the information theoretic feature which gives sense to the notion of endogenously initiated hypothesis testing in the first place: it is just *because* the sensory maps and association areas contain information about the organism's changes within an environment that it is possible to substitute endogenously generated mock environmental signals and generate meaningful hypotheses about them. The crucial point for present purposes, however, is that an organism's actually implementing the systems specifically to subserve endogenously initiated hypothesis testing may permit what already amounts to a functional representation of the organism to *function as* a representation of the organism within its own cognitive framework.

In particular, when such systems are available for hypothesis testing using an internal functional representation of the organism, some control of behaviour may be passed on from nerve tissue responsible for basic processing of sensory information earlier in the afferent signal pathway to 'higher level' tissue implementing the broader representation and hypothesis testing systems later in the signal pathway. Appropriate feedback pathways may also pass control of what endogenously generated signals are to be submitted for hypothesis testing over to the representation and hypothesis testing system itself, yielding a system suitable both for hypothesis testing *and* for generating scenarios about which to test them. It's easy to paint in broad strokes straightforward examples of the sorts of signals we should then expect to discover in an organism. For instance, perhaps the activity pattern in a sensory sheet representing visual stimulus of a food source first triggers hypothesis testing about the identity of the food. Once the food is recognised through a stable LTM resonance, some ramifications of an endogenously generated candidate motor action for acquiring it might be tested, and eventual behaviour in terms of the motor action ultimately taken might then depend on the outcome of that hypothesis testing, and so on. (Of course, this dependence of the action taken on the outcome of the hypothesis testing demands no 'supervisor' to evaluate or compare outcomes; this function may be subsumed entirely under the interactions of competing motivational and cognitive networks. See Grossberg 1984 on the place of ART in motivational contexts.) Like the stack example from section 5.3 of Chapter 6, information instantiated by the sensory maps and association areas thus acquires a functionally active rôle *as* a functional representation of the organism.

In addition to Grossberg's VAM models mentioned above in note 23, some examples strongly indicative of just such internal functionally active representation specifically in human motor control systems may be found for instance in Imamizu *et al.* (1995), Wolpert *et al.* (1995), Miall and Wolpert (1996), and especially Gerdes and Happee (1994)[26]. These studies address one central puzzle which arises when attempting to understand how motor control systems correct trajectories during visually or proprioceptively guided limb movements; in particular, the time delay introduced by signal propagation through the nervous system's sensory feedback pathways exceeds the empirically observed time scale within which control systems introduce trajectory corrections. In other words, sensory information about the real effects of given motor system efferents simply cannot be returned to the motor control system rapidly enough to figure in the sorts of rapid corrections human subjects display. This suggests the nervous system somehow *models* the effects of particular efferents internally, initiating corrections on the basis of comparatively speedier modelling long before the real exogenous signals have been adequately processed.

In any case, the simple evolutionary trick of shifting behavioural control from basic 'here and now' systems processing current input signals to higher level systems capable of running 'simulations' on internal dynamic representations offers perhaps as much selective advantage over 'here and now' organisms as the latter display over precursors controlled by systems with no top-down expectancies or hypothesis testing capabilities at all. On one hand, compared to earlier unimodal stages, processing stages later in the afferent signal pathway naturally offer greater scope for perceptual discrimination and the opportunity for integration of sensory information from multiple modalities. And on the other hand, the emergence of a coherent global model of the organism situated within its environment creates a 'centre of cognitive action', a dynamic data structure able to 'rehearse' the effects of various actions or sensory inputs even when not performing those actions or sampling those inputs exogenously. That is, the data structure makes possible hypothesis testing not only about the perceptual category of a stimulus but also about the effects of non-actual stimuli or motor actions on the organism or its environment. I suggest that the emergence of such a dynamic data structure creates a *self*.

Just as simple ART style systems may represent globally invariant properties of sensory input (see note 21 on page 195), I suggest the sort of overall system for representation and 'scenario' testing described here effectively codes invariant properties of the organism itself, as situated in an environment. (See 4.3 for some possible limitations on such invariant coding in nonhuman primates.) The organism's representation of itself—*its self*—is that which remains when sensory inputs change or hypotheses are tested; it is that embedded entity about which hypotheses get tested and with respect to which LTM codes experience. I suggest this self arises implicitly just as such an invariant feature; nowhere does the organism acquire some direct multi-modal perception from the outside of a 'thing' which it is. Instead, the 'thing' which it is never actually appears as an object of direct perception: the self as such an invariant feature is effectively a creation of the coherent relationships between information instantiated in sensory arrays, motor outputs, and global interconnections.

This self is of course the 'self model' data structure of Chapter 6 and section 1 above: the self model is, once more, something like an active representational model of the situated sensory and motor system as a whole—as 'seen' from the inside—on which, given appropriate uncoupling of motor outputs and sensory input pathways, by way of selective inhibition, and endogenous initiation of mock sensory inputs, 'off-line simulations' can be run. Instantiated by disparate modules of a functional system, the self model understood as a data structure carries primary responsibility for directing the organism's behaviour, yet (like a stack) it is not *itself* a functional module. From an evolutionary perspective, it is crucial that the hypothesis generation and testing of the self model is *not* performed by some extra representational or computational layer bolted on to an existing cognitive framework; nor does the self model develop as a selectively extraneous biological appendage until its representational capacities become sufficiently robust to deliver behaviourally useful and selectively advantageous 'simulations'. Instead, hypothesis generation and testing is performed by the very same representational systems which emerge naturally from the bottom up to subserve initially primitive but progressively more sophisticated perceptual categorisation and related cognitive processes. Although the actual neural circuitry implementing the self model must of course be built up in a discrete fashion (either functionally relevant reentrant connections between dipole fields exist or they do not), the general picture is one of relatively smooth progression from primitive sensory arrays to ART style hypothesis testing to cross modal association and hypothesis testing right up through to full-blown global functional representation by a self model conditionally coupled to the organism.

The progression requires no sudden appearance of categorically unique structures or capabilities. While the structures ultimately required to support a self model amount to some very sophisticated, massively parallel and intricately interconnected 'wiring', they are no different in general flavour to the basic networks of 3.3; no radical developmental discontinuities threaten their evolutionary plausibility. This overall approach, I believe, complements nicely Dennett's view of 'Popperian' creatures (Dennett 1995c, especially Chapter 13) as mentioned above—i.e., those which may allow their hypotheses to die in their stead—as well as his 'multiple drafts' perspective on the cognitive aspects of conscious experience (Dennett 1991). A rich and growing further literature exists on the broad benefits and capabilities of cognitive mapping and likely ways it might be implemented in real biological organisms.[27]

4.3 Self Models, Language, and Self Awareness

Notice that nowhere in this story does language emerge to play a necessary rôle, and nothing about the picture sketched here demands joint appearance of both consciousness *and* self consciousness, or self awareness, of the sort apparently lacking in all but a few primates. (I am told, for instance, that members of most species, with chimpanzees a notable exception, fail the 'acid test' of 'self awareness' in that they seem unable to recognise themselves in mirrors). In other words, being a self apparently does not imply possessing a *concept* of self, nor does it appear to imply the capacity to form linguistic expressions about the self. Being phenomenally aware that I am in pain depends only on my being able to *experience* pain and does not require any ability to *reflect* on the fact that *I* am in pain. To be sure, pain might take on a different quality for organisms enjoying such a reflective capacity, perhaps because such reflection can automatically induce other feelings (self pity, regret, longing for offset, etc.); but that suggests only that organisms with different cognitive structures may phenomenally experience the same stimulus in different ways. This is to be expected, of course, on the view that experience is an immediate feature of being a particular sort of changing data structure embedded in a cognitive framework. (It also echoes the observation from section 5 of Chapter 4 that there can be no objective description of a rose smell solely in terms of physical rose properties, since phenomenally experienced smell is a product of both the rose's properties and the smeller's properties.)

Nonetheless, the eventual emergence of language remains anatomically consistent with the particular architectures set out above for the self model. For instance, high level cross modal reentrant connections exist in the parieto-occipito-temporal (POT) area (Sanides 1975), a brain region including Wernicke's area which integrates afferents from the three primary neocortical association areas dedicated to audio-visual convergence, audio-somatic convergence, and visual-somatic convergence.[28] For this reason, Geschwind (1965) calls the POT the "association area of association areas". The broad cross modal connectivity of the POT, when coupled with reciprocal feedback to Broca's area may, according to one influential proposal (Wilkins and Wakefield 1995), constitute the evolutionary precondition for language acquisition in the hominid lineage. Notably, the same anatomical features which Wilkins and Wakefield believe promote 'modality free' representation and conceptual structure (Jackendoff 1983, 1987, 1990) and which were on their view ultimately exapted[29] for language development would also, under the picture sketched here, contribute to the appearance of a particularly robust self model data structure. This suggests that while language is by no means necessary for the emergence of the self model and conscious experience, development of a robust self, including, probably, the capacity to conceptualise the self, goes hand in hand with the development of some of the likely neuroanatomical preconditions for language use.

Interestingly, in discussing cytoarchitectonic differences between humans and nonhuman primates, and in particular the lack of a POT area in nonhuman primates, Wilkins and Wakefield refer to research by Ettlinger (1981) and Wilson (Ettlinger and Wilson 1990) which indicates that chimpanzees, while capable of cross modal recognition (such as visual recognition of an object previously presented for tactile investigation), apparently cannot form what they call the 'amodal' representations required to perform so-called 'metaphorical matching'. In other words, they lack the capacity to match stimuli metaphorically—i.e., by means of common amodal properties (temporal pattern, stimulus strength, etc.)—after the fashion of even very young human infants (Humphrey, *et al.* 1979; Wagner, *et al.* 1981). While I would distance myself from the speculation of Wilkins and Wakefield that it is specifically the POT/Broca's area pair which makes possible such 'amodal' representation, their observations underscore the notion that *whatever* the neuroanatomical preconditions of language, its emergence is very closely linked to development of the kinds of robust associations which underlie the self model.

The apparent lack of *self consciousness* in most nonhuman primates is consistent with the absence of appropriate cortical association areas for such 'modality free' representations. In fact, on the view that a sense of self is created implicitly by the global coherence between bodies of information represented by the self model, awareness of a self may require little more than the addition of an *explicit* and functionally active representation of the self as an entity. (Making such a representation functionally active is nontrivial: after all, there must be something for it to do, and a language facility seems one of the obvious candidates for providing it a rôle.) The absence of cortical association areas for 'modality free' representations in most nonhuman primates suggests an inability to form such an explicit representation. Yet, such animals plausibly implement, to a first approximation, something roughly like the self model as described here, including the implicit self, and are thus, on that assumption, conscious in the broader sense of being subjects of phenomenal experience.

4.4 The Symbol Grounding Non-Problem

On the view outlined here, the 'symbol grounding problem', the puzzle of understanding how a symbol or a representation within a cognizer acquires semantic content connecting it to the object of representation,[30] simply does not arise. Harnad's preferred solution (in Harnad 1990, for instance) grounds symbolic representations in either non-symbolic 'iconic' representations (projections from sensory arrays) or 'categorical' representations (indicating invariant features picked out directly from iconic representations). 'Elementary symbols' are names, or placeholders, for the categories picked out in categorical representations, and 'higher order' symbolic representations are symbol strings describing category membership relations. The framework presented here subsumes Harnad's strategy with only very slight modifications.

Harnad's iconic representations correspond straightforwardly to activity patterns in neuronal sensory receptor sheets. Indeed, this is approximately how he explicates the term as well. Categorical representations, in turn, correspond to activity patterns in higher level populations of neurons, which, although explored here principally in terms of adaptive resonance theory, might also be the sort of primitive feature extractors of section 2. The definition of representation used throughout this book automatically grounds such activity patterns, when taken in the context of the system of which they are a part, in the objects of representation: as in section 2, activity patterns understood within the context of the neural system instantiate mutual information content with the item represented. Finally, actual symbols naming categories correspond to the kinds of 'modality free' representations mentioned above in relation to language, while strings of symbols describing relationships between categories constitute language itself. (And again, in the spirit of Chapter 4's observation about rose smell being fixed by information both in the rose *and* in the smeller, when we understand a string of symbols as *meaning* something, we bring with us to the language game an enormous quantity of our own information in the form of representational structures we link to those symbols.)

While questions about finding semantics in the mere syntax of computational symbol manipulations have enjoyed a distinguished history in computation theory itself, the application to cognition, which only recently received the name 'symbol grounding problem' (Harnad 1987b), arguably cashes out as little more than an artefact of years of cognitive science twisted and crippled by an unduly linguistic, symbolic slant. That slant, due in no small part to Fodor's (1975) language of thought hypothesis and the symbolic model of mind he champions (Fodor 1980), very naturally gives rise to questions about how symbols can come to stand for anything particular in the first place. Exponents of the approach, such as Pylyshyn (1980a, 1984) and many others, maintain that mental processes just *are* symbol manipulations and that the appropriate level for analysing mentality and function *must* be a symbolic one. Yet, for reasons canvassed in section 4.3 of Chapter 5, little indication exists that the sort of objective functional decomposition explored in that chapter should yield transactions in a language of thought, and the neural architectures outlined here likewise provide the idea scant support. There seems little reason to think that activity patterns in neural networks handling jobs as diverse as contrast enhancement in my lateral geniculate nucleus and crafting the words I will write in my next book and mapping the tactile sensations of my left little toe—and the diverse interactions between those activity patterns—could all be brought together under the umbrella of one grand language-like construction sufficiently simple and ungerrymandered to shed any significant light on questions about cognition.

Obviously, *some* mental processes do rather straightforwardly look like processes of symbol manipulation, but I believe a cognitive science focusing *strictly* on such processes holds little promise of explaining broader problems of cognition or of overcoming the symbol grounding problem itself. Yet that problem may be avoided and, hopefully, real progress made on such broader questions by adopting a different approach entirely, one approximately like that offered here, at least in spirit even if not in fine detail.

In any case, having explored neural aspects of the self model approach to consciousness, the following chapter outlines an interesting feature which may be displayed by some neural systems relevant to cognitive science. In particular, the chapter summarises theoretical evidence suggesting that some analogue neural networks may not behave in ways strictly in the same class as what may be simulated by a Turing Machine.

NOTES

[1] Note that in the general case, the lion's share of relevant mutual information content obtains between the self model, the environment, and the sensory and motor systems of the organism (rather than the entire organism); we would not in general expect to find within the self model detailed representations of things like internal organs or bones, even though its instantiating components (neurons) may well contain a great deal of information about such structures (including information about DNA configuration, cellular metabolism, and so on).

[2] Thanks to Greg Chaitin for convincing me (personal communication) that this observation remains essentially unchanged even when information content is evaluated in lossy terms—i.e., even when absolute information content is understood as the length of a minimal program for generating a string which may deviate from the string in question at a certain number of bit locations. Permitting errors in compression bears only a comparatively little on the overall proportion of string pairs with high mutual information content. Quantifying the change precisely requires a good estimate of the logarithm of binomial coefficients, which starts with Stirling's approximation to factorial. (See Feller 1950, pp. 41-44.)

3 For general introductions, good starting places include Kuffler *et al.* (1984) and Shepherd (1990) or the classic Eccles (1964); also see Hille (1984). Many new to the area, including myself, are surprised to learn that 'cognitive neuroscience' as a discipline, created by a blend-

ing of the methodologies and interests of neuroscience and cognitive psychology, is a remarkably new entry. Gazzaniga (1984) and LeDoux and Hirst (1986) are the first collective volumes marking its birth. (Also see LeDoux *et al.* 1980.) For an indication of the dizzying rate of development, compare Nadel *et al.* (1989). In many ways, Stephen Grossberg (see section 3), cited not once in LeDoux and Hirst, was there first—beating the rest of the field to the mark by at least a decade. Horgan and Tienson (1991), Churchland (1986), and Churchland and Sejnowski (1992) blend more philosophical concerns with neuroscience or connectionism.

[4] For the purposes of this chapter, particularly with respect to the discussion below on synaptic modification rules, I mean specifically *vertebrate* neurons.

[5] The qualifier 'theoretical' appears because, as it will emerge in Chapter 9, strong evidence suggests that any real physical system is limited in power to the level of finite automata.

[6] See Wigstrom *et al.* (1982) or Carew *et al.* (1984), but compare Kelso and Brown (1986).

[7] The 'backpropagation' of Rumelhart and McClelland refers to virtual backward propagation of a signal quantifying the 'error' between a cell's actual response and the 'desired' response; this corresponds to the Widrow-Hoff rule's 'delta', reflecting the change in weight between two cells i and j as a function of the output of i and the difference between the actual and desired output of j. Similarly for the Rescorla-Wagner rule, which links the change in an expectancy of a particular conditioned stimulus (CS) for an unconditioned stimulus (US) with the difference between the actual occurrence of the US and the expectation of the US generated by all the CSs. See Lynch *et al.* (1989, pp. 182-183) for a quick summary of the second two.

[8] Formally, this simplifies the matter unreasonably, since some coding cost is incurred by the process of *extracting* the implication. This may safely be ignored, however, because any deviations in mutual information content so brought about nonspecifically affect *all* the representations, without bearing on the fact that mutual information content remains. Likewise for the case of comparatively small neural populations, where both mutual information content and total information content may be swamped by the $O(1)$ overhead of a particular Turing Machine; this can be remedied simply by choosing one better suited to the task. In either instance, the relevant information theoretic properties meaningfully contrast representations and non-representations *relative to each other* (as intended) rather than reflecting features of particular objects in isolation.

[9] As this book was about to go to press, I encountered a paper by Shimon Edelman (forthcoming) which dresses up this standard idea (which lurks in the background of most if not all research into connectionist classifier systems) in a mathematical framework intended to formalise the notion of representation by second order isomorphism. (See Chapter 3's endnote 12 on page 1 on first and second order isomorphism.) Although I'm sceptical that the paper, slated to appear in *Behavioral and Brain Sciences*, adds very much of significance to debates about representation, some readers may nonetheless find it valuable as one way of fleshing out the details underlying what is here glossed over very nonrigorously in just a few sentences.

[10] Publications of Edelman (1978, 1981, 1989) and colleagues (Edelman and Reeke 1982, Edelman and Finkel 1984) offer entry into the literature.

[11] The group context, conversely, remains important for understanding the operation of lower level mechanisms. For instance, Edelman explicitly predicts that Hebb's rule will be found to be incorrect: "*In general*, synaptic change will operate effectively only in populations. Pre- and postsynaptic modifications will occur independently through separate mechanisms; in no case will they be found to be contingent *only* upon correlated firing across individual synapses" (Edelman 1989, p. 324, emphasis original).

[12] Although some conceptual overlap exists, 'reconstructionist' does *not* here refer to the same body of thought as that underlying reconstructionist approaches to computer vision.

[13] A more comprehensive treatment of representation use than what I offer here would certainly include a look at the information processing literature from within psychology and, in particular, the notion that the best internal representations are those which effectively compress sensory information optimally (Pomerantz and Kubovy 1986).

[14] It's also interesting to note in passing the bearing of this observation on the *binding problem*, the puzzle of how pieces of information represented in physically discontiguous parts of the brain can be 'brought together' to form a coherent experience. The solution to the binding problem assumed here is that representational structures active at the level of the self model may—exactly like the stack, which is also implemented in physically discontiguous locations—be understood at a slightly higher level than that of functional organisation. (See section 5.2 in the chapter on self models.)

[15] Inexplicably, Grossberg's work remains less well known than that of many of his contemporaries, some of whom are popularly accorded credit for his discoveries. For instance, Grossberg did the 'Hopfield model' well before Hopfield; Kohonen apparently used Grossberg's equations, unmodified and without citation (Kohonen 1984), for what became 'Kohonen maps'; and his work on competitive learning predated Rumelhart and Zipser by almost a decade. For revealing historical reviews which, in my opinion, reflect very poorly on the competitive contemporary culture of cognitive science and neural net research, see Grossberg (1987b, 1988).

[16] Grossberg's remarkably prolific research career makes selecting representative samples of his work difficult; some good places to start include Grossberg (1976a, 1976b, 1978, 1980, 1982a, 1982b, 1984, 1987a) and Grossberg and Kuperstein (1989). Grossberg (1982d) provides a stunning historical perspective.

[17] The gated dipole can also be implemented with the slow transmitters in inhibitory stages of a disinhibitory pathway, and even a single cell can act as a dipole. (See the Appendix of Grossberg 1984.) The latter are thought to exist in, for instance, rod photoreceptors of *Gekko gekko* (Kleinschmidt and Dowling 1975, Carpenter and Grossberg 1981).

[18] Neuroscientific use of the word 'phasic' has confused many outside the field, myself included. In this context, 'phasic' contrasts with 'tonic', where the latter describes a cell which is internally and persistently activated and the former indicates a cell (or signal) with a lower basic frequency and transient higher frequency induced by excitatory afferents.

[19] For overviews of other approaches to long term potentiation, see Swanson *et al.* (1982), and Lynch and Baudry (1984). Also see Lynch (1986) and Lynch *et al.* (1989).

[20] Of course, the nonspecific arousal might pass along other pathways as well. For instance, it might adjust the QT of a motor control network partially affecting the supply of afferent signals to F_1—the neural equivalent of a message like "a surprising event has occurred, prepare to get more information!".

[21] Successive filtering stages like those at the heart of the ART architecture are also especially adept at coding global invariants of spatially distributed data and recognising patterns independently of spatial translation (Fukushima 1980).

[22] I suggest (somewhat speculatively) that the mechanisms enabling resonance between matching STM traces in different modalities also underlie the feature of perception which neuroscientist George Reeke calls 'zoomability': the capacity to 'zoom in on' an object of attention and extract perceptual details about it which might be unavailable were it still an object of perception but not of active attention. In particular, the resonance which accompanies attention in a given modality may also prime receptors in other modalities (or in distinct pathways of the

same modality) to expect signals normally correlated with those of the first, effectively lowering their QT and raising their sensitivity to such signals.

[23] More generally, related issues about motor control are better addressed by Grossberg's architecturally complementary vector associative maps, or VAMs (see Gaudiano and Grossberg 1991 for a particularly elegant example). While this chapter focuses on ART for the sake of exposition, it is actually the combination of ART and VAM models, taken together, which we should expect to feature in real examples of cognitive systems supporting self models.

[24] After all, characteristics inherited from a successful progenitor would be of little use if past environmental conditions were *entirely* dissimilar.

[25] See section 3 of Chapter 10 for a brief note on the connection between optimal representational compression and Bayesianism and the second half of Godfrey-Smith (1996) for some formal models of the selective advantage of good mental modelling of the environment.

[26] Thanks to Andy Clark for bringing some of this literature to my attention. Also see Tani (1996) for related work in robotics and Nolfi and Miglino's paper 'Studying the Emergence of Grounded Representations' in Mulhauser (forthcoming) for examples suggesting general caution when it comes to inferring the presence of internal maps even in the case of quite sophisticated behaviour which clearly *seems* to require such maps.

[27] Holland (1992) offers general comments; O'Keefe and Nadel (1978), Nadel and MacDonald (1979), McNaughton and Morris (1987), McNaughton (1989), McNaughton *et al.* (1989), and O'Keefe (1989) discuss spatial mapping specifically in the hippocampus. Also see Schmajuk and Thieme (1992), Schmajuk *et al.* (1993), and Schmajuk and Buhusi (1997). See McNaughton and Nadel (1990) for a related view on incorporating behavioural response coding.

[28] See Pandya and Yeterian (1985); for related work on temporal sequencing, see Tallal and Schwartz (1980), and on hierarchical organisation of behaviour, Greenfield (1992). Ingvar (1985) is similar, with an emphasis on prefrontal cortex. See Lynch (1980) on association in posterior parietal cortex.

[29] Favourites vary for a term to supplant Darwin's (1871) use of 'preadaptation', denoting a feature which emerges under selective pressure of one kind but is ultimately taken over for a distinct use in adapting to selective pressures of a different kind. Darwin's term has acquired unfortunate connotations of direction, or planning, in the evolutionary process; Wilkins and Wakefield prefer 'reappropriation', while 'exaptation' follows Gould and Vrba (1981).

[30] For entry into the symbol grounding literature, Harnad's work (1987a, 1990, 1992, 1993a, 1995, and Harnad *et al.* 1991) is the best start. Ford and Hayes (1991) address the frame problem, which Harnad (1993b) argues is a particular symptom of the broader grounding problem. Searle's (1980) classic Chinese Room can also be read as an instance of the symbol grounding problem. The general problem was probably first expressed by John Locke in 1690, in Chapter IV of *An Enquiry Concerning Human Understanding*.

Chapter Nine

Chaos, Computability, and Real Life

Many interesting differences distinguish the cognitive capacities of real biological organisms from those of virtually all existing computer technology. One of the most intriguing is that in the case of real biological cognizers, hardware ('wetware') architecture implements software *directly*. No principled distinction separates a stored code space from a stored data space and both from a processing unit; all is simply thrown together in a biochemical swamp of a brain which nonetheless manages to transform potentials and ion concentrations in such a way that behaviour can be adaptively and advantageously mediated. Adjustments in the functioning of that biochemical swamp are effected not by altering code or data stored atop a general purpose and invariant underlying architecture, but by transforming the underlying wetware itself. To phrase it entirely oxymoronically, the brain is something like a 'hard wired' application specific integrated circuit which is constantly undergoing transmutation for a new specific application. As Gerald Edelman (see section 2.2 of Chapter 8) puts it, somewhat more succinctly,

> In contrast to computers or Turing machines, there is no general-purpose animal—only the adaptive evolution of particular sensory sheets and adaptive motor ensembles and of the somatic selection principle itself evinced by particular mechanisms within the phenotype. (Edelman 1989, p. 19)

General purpose digital computers, by contrast, perform tasks as diverse as calculating taxes and portraying marauding baddies in the latest 3D shoot-'em-up game just by executing different software stored by their fixed architecture hardware. But the software flexibility of general purpose digital computers also contributes to a great weakness: in many cases, it is vastly more efficient simply to *perform* a task physically than to *calculate* or simulate the results of performing the task. Where a biological system may be 'wet wired' to perform some function directly, a digital computer with a hard wired architecture must resort to less efficient software tricks. It is much easier for a particular batch of neurons to fire in response to some afferents, for instance, than for a digital computer to calculate the precise details of such a batch of neurons firing. (Alternatively, consider how much easier it is to play with a yo-yo than to calculate the trajectory of a yo-yo being played with.) As a result, even the world's most powerful general purpose supercomputers, which can only *simulate* the underlying physics which their biological counterparts exploit directly, lag far behind in the performance of biologically common tasks like pattern recognition.

Given this marked difference between hardware and software on the one hand and the biological wetware/software unity on the other, questions about the inherent physical properties of biological flesh naturally become more urgent for the cognitive scientist than analogous questions about the inherent physical properties of silicon, gallium arsenide, or buckytubes are for the computer scientist. The inherent physical properties of real cells dictate what 'programs' real organisms may run in a way strongly disanalogous to the case of silicon chips and general purpose digital computers. The question which seems to have attracted the greatest interest among philosophers and cognitive scientists, however, is a comparative one, picking out only a thin slice from a whole range of questions: how do the computational capabilities of wet wired biological organisms (such as humans or goldfish) relate to those of various theoretical classes of general purpose computers (such as Turing Machines or finite automata)?

This chapter explores one small aspect of the wetware/software whole with respect to the question of what mathematical framework is most appropriate for modelling the capabilities of real biological systems. In particular, it outlines some interesting computational features of models which are at once analogue[1] and chaotic and compares them to the digital Turing Machine. In general, neither chaos theory nor computability theory are as relevant to cognitive science or philosophy of mind as the burgeoning literature on the topics might suggest. But recent work on a special class of chaotic analogue models, exhibiting behaviour which may plausibly reflect activity in human and other biological brains, suggests such systems might surpass the computational power of Turing Machines. Since considerable empirical evidence suggests areas of the human brain can be described as analogue chaotic systems, the special computational power of the new models may bear on cognitive scientific and philosophical questions about cognition.

The particular focus of attention here is the *match* (or the lack of a match) between a particular model or computational framework and the physical world being modelled. Of central concern in the chapter is the extent to which traditional functionalism and other approaches to cognitive science which take inspiration from the classic Turing Machine model of computation may risk overlooking features of real physical systems. An alternative analogue model of computation *may* better capture relevant features of real biological systems. I would stress at the outset that I *do not* necessarily take the view that special capabilities of *abstract* chaotic analogue models of computation bear directly on the capabilities of apparently chaotic analogue systems in *real* brains. Likewise, I certainly do not advocate attempting to build general purpose analogue computing machines in hopes of solving 'super-Turing' problems with them.[2] On the contrary, there are strong reasons for thinking that all real physical systems are constrained to the power of finite automata—see page 212 below. I *do*, however, think these aspects of the relationship between models and the physical world deserve more exploration and debate than is usual in the literature.

1. TOO MUCH OF A GOOD THING?

Escaping the word 'chaos' in contemporary cognitive science is a challenge. The recent onslaught of a huge number of popular books on chaos theory probably fuels speculation about the rôle of chaotic dynamics in human cognition as much or more than decades of pioneering work by high profile brain researchers such as W.J. Freeman and colleagues[3] and recent technical debates on the topic.[4] While many older mathematical models of cognition fall short of the remarkable richness of real human behaviour, it might be thought that new models featuring chaos, strange attractors, sensitive dependence and the like could better match the complexities of biological reality. Like quantum mechanics, whose connections with cognitive science were the target of Chapter 7, chaos theory offers a candidate mathematical framework, exhibiting properties which are in some ways unique, within which it is tempting to model the admittedly remarkable capabilities of human cognizers. But again as in the case of quantum mechanics, it is all too easy, in the absence of empirical or theoretical evidence establishing clearly relevant links between the two, to oversell the analogy between observed higher level features of human cognition and properties of the candidate framework.

This chapter explores aspects of the relationship between dynamical models in cognitive science, with a special focus on analogue ones which also happen to be chaotic, and the actual neurophysiological processes which those models describe. In particular, it examines the relationship between theoretical capabilities of the models and real capabilities of the modelled. It turns out that some dynamical models of potential relevance to cognitive science, chaotic models which are specifically *analogue*, exhibit computational capabilities different from those exhibited by *discrete* models based on the Turing notion of computation. The particular cases in which such differences occur call into question common interpretations of the Church-Turing thesis and suggest that whether a particular neurophysiological process is best described with an analogue or a discrete model amounts to a significant empirical question. Contrary to the received wisdom of half a century, simply *assuming* discrete models suffice to capture the full repertoire of behaviour observed in Nature and especially in cognizers is justified neither by theoretical computer science nor by mathematics.

Following the method of Chapter 7, this chapter includes condensed introductions to relevant material on computational aspects of modelling, dynamical systems, chaos theory, and some extensions to classical computability. These appear in sections 2, 3, 4, and 5, respectively, and readers familiar with these topics, or bored by technical details, may wish to skip them altogether. (Observations in each about their bearing on cognitive science and philosophy of mind, however, may help keep the remainder of the chapter in focus.) Section 6, starting on page 223, explores two models, due to Hava Siegelmann (1995) and Eduardo Sontag (Siegelmann and Sontag 1994), which display super-Turing computational powers. Finally, section 7 relates the development of super-Turing computation as a theoretical possibility to dynamical modelling as practised in the real world.

2. MODELS AND COMPUTATION

For the most part, the relationship between cognitive neuroscience and theoretical computer science is one of peaceful indifference. Very roughly, researchers in the first field concern themselves with the *actual*—what really happens in a brain, and how does it figure in cognition?—while those in the second attend to the logically *possible*—what sorts of things can we in principle accomplish with particular computational processes if we ignore the messy details of the real world? Yet the two fields are linked by the fact that for all its theoretical emphasis, computer science is based ultimately upon idealised models of the real physical world. As David Deutsch suggests,

> There is no *a priori* reason why physical laws should respect the limitations of the mathematical processes we call 'algorithms'...there is nothing paradoxical or inconsistent in postulating physical systems which compute functions not in [the set of recursive functions]... Nor, conversely, is it obvious *a priori* that any of the familiar recursive functions is in physical reality computable. The reason why we find it possible to construct, say, electronic calculators, and indeed why we can perform mental arithmetic, cannot be found in mathematics or logic. The reason is that the laws of physics 'happen to' permit the existence of physical models for the operations of arithmetic such as addition, subtraction and multiplication. If they did not, these familiar operations would be non-computable functions. We might still know of them and invoke them in mathematical proofs (which would presumably be called 'non-constructive') but we could not perform them. (Deutsch 1985, p. 101)

Hava Siegelmann says simply, "Computer models are ultimately based on idealized physical systems, called 'realizable' or 'natural' models" (Siegelmann 1995, p. 545). This section outlines some aspects of the relationship between model and modelled.

2.1 Computational Limits and Physical Limits

To the extent that models of computation are based on physical systems, explorations of those models from the perspective of theoretical computer science suggest boundaries or constraints on what we should expect to find in the real world; these constraints are no less important to cognitive neuroscience or other empirical fields than analogous discoveries of physics. In the case of computability, for instance, the boundaries dictated by computer science—effectively, the boundaries of idealised physics—constrain the range of functions which we generally consider relevant for describing the behaviour of things like neurons.[5] In general, if a candidate neurophysiological model describes a neuron or any other brain structure with a noncomputable function, that is taken as good reason for rejecting the model.

Despite their mutual link to physical reality, however, one occasionally hears in defence of happy indifference the naïve objection that the limits proclaimed by computer science are in fact *too liberal* for the real world: don't Turing Machines, for instance, use infinitely long tapes and huge amounts of time while performing their computations? How can the limitations of such a logical construction be at all relevant to neural events which finish in a fraction of a second (and which obviously can't make use of anything infinitely long)? But, of course, suggesting the Turing Machine model is irrelevant to understanding finite bounded things like neurons is exactly like suggesting it is irrelevant to understanding finite bounded things like my Macintosh computer! No normal Turing Machine which computes a function and halts ever actually *uses* an infinitely long tape. The unbounded tape merely permits a *general* model which accommodates different computations requiring specific *finite* amounts of it. (Moreover, we may usefully distinguish classes of computations according to the resources a Turing Machine requires to complete them.)

In fact, a bound on the computational power of real physical systems limiting them to the capabilities of finite automata appears to follow straightforwardly from research on the thermodynamics of black holes.[6] In particular, Bekenstein (1981b, p. 624) gives the precise upper bound on the information capacity in bits N of a system with mean energy E_0 and which can just be enclosed in a sphere of radius R (all in suitable units, which render the numerator enormous) in terms of the following inequality:

$$N < 2\pi \frac{E_0 R}{\hbar c \ln 2}$$

(The cited article is quite short and relies on a series of previous papers; for the fuller story, see Bekenstein 1973, 1974, 1981a; or, more recently and with a revised proof structure, Bekenstein 1994.)

I would note, incidentally, that I do not find the conclusion that all physical systems, including human beings, are limited in power to that of finite automata to be as disconcerting or depressing as some have suggested it should be. First, the number of accessible states for systems of human temperature and size remains astronomically *vast* and continuously changing: it is not as if we are ever actually going to discover ourselves to be stuck in a limit cycle! Second, given that the notion of universal computation is defined over an infinite space of functions, it would appear that no finite series of measurements could ever reject the hypothesis that a real physical system was only computationally equivalent to a finite automaton as opposed to a universal computer such as a Turing Machine. In other words, the hypothesis that a real system such as a human being is capable of Turing power does not appear to be scientifically meaningful.

In any case, however, interest here centres on the match between models of neural or other systems and the underlying physics dictating their behaviour. That all physical systems are in fact limited in power needn't necessarily bear on the project of evaluating which computational framework provides the best general model of their behaviour.

Back to the substantial relationship: to the extent we are confident in the physical reasonableness of a model of computation, we can in general be assured that whatever the model says can't be done *really* can't be done—and *won't* be done in any brain which doesn't violate the laws of physics. What might be cause for some concern, however, would be evidence that real physical systems might, ideally, do things which a particular choice of computational model (such as the Turing Machine) cannot. Such evidence might come either empirically—here is a physical object which can do something Turing Machines cannot—or theoretically—here are two distinct models of computation which differ in what they suggest may be accomplished by physical systems under reasonable constraints on resources. This chapter explores the second alternative.

2.2 Numbers, Models, and the Real World

Of central concern in this exploration is whether real physical processes are best modelled with real numbers or with rational number approximations. The present aim is not to argue positively for one choice or another, but simply to illustrate that the choice does, contrary to the received wisdom of half a century of computer science, make a difference. It is helpful to clarify briefly the available alternatives.

The basic system of numbers is the *naturals*, denoted **N**; the natural numbers are the members of the set $\{0, 1, 2, \ldots\}$. Including negatives of the naturals after 0 yields the *integers* **Z**. Numbers which can be expressed as a fraction, or ratio, of integers, as in m/n, where $m \in \mathbf{Z}$, $n \in \mathbf{Z}$, and $n \neq 0$, are called *rational*;

the set of rationals is denoted **Q**. Of course, not all numbers can be expressed in this form; those which cannot (such as $\sqrt{2}$, π, and so on) are *irrational.* Finally, the set of all *real* numbers, rational or irrational, is denoted **R**. In general, we associate rational numbers with *discrete* processes and real numbers with *analogue*, or *continuous*, processes. The calculations of a digital computer offer a paradigm example of a discrete process, while a good example of an analogue process is the rotation of a wheel. Digital computers calculate exclusively with rational numbers, while a turning wheel passes through a range of angles of rotation described by a continuous real number interval.

Making sense of the debate over the choice of number systems and computational frameworks for modelling the real physical world requires acknowledging that the set of real numbers is 'bigger' than **N**, **Z**, or **Q**. Any set which may be placed in one-one correspondence with the positive integers, or $\mathbf{Z}^+$, we dub *countably* (or *denumerably*) infinite, while any other infinite set is called *uncountably* infinite. The naturals, the integers, and the rationals are each only countably infinite, while **R** is uncountable. In other words, the members of the other sets of numbers all correspond one-to-one with $\mathbf{Z}^+$, but any correspondence for the members of **R** invariably misses out some members. Indeed, it misses out *infinitely many*.[7]

Despite the fact almost all of physics is framed in terms of the real numbers, it is commonly assumed that **Q** suffices for all practical purposes of modelling and scientific explanation. The underlying rationale is *not* merely the banal observation that, as every experimenter knows, close approximations will always have to do for describing particular real physical entities (on account of noise and other messy details of reality). Instead, the assumption rests on the belief that, over the long term, numerically tiny deviations introduced into descriptions of the real physical world cannot make any difference to the *qualitative* likeness between our descriptions and reality. That is, even if tiny deviations might lead to large *quantitative* disparities between a model and physical reality, such as in the case of a system highly sensitive to its initial conditions, it is thought the general features of a model's behaviour won't be compromised. A weaker justification for assuming the adequacy of **Q** derives from the doctrine of theoretical computer scientists that the set of Turing Machine programs, operating with rational numbers (or, more precisely, at the lowest level with **N**), exhausts the set of physically realisable algorithms. Shortly we explore factors which render both reasons unconvincing.

3. DYNAMICAL SYSTEMS

As used here, the term *dynamical system* applies to a type of mathematical model specified by a set of coupled state variables and a set of equations which indicate how those variables change over time. By *state variables*, I mean a set of n scalar or vector variables which together describe the state of the entity being modelled; state variables often represent some subset of the observable properties of a physical object, such as its position, momentum, or temperature. The set of equations specifies how change in the values of state variables depends on time (making them *dynamical*) and on each other (making them *coupled*). We require that at least some of the interdependencies specified by the equations depend upon either differences between variable values or upon rates of change in variable values; the former type of dependence is normally described by *difference* equations, while the latter is described by *differential* equations. This requirement ensures that the metric properties of the n-dimensional metric space defined by the state variables, called *phase space* or *state space*, are relevant to the system's dynamics. In other words, *distances* between points in the space are relevant to the system's dynamics. (While it often is, the time set itself need not be a metric space, since discrete merely ordered time steps suffice for describing paradigmatically dynamical systems like shift maps. For the same reason, temporally sequential points need not be contiguous within the metric space.) The system's temporal evolution plotted in phase space—that is, the sequence of points picked out as the n state variables change over time—is called its *phase trajectory*. The trajectory of a point under the action of an *iterated function system*, to use Barnsley's (1988) preferred terminology for a class of iterative maps such as the logistic map or that used to generate the Mandelbrot set, sometimes takes the alternative label *orbit*. There is no requirement that state variables actually do describe observable properties of anything in the physical world; alternatively, there is no requirement that a dynamical system be a model *of* anything. Thus, under this working definition, a dynamical system may be purely formal, such as a shift map defined over the integers, or it may describe something physical, such as a pendulum or an apple falling from a tree.

As Chapter 7 illustrated, in the limiting case, the temporal evolution of *any* physical object may in principle be described in the paradigmatically dynamical terms of quantum mechanics. That is, there is in principle a model which describes the dynamics of any physical object in terms of quantum mechanical variables. (There is some trickiness here, in that Hilbert space is not phase space, but the details do not compromise the general point.)

Some of the confusion which infects much discussion on the topic and much use of the term *dynamical* comes from conflating properties of mathematical models and properties of the physical world. Whether a particular physical object may be described dynamically depends on our choice of state variables, and state variables refer to properties of either real or abstract entities at particular levels of description. Often we may choose from one or more different and nonequivalent sets of state variables to describe the same entity, and while some of those sets and associated equations may yield dynamical systems, others may not. Reiterating section 2 of Chapter 1, *in each case, we may speak of different dynamical models but the same entity.* For instance, a model of a real physical digital computer which describes its evolution in terms of CPU state and the content of its memory does not yield a dynamical system because the evolution of CPU states and memory bits does not take place in a relevant metric space.[8] Yet, a lower level description of the very same computer in terms of precise quantum states of free electrons and silicon yields a paradigmatically dynamical quantum system. And again at a higher level, consider the same digital computer plummeting from atop an office building. Here an appropriate choice of state variables again yields a dynamical model of a mass tumbling down under the influence of gravity and fluid mechanics.

In a perfectly straightforward sense, the descriptions above refer to the *very same (physically identical) computer* with state variables in terms of memory content and CPU state in the first case, quantum state in the second case, and (for example) orientation and relative velocity in the third case. We simply describe, or model, the same computer in a different way: a computer, by any other name (or description), is still a computer. (This observation relates to the standard distinction in philosophy of language between what Frege called 'sense', or *Sinn*, and 'reference', or *Bedeutung*. The same referent may be picked out by many different senses, or intensions.)

The term *dynamical system*, then, properly applies to *models* which describe real or abstract entities, and its appropriateness depends on the choice of state variables used to formulate the description. Recall from Chapter 5 that infinitely many mathematical descriptions fit the behaviour of any finite physical system. Thus it is not only a category error to suggest, on the basis of a model *describing* it, that any real physical object or collection of objects actually *is* a dynamical system; it also clashes with the fact that the *same* physical object may be described by a host of *different* dynamical (or non-dynamical) models. Making sense of such a suggestion demands a different definition of 'dynamical system'; preserving the general flavour of the present one while also covering real physical objects requires untrivialising the sense in which values of state variables and the equations governing them may be said to 'correspond' to features of the physical world. (Chapter 5 again.) Thus, while this chapter occasionally refers to 'physical systems' as the physical entities modelled by dynamical systems—or treats chaos as a feature of physical entities (rather than of dynamical models)—every such reference should be read as carrying implicit acknowledgement that the jump from model to modelled crosses a huge chasm well explored (and as yet largely unbridged) by philosophers of science.

Finally, while some of the many different models of physical entities may be dynamical and some not, each may afford the possibility of observing different kinds of features and of making different predictions. Useful modelling requires a careful match between the level of description of the model and the level (or levels) of description at which questions of interest are posed or explanations are sought.[9] A quantum mechanical description of a digital computer is not very useful if we want to debug a program; we're better served by an appropriate non-dynamical description of the thing as a computational system. Likewise, tracing orientation and relative velocity of a falling mass of electronics won't help us understand quantum tunnelling. Section 7 returns to the issue of alternative descriptions of the same thing with respect to the question of whether functional descriptions of human cognizers involve chaotic dynamical systems, but for now the task is to explore what it is that makes a system which is dynamical specifically *chaotic*.

4. CHAOTIC SYSTEMS

Frequently chaos is defined incorrectly just in terms of *sensitive dependence on initial conditions*. Such dependence is a necessary but not sufficient condition for chaos, and examples of non-chaotic but sensitively dependent systems abound. Consider, for instance, an idealised system of a rigid sphere left to roll, under the influence of a uniform gravitational force, from the apex down the side of a cone intersecting a plane. As in Figure 19, the point where the path of the sphere intersects the plane depends critically on the angle of a line from the cone's apex through the initial centre of the sphere, yet the system is in no way chaotic.

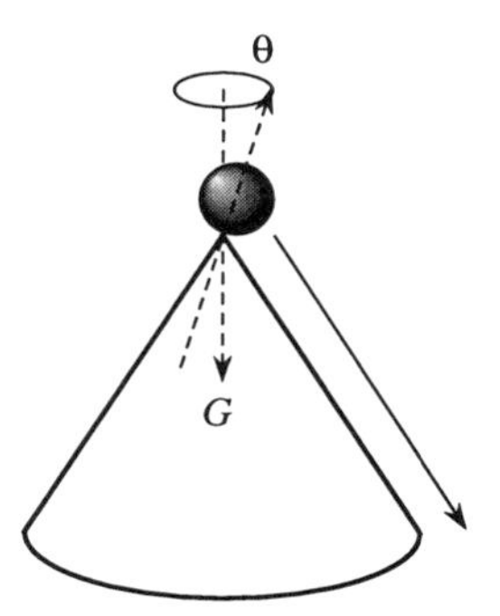

Figure 19. Sensitively Dependent Sphere on a Cone

4.1 Defining Chaos

Here I adopt a standard definition, due to Devaney (1988), which can be applied easily either to continuous systems described typically by differential equations or to iterated mappings. This definition requires *topological transitivity* and a *dense covering* of state space with periodic points. The second property is the most straightforward: the set of periodic points in the state space of a chaotic system at any given time slice, or Poincaré section, is at least countably infinite, and between any two points in the set there exists another also in the set. *Periodic points* are just those which lie on closed phase trajectories.

Topological transitivity indicates that any particular neighbourhood in the phase space of a chaotic system is eventually visited by the trajectory of some point lying in any other given neighbourhood. Consider a system mapping a metric space to itself, such as

$$C: J \to J,$$

where J is a metric space and C is a system of equations describing interrelationships between variables which pick out points in J. Topological transitivity means that for any two open bounded sets in J

$$U, V \subseteq J,$$

there exists a $t \geq 0$ such that

$$C_t(U) \cap V \neq \varnothing,$$

where $\varnothing$ is the empty set and the subscript is taken either as $t \in \mathbf{R}$, reflecting the time for which the system has been applied to the set or as $t \in \mathbf{N}$ indicating the number of iterations of a mapping. As in Figure 20, this means in other words that the phase trajectory under C of at least one point in U intersects V in finite time, regardless of how small or how far apart the two sets are to start. Alternatively, given any two such sets, there always exists some finite phase trajectory connecting a point in one to a point in the other.

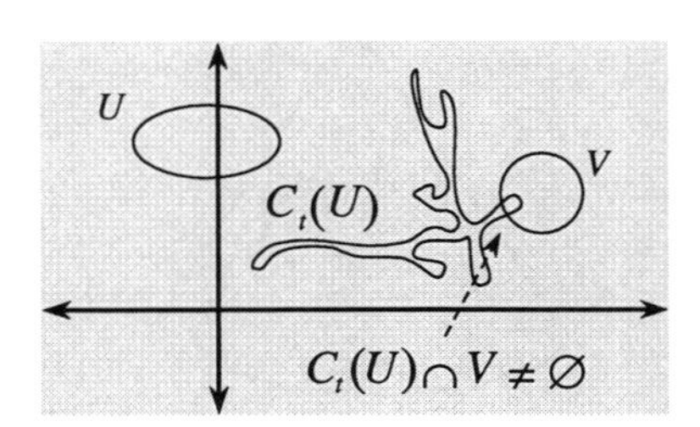

Figure 20. Topological Transitivity

Although these two properties together logically entail sensitive dependence on initial conditions (Banks *et al.* 1992), it is worth having in hand a formal definition of the latter as well. Sensitive dependence describes the case in which for any point in the state space of a system, another point within an arbitrarily small distance lies on a phase trajectory diverging from that of the first. A system C as above is sensitively dependent on initial conditions when there exists a distance $\Delta > 0$ such that for any point in the metric space—i.e., $\forall x \in J$—and for any closed neighbourhood N of x, there exists at least one point $y \in N$ and a real number $t \geq 0$ such that

$$|C_t(x) - C_t(y)| > \Delta$$

Note especially that this property occurs for every x in J, and a straightforward proof shows every neighbourhood N of x includes infinitely many such diverging points. The property appears schematically in Figure 21.

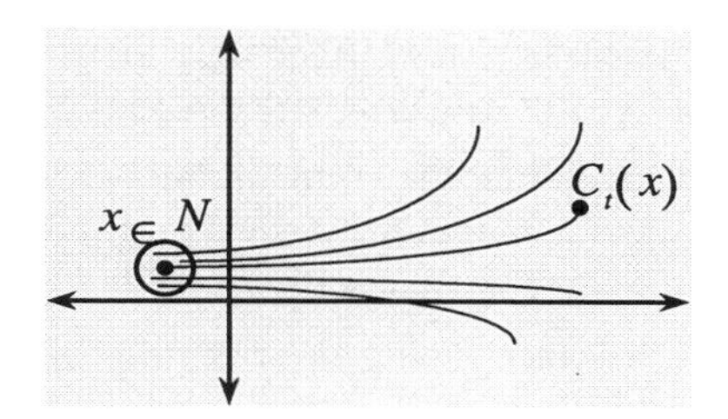

Figure 21. Sensitive Dependence in a Neighbourhood of x

4.2 Chaos in the Real World

As in the case of dynamical systems, we define chaos in terms of mathematical models, and a word of caution is in order. Properly speaking, chaos is a property of dynamical models, of particular ways of *describing* things—and whether a system 'is' chaotic is as much determined by the level of description as the question of whether it is dynamical. Moreover, while it may be rigorously *proven* that a given equation or set of equations defines a chaotic system, that certainty cannot be extended from the chaotic dynamical system to a physical entity being modelled. Starting with empirical measurements, we might search for *signs* of characteristics like sensitive dependence or quasi-periodicity or period doubling, but finite measurements can never demonstrate characteristics which mathematically demand infinite precision. For instance, we can never show empirically that the temporal evolution of a real physical system is periodic for infinitely many distinct initial states or that for any initial state there will always be another arbitrarily similar to it which diverges from it over time. Likewise, finite empirical measurements can never reject the hypothesis that a particular trajectory is simply a closed orbit, rather than lying on a strange attractor. In general, it is no more possible to prove from observation that a real physical system is chaotic than it is possible to prove it is governed by *exactly* such and such a set of equations. (Indeed, as defined here, it is doubtful whether the word 'chaotic' even applies to the real physical world. In a forthcoming book, Peter Smith explores the incongruity between the infinite intricacy necessarily exhibited by chaotic models and the fuzzy physical world, in which infinite intricacy is apparently impossible.) This should go without saying, but all too often one encounters misinformed assertions that the weather *is chaotic* or that a particular area of the brain *is chaotic*.

It is also frequently supposed that their sensitivity renders chaotic systems somehow utterly unpredictable, that there is some time t after which all predictions of a system's behaviour become useless. Yet the order of the quantifiers in the definition of sensitive dependence reveals the error: it *is* the case that for any point in some neighbourhood in the metric space, there is some time t by which the trajectory of at least one other point in the neighbourhood will have diverged from that of the first by $\Delta > 0$. But it is *not* the case that there is some time t which guarantees divergence of all the points in the neighbourhood! Indeed, the magnitude of divergence in each dimension for nearby typical phase trajectories is proportional to $e^{\lambda t}$, where λ is the system's *Lyapunov exponent* (and $e \approx 2.71828$ is the natural constant). Using the simple proportion, we can always calculate for any specific finite horizon of prediction t and any eventual error bound δ the minimal precision needed in an initial measurement to guarantee that after t our error will be less than δ. This feature is known formally as the *Shadowing Theorem*.

The theorem guarantees that for a certain cost in initial precision, we can 'shadow' any particular trajectory as closely as we wish and for as long a finite time as we wish. Therefore, the presence of chaos does not, broadly speaking, render a system impossible to model or to predict. Of course, placing any model in exact one-to-one correspondence with a real physical system's low level properties remains impossible, but while the specific behaviour of a model might eventually diverge greatly from that of a particular physical system because of this fact, its general character remains utterly unaffected. Nor does chaotic behaviour in general somehow make a system noncomputable or endow it with special non-algorithmic powers. Below I explain the relevant meaning of computability and explore some exceptions which do exist to the Turing-computability of chaotic systems, but in general there is very little mileage in the idea that seemingly chaotic behaviour in a brain, for instance, could *by itself* give a cognizer non-algorithmic powers. Likewise, while superficial links between sensitive dependence or topological transitivity and deep problems of cognitive science are occasionally entertained in the literature (including, in years past, by me!), the mere presence of apparently chaotic behaviour in a brain or other system of a cognizer does not appear to bear in any straightforward way on the frame problem, free will, creativity, or the content addressability of memory. Chaos is just an interesting mathematical feature of some dynamical models of some physical systems.

5. EXTENDED COMPUTABILITY

Underlying modern approaches to computability is the classic tradition of Turing computation over the natural numbers pioneered by Turing, Gödel, Church, and others. In the classic tradition, the most important boundary limiting what may be computed by a Turing Machine is *completeness*, or decidability, summarised in the section 'Gödel De-Mystified' in Chapter 3. As explored in that chapter, the basic problem of incompleteness has been extended to its most general form by Greg Chaitin, and the theory of Turing computation over the naturals can be usefully developed to include algebra and fields (Rabin 1960). But present purposes require a consideration of more than the raw limits of logical proof as formalised by Turing Machines operating over the naturals. The fact that physical systems are normally described by functions which take on real values demands some extension of Turing computation to cover the real numbers **R**.

5.1 Recursion Theory

The field of *recursive analysis* develops natural number computation into a framework appropriate for the real numbers. Here we use the standard recursion theoretic definitions of Pour-El and Richards (1989). A real number $x \in \mathbf{R}$ is said to be computable when there is a *computable sequence* of rationals which *converges effectively* to it. A sequence of rationals $\{r_k \in \mathbf{Q}\}$ is computable when there exist three recursive functions over the naturals[10] a, b, s: $\mathbf{N} \rightarrow \mathbf{N}$ such that for all k,

$$b(k) \neq 0 \qquad \text{and} \qquad r_k = (-1)^{s(k)} \frac{a(k)}{b(k)}$$

Such a sequence *converges effectively* to x when there exists a recursive 'error bound' function over the naturals e: $\mathbf{N} \rightarrow \mathbf{N}$ such that $\forall N \in \mathbf{N}$:

$$k \geq e(N) \qquad \text{implies} \qquad |r_k - x| \leq 2^{-N}$$

That is, the modulus, or absolute value, of the deviation from x of members of the sequence shrinks like 2^{-N}. Thus, for any desired level of accuracy, the recursive function e indicates exactly how many steps through the sequence we must progress to achieve that precision. (Obviously, any real number which happens to be rational is straightforwardly computable, but not every computable real need be rational.) Intuitively, a real number is computable if it can be approximated to an arbitrary degree of accuracy by an algorithmic method. It is interesting to note that the set of all *computable* reals, like the set of rationals, is of course only *denumerably* infinite, while the set of all reals is *uncountably* infinite. Since all reals are either computable or noncomputable, this implies that 'most' numbers are noncomputable (Minsky 1967).

Computability for a real-valued *function,* first formulated nearly four decades ago (Grzegorczyk 1955, 1957; Lacombe 1955a, 1955b), requires both *sequential computability* and *effective uniform continuity*. Consider a function f defined on a closed bounded rectangle I^q in $\mathbf{R}^q$, where

$$I^q = \{a_i \leq x_i \leq b_i,\ 1 \leq i \leq q\},$$

a_i and b_i (the 'corners' of the rectangle) are computable reals, and $\mathbf{R}^q$ denotes q-dimensional real space. The function is *sequentially computable* when it maps computable sequences to computable sequences: f maps every computable sequence of points $\{x_k\} \in I^q$ into a computable sequence $\{f(x_k)\}$ of reals. The requirement of *effective uniform continuity* is met when a particular algorithmic relationship obtains between the Euclidean distance separating points in the domain of the function and the distance between corresponding points in the range. Specifically, the condition is satisfied when there exists a recursive function d: $\mathbf{N} \to \mathbf{N}$ such that $\forall x, y \in I^q$ and $\forall N \in \mathbf{N}$:

$$|x - y| \leq \frac{1}{d(N)} \qquad \text{implies} \qquad |f(x) - f(y)| \leq 2^{-N}$$

In other words, the sizes of neighbourhoods before and after the action of f are computably related by the natural number function d.

No immediate inconsistency threatens the notion of such a real-valued function which meets the requirements of both sequential computability and effective uniform continuity and which *also* serves as the basis of a topologically transitive dynamical system with a state space densely covered with periodic points. The definitions of computability and of chaos refer to properties as different as apples and oranges, and apparently little or no scope exists for establishing direct logical implications between them: indeed, there is utterly no reason why any chaotic system should—*solely by virtue of being chaotic*—be noncomputable.

To be sure, we might contrive a noncomputable chaotic system from a computable one simply by introducing noncomputability into a coefficient of one of its equations. There are several examples of other functions which might similarly be worked into chaotic systems (Pour-El and Richards 1979, 1981; also see Aberth 1971). But we might just as easily contrive a noncomputable *non-chaotic* system in the same fashion. The point is that *in general*, chaotic behaviour has nothing to do with computability, either in the classic framework of computation over the naturals or in the extended recursion theoretic framework of real-valued functions. Neither topological transitivity nor dense coverings of state space with periodic points (nor sensitive dependence, for that matter) threaten, singly or jointly, sequential computability or effective uniform continuity. Thus, with an important exception discussed below, the presence of chaotic behaviour in some part of a cognitive system has no bearing on the Turing-computability of that system's behaviour, and it has no bearing on the question of whether that system mightn't be effectively simulated by (or be computationally equivalent to) a Turing Machine. Chaos is an interesting trait of some kinds of dynamical systems which may be relevant for explaining observed behaviour of some cognizers, but it generally affords no opportunity to import noncomputability into that behaviour.

6. COMPUTABILITY AND CHAOTIC ANALOGUE SYSTEMS

Although the last point above is true in the general case, recently a curious exception has emerged which bears not only on cognitive scientific and philosophical questions about cognitive systems but also on the theoretical foundations of computer science. Hava Siegelmann (1995) and Eduardo Sontag (Siegelmann and Sontag 1994) have recently introduced ideally realisable, highly chaotic dynamical systems—an analogue recurrent neural network and an analogue shift map—the computational powers of which exceed those of the standard Turing Machine model.

6.1 Super-Turing Computation

Both the analogue recurrent neural net and the analogue shift map compute in polynomial time[11] the class of functions denoted *P/poly* (Karp and Lipton 1980; Balcázar *et al.* 1988). This class includes all Turing-computable functions plus some (comparatively few, in fact) nonrecursive super-Turing functions such as the so-called *unary halting function*, which determines whether a given program encoded in unary halts on a particular unary input sequence. The two models are computationally equivalent to the family of nonuniform circuits of Boolean (AND, OR, NOT, etc.) gates, where *nonuniform* means the actual circuit layouts need not be recursive sets—i.e., where the circuits may be noncomputable. The family of nonuniform circuits is also formally equivalent to nonuniform Turing Machines, such as those nonrealisable models which, in addition to their input string, 'take advice' (Turing 1939) to assist in a computation or consult sparse oracles for help during the course of computation. (Here *sparse* means the number of words of any particular length in the oracle set is polynomially bounded.) All three families—analogue neural nets and analogue shift maps, each of which is *uniform*, nonuniform Boolean circuits, and nonuniform oracle machines—compute P/poly. (The models of Siegelmann and Sontag are the first to be discovered which are uniform but display super-Turing properties.) This section explores both the neural net model and the analogue shift map with a view to revealing the mathematically beautiful simplicity of each and uncovering their relationship to dynamical systems which might be relevant to cognitive science.

6.2 Super-Turing Neural Networks

The remarkably simple neural net consists of a finite number N of processors with activations updated synchronously at discrete time steps by the scaled output of a piecewise affine function (i.e., a linear combination) of the previous activations x_j and M external input signals u_j with *real* weights a_{ij}, b_{ij}, c_i:

$$x_i(t+1) = \sigma\left(\sum_{j=1}^{N} a_{ij}x_j(t) + \sum_{j=1}^{M} b_{ij}u_j(t) + c_i\right), \qquad i = 1,\ldots,N,$$

That is, each network node activates according to the sum of its x_j inputs from both other nodes and direct external signals u_j, with each input weighted according to a connection strength a_{ij} or b_{ij}, in addition to any extra bias term c_i (which varies according to the node). Bias terms are typically understood interchangeably either as artificial neural network models' sanitised rendition of a real neuron's inherent susceptibility to fire or as a continuous tonic signal of a given strength. (See section 2.1 of Chapter 8 for more on basic neural models.) This linear combination appears schematically for 'neuron' 3 in Figure 22.

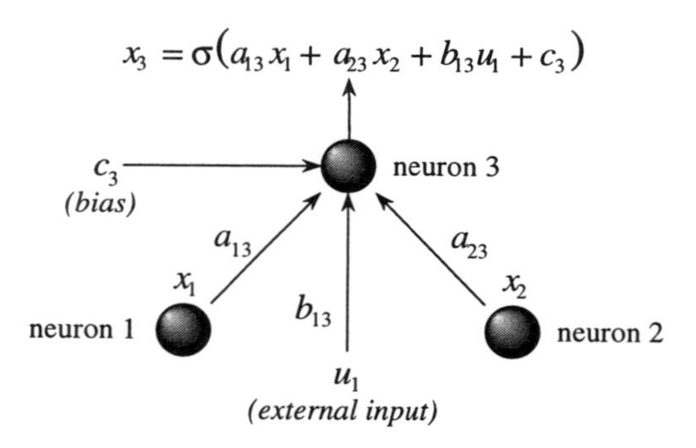

Figure 22. Calculating Node Activation From Weighted Inputs

(Interestingly, higher-order nets with more complex versions of this function, using polynomial rather than linear combinations of inputs, offer no increase in power apart from polynomial improvements in processing time.) The output of each node is constrained to the range [0...1] by a scaling term σ, the simplest possible sigmoidal ('S'-shaped) nonlinearity, shown in Figure 23.

$$\sigma(y) := \begin{cases} 0 \text{ if } y < 0 \\ y \text{ if } 0 \leq y \leq 1 \\ 1 \text{ if } y > 1 \end{cases}$$

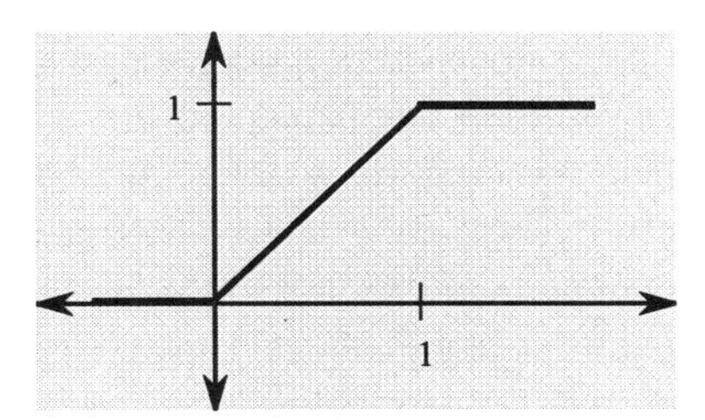

Figure 23. 'Saturated-Linear' Output Scaling Function

Just as more complex polynomial combinations of weighted inputs offer no increase in power, this 'saturated-linear' output scaling function also contributes exactly as much computational power as the more usual differentiable sigmoids like that depicted in part (b) of Figure 13 in Chapter 8.

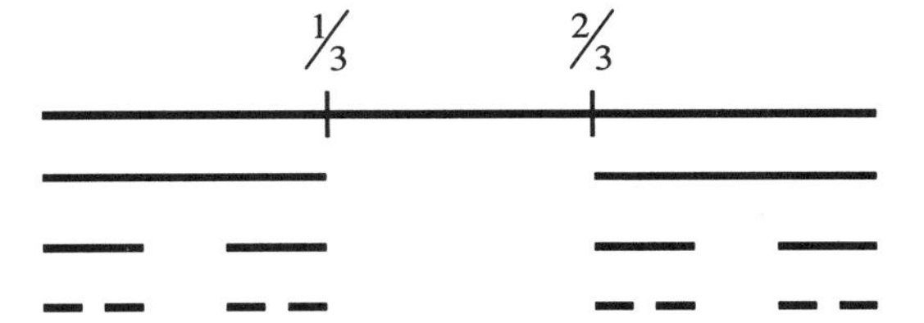

Figure 24. Constructing a Cantor Set

The key feature which separates this model from Turing-computable relatives is the presence of weights, representing synaptic efficacy, described by *real* numbers. With weights constrained to the rationals, such networks exhibit exactly the same power as Turing Machines (Siegelmann and Sontag 1992). In fact, the formal proof of the analogue net's super-Turing power, achieved by showing effective simulation of the family of nonuniform circuits, rests on the presence of *just one* irrational (but not necessarily noncomputable) weight. In the proof, the single irrational weight encodes, in a style inspired by the Cantor set, a specification of the particular nonuniform circuit being simulated (Siegelmann and Sontag 1994). In general, Cantor sets are constructed by recursively deleting segments in a continuous interval. Figure 24 shows the first few steps in the construction of a Cantor set over the unit interval [0...1] by deleting the middle third of each remaining segment of the interval.

Since the Cantor set construction requires that every continuous interval have its middle third deleted, it's easy to understand the set's most relevant feature for present purposes: any two points in the set are always separated by a finite interval not in the set. By coding nonuniform circuit descriptions as points in a related set (with the interval divided instead into ninths, and segments in alternate ninths deleted), Siegelmann and Sontag's networks can thus manage to 'read' an encoding with only finite precision. Indeed, while it is nearly universally assumed that analogue computation with real numbers could exceed that of standard computation only by exploiting physically unreasonable infinite precision, each step of the model's computation relies only on finite linear precision. That is, the precision required at any given step relates in a linear fashion to the desired precision in the result. (This is analytically more straightforward when it comes to the finite domains of dependence for the transfer functions of the computationally equivalent analogue shift map, discussed below.)

Curiously, the network's activation function, including the scaling term σ, is a computable equation, and the actual layout of the finite network can be described recursively. Yet obviously but somewhat paradoxically, because the network has super-Turing power, not all features of its behaviour can be simulated by a Turing Machine. (In fact, this really is little more paradoxical than the fact that some r.e. sets are not themselves decidable; see Chapter 3.) There is not only a quantitative mismatch but also a *qualitative* mismatch between the behaviour of such a network and its nearest Turing simulation. As early as 1992 (Mulhauser 1992, 1993c), I suggested on much less rigorous theoretical grounds that just such a situation—computable governing equations but elements of noncomputable behaviour—might be observed for a class of specifically analogue chaotic dynamical systems. This suggestion was based on the following idea, given an analogue system described by computable equations. If such an analogue system is appropriately sensitive to continuous parameters, as in the case of chaotic systems, there is no *a priori* reason to think it mightn't do something 'interesting' within the approximation neighbourhood with respect to which any rational simulation must be limited. In other words, there is no immediately obvious reason to think the Shadowing Theorem guarantees a qualitative correspondence between members of the *uncountably* infinite set of phase trajectories described by real numbers and the *countably* infinite set of rational approximations. Just as there is no one-to-one mapping of members of **R** to members of **Q**, perhaps there is no one-to-one mapping of phase trajectories described with **R** to qualitatively similar ones described with **Q**.

Such a view took support from the emergence of the pathologically 'riddled' attractor basins of the Sommerer and Ott model (1993), where computable equations describe a chaotic system in which, although we may shadow individual trajectories arbitrarily closely, the question of which of two attractors they (or any of the others within the shadowing neighbourhood) will eventually converge to remains undecidable. That is, the 'destiny' of individual trajectories never becomes clear: we can never determine with certainty whether any particular trajectory lies within one basin of attraction or another (except, obviously, for those which actually begin at zero distance from an attractor). Hence, there is no guarantee that *any* of the rational approximations within some neighbourhood of an analogue number trajectory share its ultimate destination.

The neural networks of Siegelmann and Sontag make the point even more persuasively. They are *essentially* analogue and *essentially* chaotic; while Cantor sets are the fractal invariant sets of a family of well known chaotic iterated maps (Barnsley 1988), the actual *dynamics* of the networks while computing super-Turing functions are also chaotic (Siegelmann, personal communication). Most importantly, they may follow phase trajectories, while computing non-Turing problems in P/poly, which differ both quantitatively and qualitatively from those of any Turing-computable rational number approximation. And they do so in the *polynomial time* relevant to computation in the real world.

6.3 Super-Turing Shift Map Model

Similarly chaotic is Siegelmann's (1995) computationally equivalent and ideally realisable dynamical system, the analogue shift map (which differs slightly from the 'generalised shift' of Moore—see below). This model has the advantage of admitting slightly easier analysis while remaining relevant to a host of real systems. (Shift maps are widely applied to model all kinds of physical systems.) The shift map acts over a dotted sequence in the space denoted $\dot{E}$, constructed from the finite alphabet E—i.e., two sequences of letters from E separated by a dot. (For instance, given the alphabet $E = \{1, 3, 4, 5, 9\}$, 3.14159 is a dotted sequence in $\dot{E}$.) Given such a sequence a and an integer $k \in \mathbf{Z}$, Siegelmann defines the shift map

$$S^k : \dot{E} \rightarrow \dot{E}$$

with shifts of the individual elements of the sequence, denoted as

$$S^k : (a)_i \rightarrow (a)_{i+k}$$

We may think of the index i as indicating the element i places to the left of the dot for positive i or i places to the right of the dot for negative i. Thus we may interpret the shift map either as moving each element in a to a new position while leaving the dot in place or as shifting the dot k places to the right for positive k and similarly to the left for negative k. For instance, Figure 25 shows the shift on 3.14159 for $k = 2$.

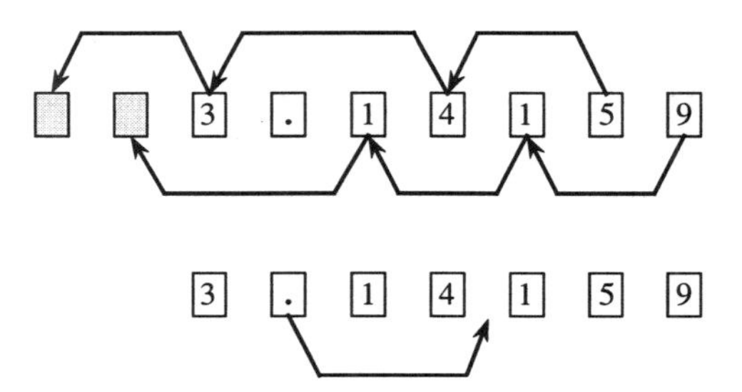

Figure 25. Interpreting the Shift Operation on a Dotted Sequence

The analogue shift is then defined as a process of first replacing a dotted substring with another of equal length according to a function G and subsequently shifting the new sequence an integer number of places according to a function F. Overall, the analogue shift map is thus

$$\Phi : a \rightarrow S^{F(a)}(a \oplus G(a))$$

Here G, operating over the dotted sequence space, modifies the sequence, while the function F, from the dotted sequence space to the integers $\mathbf{Z}$, indicates the amount and direction of the shift under S. The operation $\oplus$ replaces the elements of the first dotted sequence with the matching element of the second dotted sequence if that element exists in the second sequence, or leaves it untouched otherwise. Where g, like a, is a particular dotted sequence,

$$(a \oplus g)_i = \begin{cases} g_i \text{ if } g_i \in E \\ a_i \text{ if } g_i = \varepsilon \end{cases}$$

and ε is the empty element. (At first glance awkward, this two-step process for modifying the dotted sequence a simplifies the job of specifying the function G.)

The analogue shift is iterated, and the associated computation is complete when and if the system reaches a fixed point—i.e., when $F = 0$ and a remains unmodified by the action of G. If the process does halt, the input map is then from the initial dotted sequence to the final subsequence to the right of the dot. Siegelmann demonstrates that even when attention is restricted to systems offered finite dotted sequences for input and halting with either finite or left-infinite dotted sequences, the analogue shift displays super-Turing power. Her proof establishes computational equivalence between the (uniform) analogue shift and the (uniform) analogue recurrent neural net, previously proven exactly as powerful as the family of nonuniform Boolean circuits.

Recall that results of computations from the analogue recurrent neural net demand only linear precision at each step, even though the actual values internal to the network during computation are continuous reals in the general case. This feature is matched in the analogue shift: both F and G have a finite *domain of dependence* (DoD). That is, their values depend only on a finite dotted substring of the dotted sequence they take as input. The *domain of effect* (DoE) of G may be finite, infinite on one side of the sequence, or infinite on both sides of the sequence (bi-infinite). And similarly again, just as analogue networks are Turing-equivalent with weights constrained to the rationals, so too is the shift map Turing-equivalent when the DoE of G is finite (and thus reducing to the 'generalised shift' due to Moore 1990, 1991)—i.e., when G outputs only rationals (Siegelmann 1995). Once more, although the super-Turing power of the analogue system depends on the use of real number values, each step of process requires finite precision only.

6.4 Is the Church-Turing Thesis False?

These models call into question, for a class of analogue systems, typical interpretations of the Church-Turing thesis (a label which originated, not with Church or Turing, but with Kleene 1952 as *Church's Thesis*)—sometimes stated as the principle that the Turing Machine and equivalent models formalise the notion of *algorithm* or the principle that no ideally realisable computer can have more power (apart from polynomial speed improvements) than the Turing Machine.[12] In one of the most important recent papers on analogue computation, Vergis and colleagues (Vergis *et al.* 1986) proved a restricted form of the thesis for a class of well behaved linear analogue systems and concluded that "any interesting analog computer should rely on some strongly nonlinear behavior" (p. 93). We now have an example of such an interesting 'computer', and Siegelmann and Sontag (1994) have proven for a large class of dynamical systems an analogue version of the Church-Turing thesis, the conjecture that "any reasonable analog computer will have no more power (up to polynomial time) than first-order recurrent networks" (p. 355), where *first-order* denotes networks whose node activations depend only upon linear combinations of inputs and where we may take *reasonable* to mean *ideally realisable*.

The apparent failure—or, at the least, weakening—of the Church-Turing thesis as commonly understood carries profound implications for computer science as well as for approaches to mind inspired by it. The Turing model, which has served as the foundation of theoretical computer science for more than half a century, does not adequately describe the class of all algorithms; a full description demands in addition something akin to the analogue framework of Siegelmann and Sontag.[13]

To the extent that cognitive scientific approaches such as functionalism rely on the Turing model primarily just to supply the class of algorithms with which, it is hypothesised, human cognitive processes may be completely described and/or simulated, it might seem the emergence of super-Turing computation offers little new except a reminder that attention should be paid to the analogue nature of the brains or other systems with which cognizers perform.[14] However, the super-Turing capabilities of some chaotic analogue systems also suggest the basic metaphor of how a computation comes to be performed, presently a metaphor supplied by the Turing Machine, may benefit from closer alignment with the framework of continuous dynamical systems such as the analogue recurrent neural network. The difference is roughly one between discrete symbol manipulation on the one hand and (not necessarily symbolic) manipulation of real-valued signals on the other.[15] The theoretical picture suggested by a functionalism based on the former may differ in significant respects from the picture suggested by a functionalism embracing the latter. However, a wide range of questions, both in cognitive science in particular and about the project of naturalising philosophy of mind in general, remains essentially unchanged after adapting functionalism or other approaches to a broader theoretical base including super-Turing analogue computation. An adapted functionalism or similar approach leaves unaltered basic philosophical questions about whether cognizers such as humans could at some level be equivalent to computers—except that the question *what kind of computer?* (digital or analogue) becomes rather more pressing.

7. CHAOS AND COGNITION IN REAL LIFE

In this context, it is helpful to recall the note at the end of section 2.2 concerning the common assumption that rational number models suffice 'for all practical purposes' (or 'FAPP', following the quantum mechanical tradition described in section 6 of Chapter 7). Building functionalist theories of cognition and building dynamical models of neurophysiological processes are closely related but distinct enterprises. While the former might succeed as an explanatory strategy largely on the basis of theoretical considerations (and even without very much concern over particular models of computation), the latter demands attention to the match between particular dynamical models and particular brain structures. Sidestepping for the moment Peter Smith's intriguing problem, mentioned in section 4.2, about the relationship between chaotic models which *must* exhibit infinite intricacy and a real world which apparently *cannot* exhibit infinite intricacy, it is clear that using dynamical models at all requires facing up to questions about matching model and modelled.[16]

(In an unpublished manuscript, Maass and Orponen 1996 propose a proof that the power of *any* discrete time computational system with a finite dimensional, bounded state space which is affected by piecewise uniformly continuous noise is bounded by that of deterministic finite state automata. Perhaps the very strength of their conclusion hints something might not be *quite* right;

moreover, their assumptions differ considerably to those of Siegelmann 1994, who previously proved that the unusual powers of analogue recurrent networks are robust to certain kinds of noise. The project of fully assessing and comparing the two proofs awaits an ambitious theoretician with greater command of the material than I can offer! Of course, page 212 noted already that *all actual* physical systems are apparently limited to the power of finite automata. Questions about capabilities of the underlying model of computation and its match with reality persist, however.[17])

7.1 Model Differences and 'Computation'

Models based on **R** and those based on **Q** need *not* behave the same. This difference in behaviour is at once quantitative and qualitative. On one hand, the behaviour of *any* rational number approximation of a chaotic analogue system must differ from it quantitatively, simply because of sensitive dependence. As in section 4.2, however, the difference in behaviour is quantified precisely (and computably) by the Shadowing Theorem. But as in section 6.2, the behaviour of a rational number approximation of a chaotic analogue system also may differ qualitatively, in the sense that analogue systems apparently may solve 'problems' which rational approximations cannot. This kind of difference in behaviour is not quantified precisely by any function which computably relates the 'answers' of the approximation to those of the analogue system.[18] Given such a difference in behaviour between a rational number model and a real number model, there is some tension in the notion that they could *both* be the best fit to the physical reality being modelled. Moreover, the way in which two such models differ may actually be rather important, since the distinction happens to mark out a set of computations which Turing-based models not only *don't* perform (in some particular example model) but *cannot possibly* perform (in any model).

Clearly, all this is independent of whether we interpret neural networks as actually *computing* anything, and the substance of the debate mentioned in section 2.1 of Chapter 1 between computationalists and dynamicists (with, say, Churchland and Sejnowski 1992 or Crick 1994 exemplifying the former and Edelman 1989 the latter) is of no present concern. Although the formal proof of analogue neural nets' super-Turing power, for instance, is phrased in terms of particular outputs they may produce in response to particular inputs, this is purely a convenient formality; here we are interested in such networks' overall behaviour, with or without particular nodes selected for communicating inputs and outputs. Likewise, the Cantor-like method of encoding a nonuniform Boolean circuit is entirely an artefact of the proof structure. Generally speaking, my concern is not whether we might exploit analogue neural networks as some sort of general purpose computing devices with super-Turing power—indeed, I am inclined to think we could not—but whether the behaviour of neural networks in the real world should be understood with real or rational number models.

This also reveals the flaw in the naïve objection that real number 'computation' cannot be important to neural modelling because we can't ever instantiate particular real numbers, chosen in advance, in actual physical parameters such as position or momentum at a low level or spiking frequency at a higher level. That is, someone might object that because we cannot *in practice* encode real number values in such parameters (and certainly not with Cantor sets!), we needn't concern ourselves with what computations might be performed on them by analogue systems if we could. But the general fact that we cannot encode *any* specific numbers directly in such parameters—real, rational, or whatever—is just a basic feature of thermodynamics: specifically, we cannot by any finite process reduce to zero the temperature of the degrees of freedom in the correlation between a prepared system and its environment.[19] (It is only in the limiting case of quantum mechanics that a strict correlation between states of two systems may be prepared.) Intuitively, this just means we cannot escape noise. The fact applies universally and has nothing to do with real numbers. More importantly, that we cannot so encode real numbers has no bearing on whether such parameters nonetheless do take on real number values. Thus it doesn't matter that we can't *choose* what real numbers to feed a particular neural system; it matters only whether real numbers are there anyway.

And on this very deep question—whether real values, as distinct from rational ones, actually occur in the physical world—I can only acknowledge the fact, already mentioned in section 2.2, that nearly all of physics is framed in terms of real numbers. In the absence of an *argument*, as distinct from the usual FAPP assumption, as to why the features of the world described by physics can only take on rational values, it seems there is little more to say. (Well, there is a *little* more; see section 3 in Chapter 10.)

7.2 Applicability and Relevance of Super-Turing Potential

Significantly, these observations about the difference between rational and real models of the physical world apply quite generally; for the most part, the details of particular neurophysiological models do not bear on the main point. While it is trendy to talk of modelling 'the edge of chaos' or the dynamics of systems hopping between chaotic and non-chaotic regimes or mechanisms for 'controlling chaos', the possibility of a mismatch between the capabilities or qualitative behaviour of an analogue chaotic system and a discrete one exists in any case. Siegelmann and Sontag's analogue networks show the possibility may particularly bear on neural systems, while Siegelmann's shift map indicates its relevance to a wide range of chaotic analogue systems.

Yet, the question of whether we should care about this general applicability remains linked to the actual *rôle* of chaos in a particular model.[20] In the work of Freeman cited in section 1, for instance, experimental observation of behaviour characteristic of chaos suggests a model in which chaos is an important feature of a normally functioning olfactory cortex. Here the analogue nature of real neurons may be relevant to the match between model and real observed behaviour. In Carpenter and Grossberg's (1983, especially section 11) gated dipole model of circadian rhythms, by contrast, chaotic behaviour appears merely as a side effect for unusual parameter values in the equations of a simple oscillatory dynamical system. Since these parameter values lie outwith the normal range of interest, and chaos in this case is *not* an important feature of a normally functioning hypothalamus, differences between analogue and discrete models are apt to be of limited interest.

As recalled in section 3, the same object may usefully be modelled at many different levels of description. It would be helpful to know whether chaotic behaviour is really essential at the level (or levels) of description specified by useful overall functional models of cognition. Circuits etched into the silicon chips of modern digital computers, for instance, behave as they do partly because of quantum mechanical effects. Yet the operation of the computer *as a computer*— its overall functional behaviour at the relevant higher level of word processing and sending email—is entirely classical and does not essentially depend in any way upon 'accidental' quantum effects in semiconductors. Modern computers would work just the same way, albeit more slowly, if their millions of tiny circuits were instantiated instead by valved water pipes or, to use Ned Block's (1978) image, by a billion Chinese people exchanging coded English messages over telephone lines. Likewise, although many of course dispute the idea, it *could* be the case that the low level, possibly chaotic neural systems of cognizers like humans could in principle be replaced with non-chaotic and/or digital components without altering their higher level functional structure. Knowing something of the extent to which overall functional descriptions of human cognizers *essentially* involve analogue chaos would provide enormous clues about appropriate levels of modelling.

The basic message of functionalism as an explanatory approach survives essentially unscathed by super-Turing computation. However, the overall picture it suggests may differ significantly according to whether it takes Turing-style discrete symbol manipulation or neural network-style analogue signal manipulation as its central model of computation. *Which* model best fits biological reality remains an open empirical question.

While chaos as a feature of the mathematical framework of dynamical modelling may have little direct bearing on such properties of the cognizers being modelled as are popularly linked with it, the special capabilities of chaotic analogue systems may yet be of profound significance for cognitive science and the philosophy of mind. Questions of how important chaos may be for real cognizers and of what capacities it may bestow on them are far from being answered.

NOTES

[1] The 'analogue' models described in this chapter take continuous values but operate over discrete time. Although the analogue/digital distinction properly has two dimensions—models may be continuous or discrete in both signals and time—this subtlety is largely important in the context of this chapter.

[2] Although, as the discussion above suggests, there might be other reasons for attempting to build *special purpose* analogue machines directly inspired by biological brains.

[3] Examples include Freeman (1964, 1979, 1987, 1988, 1989), Freeman and Skarda (1985), Skarda and Freeman (1987), Yao and Freeman (1990), and Eeckman and Freeman (1991).

[4] Some examples: Horgan and Tienson (1993), Mulhauser (1993b, 1993d), Smith (1993 and forthcoming), Wright *et al.* (1993, 1994), and Tsuda (1994).

[5] Of course, there are exceptions to the computability of functions relevant to describing the real world; these feature later.

[6] Why black holes? As Bekenstein (1981a, p. 287) notes, "black holes have the maximum entropy for given mass and size which is allowed by quantum theory and general relativity": thus, any bound on the maximum number of internal configurations for a black hole applies with equal force to every other physical system.

[7] Those whose intuitions recoil at this fact will find details in any introductory text such as Ross and Wright (1988).

[8] The usual metric for discrete bit spaces is the *Hamming metric*—see note 13 on page 1. In general, Hamming distances between a digital computer's distinct memory states are entirely irrelevant to its temporal evolution. That is, the temporal evolution of a digital computer is unrelated to distances between distinct CPU and memory states.

[9] One gloss on the *purpose* of modelling is that it aims to capture the characteristic features of a system's behaviour with a small number of variables and equations, thus *compressing* a description of the system's behaviour at what is normally some lesser degree of precision. (Recall the note in section 2.1 of Chapter 3 on Solomonoff's application of algorithmic information theory to quantify the unifying power of scientific theories in terms of the degree of compression they make possible.)

[10] Here, a recursive function is one which can be computed by an effective procedure—a Turing program guaranteed to produce the required results and halt.

[11] 'Polynomial time' complexity, indicating a polynomial relationship between the size of a problem and the resources required to solve it, is generally taken to describe problems which are computationally *tractable* in the real world. By contrast, those problems which demand resources exponentially related to their size are normally considered computationally *intractable*.

[12] See, for instance, Yasuhara (1971); also see Boolos and Jeffrey (1989).

[13] Their approach is, for many reasons, preferable as a model of real computation to other analogue frameworks such as that of Blum *et al.* (1989), which is both infinite dimensional and allows infinitely precise comparisons between values.

[14] Incidentally, Siegelmann and Sontag (1994) show that the equation P = NP implies much the same unappealing consequences under their framework as in the standard Turing context and conclude it is therefore unlikely analogue nets can solve NP-hard problems (those thought to require exponential resources) in polynomial time. Nonetheless, the possibility of special analogue capabilities with respect to NP-hard problems and the prospect of a finer characterisation of NP analogous to Selman (1982a, 1982b) remains open, with, for instance, Vergis *et al.* (1986) able to predict exponential settling time only as a *worst case* for analogue spin glass models as in Barahona (1982) or Johnson (1983).

[15] The rigorously provable distinction between capabilities of analogue networks and Turing Machines puts an interesting twist on the connectionist-logicist debate. Given that analogue networks may perform super-Turing computations, it cannot be the case that their processing may be reduced to symbol manipulation *of the Turing kind* at nontrivial levels of description. Whether it may be understood as symbol manipulation of some other kind is another question.

[16] For more general observations on models, including chaotic ones, see Morton (1993); for special emphasis on the attractions of chaotic models, see Rueger and Sharp (1996).

[17] It is worthwhile recalling that in considering real biological systems, our concern is not with discrete time networks anyway. See Orponen (1996) for a survey of the extremely limited current field of research on computability and complexity aspects of analogue time computation; more generally, also see Pour-El (1974). Thanks to Dave Chalmers for searching out the Maass and Orponen (1996) paper mentioned in this paragraph.

[18] If there were such a function, then it could be used to compute the 'answer' given by the analogue system from the 'answer' given by the rational approximation, thus offering a computable method for calculating a noncomputable value.

[19] Obviously, this observation applies strictly to the physical parameters *themselves* and is independent of the fact that we regularly 'encode' numbers at higher levels in wristwatches, computers, dice, and the like.

[20] Also see Wolfram (1985) for an interesting speculative distinction between *homoplectic* processes, those whose macroscopically random appearance results simply from magnification of low level environmental noise, and *autoplectic* processes, those which persistently generate complex behaviour independently of variations in background noise. With this distinction in mind, Bennett (1990) wonders whether something like a waterfall could be a logically deep (see section 2 of Chapter 5, starting on page 1) autoplectic process containing objective internal evidence of a long dynamical history: is there any objective difference between a day old waterfall and a year old one?

CHAPTER TEN

My 'Hidden' Agenda

Exactly two months to the day before my first visit to British Telecom Laboratories where I now work in artificial life research, I found myself before a multidisciplinary panel of scientists at the offices of the Royal Society of Edinburgh. The government's Scottish Office and the Royal Society, the organisation which bills itself as "Scotland's premier learned society", were sponsoring a small number of Personal Research Fellowships, and I had been short-listed and invited to interview. If successful, my next few years of research were to focus on the development of new cognitive models, inspired by human studies, which would be especially tailored for the goal of implementing data structures like the self model in non-biological substrates such as computers. My proposal was entitled 'Engineering the First Generation of Conscious Machine'.

The palpable feeling of awkwardness and the interviewers' blank faces when I entered the room left me anxious from the start as to what sort of reception lay in store for my prepared presentation. But I struggled through a dozen or so transparencies anyway and sat down to take questions. Only a few were of the scientific or technical sort which tend to stick in my mind, and the excitement of the day has left me with little recollection of most. As I was finishing some answer, however, one Fellow leant forward and interjected, with a flash of intensity in his otherwise carefully measured tone, the single question which burned itself with perfect clarity, word for word, into my memory:

But don't you think building a conscious machine is a bit like building an atom bomb?

Reflecting on it later, I realised I had never in my life been so taken aback by an interviewer's question. But at the time I was slightly surprised—and the panel perhaps seemed disappointed, I remember thinking—by the immediacy of an answer which I would still give today if forced to address the question seriously. No, I replied, building a conscious machine is not like building an atomic bomb, at the very least because the process of building a conscious machine would teach us something about our innermost selves, something about those aspects of our personal existence with which we are so intimately familiar yet which remain so deeply puzzling—something which making atomic bombs has not taught us. In building a conscious machine, we would learn about ourselves. We would learn about our *selves*. As with the bulk of my answers that day, I don't think they were convinced.

The response accurately reflects, however, the motivations underlying my interest in theories of consciousness and cognition in the first place; it reflects the craving to understand which lurked behind my resoundingly unsuccessful application to the Royal Society, the irrepressible urge to grasp something of what we conscious cognizers are and of how we relate to each other and to the physical world around us. My aim is to search out those basic principles of organisation, structure, or dynamics (or whatever else it might be) with which Nature has built minds out of matter. Comprehending Nature's feat, by whatever research strategy—by developing the range of theoretical frameworks available to philosophy of mind, by gathering relevant empirical data, or, the ultimate prize, by demonstrating the basic principles of consciousness artificially coaxed into action in non-biological substrates—would offer unparalleled insights into the fundamental physical foundations which effectively unite all conscious organisms with each other and with the rest of the natural world. I hope this book may bring that goal of understanding consciousness, its foundations and relationships a little nearer our reach. The first section below reflects on the significance of the self model view for a few selected areas of this broader context. Later, the chapter recalls central themes of the book and notes some related open questions and possible avenues for future research.

1. WHAT GOOD IS A THEORY OF MIND AND BODY?

It seems, on the face of it, that remarkably few of life's activities depend on our adopting any particular theory about how it is that myriad cognitive processes converge in a conscious self or about how that consciousness relates to the physical world. The choice of a theoretical framework, whether it be simple mind/brain identity, interactionist dualism, epiphenomenal dualism, some flavour of agnostic mysterianism, or whatever, seems at first glance to matter not very much at all to the important businesses of sleeping, feeding, and reproducing with which the great bulk of our resources are, directly or indirectly, taken up. For the most part, questions about why it feels like it does to be us keep company with those like "so what's it all about, then?" whose rare appearances mainly coincide with evening forays to the pub. Except for a few enthusiastic students in their first or second year of undergraduate work, probably no one proclaims to their friends, "I read Descartes, and he changed my life!".

In fact, I think it is no great affront to society to say that most of us haven't *any* particularly well developed framework for thinking about relationships between conscious mind and physical world—but that if we did have such a framework, perhaps the choice *would* make a difference in our day to day lives.

A theory of mind and body provides one more element in a backdrop against which individuals may evaluate their feelings and behaviours toward themselves, other people, and the physical world around them. It amounts to one more set of tools with which individuals may think for themselves about life and its challenges. Borrowing from absurd extremes to make the point, surely an individual's behaviour and feelings would be different if she believed, for instance, that every object larger than one centimetre tall had a little conscious spirit living inside it than if she believed that no one else in the whole world ever had conscious experiences except her. While contemporary theories about minds are not quite so extreme (although a few are probably just as absurd!), I believe that seriously incorporating any of them into one's beliefs about the world must have consequences for daily life. If anything is 'hidden' about my agenda for research in cognitive science, philosophy of mind, or artificial life, it is this: I hope to help construct a theoretical framework which matters to the way individuals think about themselves, each other, and the world around them; a framework which can make a positive difference to the way people live.

(As it should become clear in due course, this aspiration needn't carry any religious overtones and, indeed, I don't think it necessarily conflicts with the aims of various theological traditions. Picking an example arbitrarily, the Christian belief in a soul is not a view about how a subject's consciousness, or soul, for that matter, manifests itself physically. It is no more a belief about how to relate consciousness itself to the physical world than the atheist conviction that God does not exist is a belief about why it feels like it does to be us. The account of consciousness developed here addresses a body of concerns which are different and largely independent from those with which these or other religious traditions principally concern themselves. Although it may be that some ideas advanced in this book do bear indirectly on scattered religious issues, disentangling potential connections is thorny business, and such delicate questions are here quite deliberately left to the side.)

1.1 Situating the Self in the Physical World

The approach to the conscious 'self' outlined in Chapter 6 brings with it a view of the world and its minds which differs in significant respects from that which probably suggests itself to many of us upon first reflection. Centrally, the self is not on this view some entirely separate entity, categorically different from both the outside world and its own physical body—*even though* it might look clearly, perhaps even obviously, that way to introspection. While mind/body dualism seems to offer the most comfortable conceptual fit with our pre-theoretic hold on the phenomenal territory available to introspection, the self model approach nonetheless suggests that the phenomenological appearance of things may be preserved without resorting to any relationship of opposition or interaction between mind and body. Understanding individual selves as dynamically changing data structures supervening on microphysics, as particular variations on a shared theme of underlying matter, prompts a view of the mind/body as a *whole*, a wholeness within which the most significant division is one of *perspective*—the distinction between first person and third.[1]

At an appropriate level of description, the individual conscious subject is just one especially robust, self organising corner of an overall pattern of physical particles changing all across the universe. And each particular pattern implementing a conscious mind shares a great deal in common with every other; if the view of a conscious mind as a specific variety of supervening data structure is remotely correct, far more unites individual examples of such data structures, particularly human examples, than separates them.[2] Placing the idea squarely in the human sphere, the self model approach suggests that each of us differs only in fine structure and not in 'substance' from both the most cowardly terrorist murderer and the bravest, most saintly peacemaker.

I suggest this not as another brand of soft hearted liberalism grounded solely in emotion (although I would not want to denigrate such feelings), but as an empirically meaningful hypothesis, deriving from the theoretical framework of self models, about structural features shared across the set of conscious human minds. Section 1.3 below returns to this point in the context of moral responsibility and value, suggesting that some fundamental degree of respect is due each member of the set of conscious minds, including both terrorist murderer and saintly peacemaker. The main point for the moment, however, is simply to draw attention to the 'embracing' and egalitarian outlook on the community of minds which the approach to consciousness advanced here appears to underwrite.

1.2 Self Determination and Free Will

Understood within the self model framework, we are all, much like the stack example in Chapter 6, just hitching a supervening ride on our physical bodies. Most consequences of this view might be basically appealing and innocuous, such as the thread of unity and commonality which it suggests runs through the community of conscious subjects. But some might understandably regard the self model approach as disturbing for other reasons. For instance, if we really do supervene on microphysics, and if the cosmos is a closed system governed entirely by the laws of physics, then when it comes to thought and action, it seems that in a sense *we* don't really move our bodies—*they move us*. The notion that it is only in virtue of the physical processes in the supervenience base that we think, feel, and act as we do appears on the surface to threaten the very foundation of our freedom as agents. Although nothing here presupposes that these underlying microphysical processes are actually deterministic, one spin on the concern places it in the near vicinity of longstanding philosophical debates over the compatibility of free will and determinism. (My own favourite in this area is van Inwagen's 1983 *An Essay on Free Will*, while Earman 1986 is frequently cited for the groundwork on determinism itself.)

Without directly engaging the huge literature on this topic, a key point borrowed from the 'compatibilist' camp illuminates a promising response to the concern about self determination in the context of self models. In particular, whatever we make of the rather misleading dichotomy between 'we move our bodies' and 'our bodies move us', it remains that there are times when *we are able to do what we want*. Many varieties of free will feature significantly in the literature, but the crux of the issue for the compatibilist is not whether situations exist where some external constraint such as a physical threat or gravity or a pair of handcuffs (or even inner psychological torment) prevents our doing what we would like to do, but whether determinism *itself* somehow compromises free will, independently of such straightforward constraints. Likewise for the present concern: the question is whether, in the absence of physical restraints such as manacles and chains or other constraints like conservation of energy, our very supervenience on microphysics somehow compromises our capacity to do what we want. But clearly supervenience does not compromise *that* sort of free will; supervenience is altogether compatible with that sort of freedom. And the freedom to do what we want—rather than, say, the 'freedom' to want what we want to want[3]—is the variety of free will which theorists worried about supervenience are probably most concerned to preserve. (Also see Dennett 1984.)

The worry can also be viewed as another rendition of the general problem of explanatory exclusion, and an appropriate response follows straightforwardly from the discussion in section 1 of Chapter 6. Namely, although an underlying microphysical process may be interpreted as the cause of some given action or thought, it remains the case that a supervening entity such as a self model data structure may *also* be interpreted as the cause—and, indeed, as a better *explanation*. Thus, recalling Hellman and Thompson's framework (from Chapter 1, on page 10) coupling ontological determination with explanatory anti-reductionism, although nothing ever happens which is not, at the lowest level of description, entirely a matter of physics, higher level supervening entities such as self models still feature crucially in explanation. And as expressed in the earlier supervenience discussion, (logically) supervening entities 'inherit' causal relevance from their supervenience base. Far from compromising the conscious subject's self determination, it is in fact supervenience on microphysics which supports the causal powers of the self in the first place!

1.3 Data Structures as Objects of Moral Scrutiny

But even if self determination or free will by themselves present little reason for worry about the supervenience of conscious minds on microphysics, the issue also raises general questions about the status of supervening minds as objects of moral scrutiny. If individual conscious subjects are data structures 'hitching a ride' on their physical bodies, what does this say about their rôle as moral agents? What does it say about our natural tendency to attribute responsibility, to parcel out punishments for nasty actions or to respect those who act admirably? Here again, the problem can be seen as another example of explanatory exclusion: if a complete account of my actions can be given in terms of microphysics, without ever mentioning *me* as a cognizer, how can I be held responsible for those actions? Even though I retain explanatory importance, and even though I inherit causal relevance from my supervenience base, it might be said that the 'real' causal relationships exist between subatomic particles and have little or nothing to do with me. Of course the saints acted saintly, one might argue, but it was all to do with their protons; their virtuous convictions and integrity were quite irrelevant!

As before, here is not the place (and I am not the author) to engage critically the enormous literature on moral responsibility, and I have only a little to say on the topic for the moment. But for my part, I am content to bite the bullet and concede that the problem of explanatory exclusion presents serious problems for the foundations of many moral frameworks and for efforts to justify rationally the everyday tendency to judge agents as noble or mean spirited in light of their actions. Yet I certainly would not give up altogether on the notion of moral responsibility just because of these problems; if nothing else, some species of moral responsibility remains attractive for purely utilitarian reasons. And the bearing on utilitarianism of two points arising from the self model view may be worth pondering. First, the counterintuitive excesses of utilitarianism as a prescription for action[4] might be significantly tempered by appealing to the 'balancing' notion that every member of the set of conscious minds is due some fundamental degree of respect simply in view of their membership in that set. The underlying similarity between organisms possessed of self models may by itself suffice to ground the extension of fundamental respect from the first person to the third. This fundamental respect would proscribe the kinds of gross abuse of individuals which undiluted utilitarianism might dictate.

In fact, the notion of transitivity to which this suggestion appeals, allowing extension of the respect attributed to the first person to other selves in virtue of architectural similarity, appears to underwrite the extension of basic respect to other objects instantiating similar data structures which don't happen to be self models. The living but nonconscious evolutionary relatives of conscious life on our planet come to mind: because of their similarity to entities which led, over the course of phylogenetic history, to physical patterns which today do implement consciousness, it would seem appropriate that they share to some degree in that same fundamental respect.

The second, closely related point arising from the self model view which bears on the utilitarian framework is precisely its capacity to ground the claim that respect is appropriate in the first place: supervenience offers a simple mechanism for exploiting the general supposition that *conscious experience has value*. One attractive view holds that it is not merely in virtue of one or another particular physical structure that an entity is worthy of respect; it is in virtue of the relationship of that physical structure to supervening *consciousness*. Recalling the example from 1.1 above, it is not because the cowardly terrorist or the brave saint each possesses knees or a nose or ear lobes that they share common fundamental worth, but because each participates in the instantiation of a *conscious mind*. The supposition that consciousness itself bears value may in turn be justified in the first instance simply by appeal to first person experience: we ourselves individually experience phenomenal consciousness, and as a result we come strongly to believe that it has value. (Relative to the perspective of those who have experienced consciousness, one might argue that it is simply a brute fact that such consciousness is, in the general case, valuable.)

Moreover, as I suggested in section 3 of Chapter 1, there being a physical account of wonderfully rich phenomenal experience renders it no less wonderfully rich phenomenal experience; in no way does it compromise or reduce the painfulness of the pain or the lustfulness of the lust. Likewise, the extent to which experiences of aesthetic rapture or bodily pleasure derive from a physical basis bears not at all on the essential quality or value of those experiences for the subjects who enjoy them. The first person *a posteriori* attribution of value to conscious experience is by itself too weak, however, for purposes of extending respect to other subjects; but that extension may yet be justified by appealing to logical supervenience itself and the third person notion that the first person feeling that phenomenal experience has value is a *logical* requirement of the instantiation of the right kinds of physical structures. In other words, I suggest that even from the third person perspective, it can be seen that first person phenomenal experience is in some sense valuable *in and of itself*—because it is valuable to the conscious subject from its own perspective. This grounds some basic concept of value which may be suitable for supporting the notion of fundamental respect or worth as well as the utilitarian goal of maximising for conscious minds some hypothetically measurable quantity (such as pleasure, happiness, etc.).

Finally, in the face of logical supervenience and the underlying similarity and inherent value shared across the community of conscious minds, one must surely be driven to take only the most sombre view of *punishment*, even in cases where its appropriateness may be dictated unequivocally by most or even all moral theories available. To my way of thinking, it seems absolutely clear that if the approach to consciousness outlined in this book bears any recognisable resemblance to the correct one, the moment of throwing the switch on the electric chair or watching a convict led away to solitary confinement should be accompanied not by joy that justice has been served, but by solemn regret that conditions necessitated its being served at all. Human emotional factors clearly do lead us sometimes to rejoice when punishment is meted out, and there is little room for doubting that punishment is often warranted and perhaps even essential to society's preservation of itself. But I am sceptical that a convincing rational basis could be contrived for any response other than regret.

I recall a recent news report which included footage of a woman campaigning in favour of the death penalty outside a Texas prison on the morning a man convicted of murder was to be executed. She sang triumphantly over a megaphone:

Hi dee, hi dee, hi dee ho,
Lethal injection is the way to go...

While I realise intellectually that such a time is unlikely, nonetheless I cannot extinguish the hope that some day people will begin to think in a different way entirely, and that no such refrain will ever again be heard.

In any case, numerous other difficult questions about the moral status of conscious subjects understood as dynamic data structures lurk in the vicinity of those mentioned here, and I cannot claim to have done any more than scratch the surface of the topic. For most of these questions, satisfying answers must await further work, and my barely skeletal grasp of the relevant literature suggests I should refrain from speculating. Nonetheless, I hope it is clear that a physicalist account of consciousness such as that advanced in this book may have as much to offer theories of value and moral responsibility as it potentially strips away: while such an account may bear critically on some traditional ways of understanding value, moral responsibility, or punishment, it offers physically grounded replacements which may serve our conceptual needs just as well.

2. A PARTIAL PICTURE IN SOFT FOCUS

In the first chapter, I described this book as a full frontal assault on the clutch of confusions which philosophers collectively refer to as the mind/body problem. Exploiting resources and methods from fields ranging from traditional philosophy to computer science, mathematics, physics, and neuroscience, I promised to develop and explore an account of how conscious minds may be implemented by matter. This section takes a step back, surveying the principal developments of the subsequent chapters and linking them together into a coherent picture without the clutter of specific details.

The initial warm-up exercise about constructing zombies illustrates why analysis of externally observable behaviour, however thorough and finely grained, just isn't enough for purposes of understanding consciousness. That chapter highlights the mistake of mathematical mysticism: the claim that even though any finite sequence of observable behaviours seemingly indicative of consciousness can in principle be produced by infinitely many distinct mechanisms, *all* of those mechanisms truly are conscious. In its weaker form, mathematical mysticism is the claim that out of all those infinitely many mechanisms, only those which really are conscious are also consistent with the laws of physics. The weak mathematical mystic presupposes that it is physically impossible for any system to display external behaviours indistinguishable from conscious behaviours without actually being conscious. The chapter shows that strong mathematical mysticism is clearly false; for all we know, the weak mathematical mystic *may* be right. But simply *assuming* such a view is correct should be too much to stomach for anyone trying to understand how it actually comes to be that so many (probably all) of those physically instantiated creatures we observe around us behaving as if they're conscious really are conscious.[5]

In preparation for a careful look at the *internal* processes which the zombie problem means any satisfying account of consciousness will have to address, Chapter 3 outlines a modern physical view of information and introduces a well established mathematical framework for discussing it, that of algorithmic information theory. While the first half of the chapter indicates why making physics work properly requires taking all information to be physically instantiated, the second shows how to quantify both information content and information relationships in terms of the lengths of minimal self delimiting Turing Machine programs. This framework brings with it an array of powerful tools from computer science and mathematics and also provides an easy grammar for discussing formal issues of decidability and completeness. Defining the basic relationship of *representation* in terms of algorithmic information theory improves the mathematical rigour which can be brought to bear on philosophical discourse on the topic and opens the door to a new look at representation *use* by cognizers later in the chapters on self models. The approach underwrites an important distinction, according to which only cognizers can *use* representations as such, but items in the world may exist in representational relationships quite independently.

Philosophical puzzlement about the perspectival nature of subjective experience emerges as the main topic of the next chapter. From the discussion about Frank Jackson's chromatically challenged neuroscientist, her botanically inclined colleague, and their computer counterpart, it turns out that the mere fact a cognizer cannot 'derive' an understanding of a particular experience simply by reading extensive physical information about it and performing some cognitive gymnastics should in no way be construed as an argument against physicalism. Like a sex-crazed red herring, Jackson's Mary has spawned reams of fishy philosophical musings to the effect that this logical barrier means physicalism must be false. But, on the contrary, Chapter 4 shows why the story about Mary turns out just the same even if we *assume* physicalism. Expunging the ludicrous underlying assumption that philosophers' cogitations should be able to achieve all the same information transfers or transformations as microphysics is one of the book's central moves, and it figures prominently in later discussions of supervenience.

The next chapter summarises a few basic mathematical pitfalls which await traditional attempts to describe complex systems in functional terms. It then extends the information theoretic framework of Chapter 3 to illuminate a new picture of functional decomposition which avoids these traps. Based on a rigorous and objective measure of process complexity, the chapter demonstrates in sketch form how the internal interactions of a composite system may be given an essentially unique functional description in terms of the simplest possible components interacting through the simplest possible pathways. Although 'computational' in its construction, this method of functional decomposition occupies a position almost entirely orthogonal to the debate between computationalism and dynamicism and may be applied equally well within either context.

So, as of the end of Chapter 5, the book has justified the focus on internal processes for the task of understanding consciousness, provided a mathematical framework for discussing the properties of those internal processes in terms of information, and extended that framework to allow these processes to be understood functionally. In addition, it has argued that there is no reason to believe a cognizer's logical manipulations of information at a high cognitive level should be able to capture all the details of the information transforming processes at work in the cognizer's own lower levels. In particular, there is no reason to believe a cognizer should be able to 'think' his or her brain into a particular state just by reading information about that state. In scrutinising the last point, I have attempted to bring the general problem of perspectives to the front, tugging it out from the background where it has been quietly lurking and spinning off apparently insoluble balls of confusion.

The next chapter rejects the traditional philosophical notion of 'mental state' as flatly incoherent, replacing it with an essentially temporal alternative. It goes on to investigate the *a priori* conceptual connection between conscious experience and its underlying cognition, showing that although *a priori* argument leads straightforwardly only to the weaker relationship of metaphysical supervenience (of consciousness on cognition), the stronger logical supervenience grounded in *a priori* intension is ultimately to be preferred. This lays the groundwork for the self model, proposed as a conceptually motivated but empirically falsifiable hypothesis linking the subjective feel of 'what it is like to be' with underlying cognition. On this view, the subject of conscious experience is a particular variety of materially instantiated dynamically changing data structure, the self model. Conscious experience itself is what it is like to be such a changing structure. The full *perspective* of any other self model except our own is not directly accessible to us for the contingent physical reason first canvassed in Chapter 4: in the general case, we cannot, through creative cogitations or otherwise, transform our own physical state into the physical state of another.

Skipping over the quantum mechanical interlude for the moment, Chapter 8 fills in more carefully some of the information theoretic features of the self model, which were initially described only roughly and on the fly. Building on the precise definition of representation offered earlier in the book, the chapter explores representation *use* in the context of a self model and shows how biological substrates such as networks of neurons can be understood in representational terms in the first place. The discussion continues by outlining some varieties of neural structures which we might expect to support self models in naturally evolved organisms, with a special focus on one architecture in particular, Stephen Grossberg's adaptive resonance theory, or ART. While there is certainly no *direct* connection between ART itself and consciousness, the architecture displays, in a biologically plausible way, many of the capabilities which would be required to support the functional features of data structures like the self model.

Between the two components of the self model exploration, I have inserted a chapter intended to short circuit the common objection that a good theory of consciousness should either depend in some essential way on special quantum effects in the brain or should at least account for the rôle some believe consciousness plays in physical processes like quantum measurement. However, consciousness *per se* needn't play any rôle whatsoever in quantum measurement, and the most popular claims about wild quantum effects in the brain, such as in microtubules, are textbook examples of bad physics.

Finally, Chapter 9 investigates interesting theoretical properties of chaotic analogue systems which *might* be relevant to theories of mind attempting to ground consciousness and cognition in analogue devices such as naturally occurring neural networks. In principle, recurrent neural networks operating with real valued signals may compute a small class of functions which discrete Turing Machines cannot. The twist this fact receives here is this: given that models operating with discrete values and with continuous values differ in their theoretical capabilities, they cannot both be the best match with the real physical world. In the context of models of cognition, I suggest that the choice of analogue or discrete variables actually does make a difference to the way we understand the world which we're attempting to model. Are there any important features of real cognizers which we risk overlooking by focusing solely on discrete models? The discussion is by no means intended as a comprehensive analysis of this or related questions, however, and the full relevance to the real world of the recently discovered special capabilities of some theoretical chaotic analogue systems remains to be seen.

3. PHILOSOPHICAL FUTURES

Oddly enough, many of the topics which have appeared in this book aren't really those which interest me the most. The book's structure is largely an artefact of my own struggle to extract from professional philosophy some bigger mind/body picture that might help me grasp what are to me the more interesting issues and broader applications—those which bear directly on questions like why it feels like it does to be us (rather than like something else), or how we relate to the physical environment around and within us, or how we should think about other conscious people or animals. In studying philosophy, I was initially after a worldview: not just an archipelago of philosophical islands grown up around individual questions, but a whole model of the way things are which would yield insights about particular questions without obscuring how they relate to all the other questions waiting to be asked. I naively turned to philosophy believing it was the place to find a philosophy of life. But it wasn't.

Neither is this book.

But the book attempts to fill a few of the gaps in a set of tools which might eventually help to construct a worthwhile model of the way things are. It seems to me that philosophy by itself is simply inadequate to answer many of the questions traditionally labelled 'philosophical'. Before launching into some of the really fascinating parts of the mind/body problem and its broader implications, I have tried first to build up a set of tools from philosophy and other fields and to lay some groundwork for what may some day be a larger scale 'frontal assault'. Perhaps some of these other questions might even have been included in this book if it had been delayed by a year, or two, or five or ten. They might be topics for future work. But probably every author has experienced that awkward moment when the decision must finally be made to stop expanding, editing, and mutating some writing project. At some point, the thing just has to be declared finished, omissions, confusions, warts and all.

No one will have much difficulty locating the warts, I suspect, and the set of unaddressed questions includes just about everything. But in this final section, with a nod to the future, I'll take note of a few of the most obvious issues left open by the discussions which did feature in the book and suggest routes for exploring them more thoroughly.

First, I wonder about the prospects for arguing from 'first principles' to defend the intuitions of weak mathematical mysticism. This might be something in line with John von Neumann's work on mathematical analyses of life or Chaitin's *d*-diameter approach to structural complexity described in Chapter 5. Both mathematicians wanted some way to characterise in a precise way the conditions which must be met in order for life to develop. The battle to justify a similar characterisation for consciousness would have to come in two parts, and I suspect both would be steep uphill struggles. First, we would have to narrow the class of physically possible mechanisms for implementing apparently conscious behaviour to all and only those displaying some particular range of architectural features. Second, we would have to show that those architectural features necessarily implement consciousness. But the class of *logically* possible mechanisms is infinite, and the class of physically possible mechanisms, while finite, is extraordinarily vast. If someone were to offer a convincing argument defending mathematical mysticism, they would have succeeded in characterising this extraordinarily vast space of physical possibility (the first part), while simultaneously crossing the difficult territory between metaphysical and logical supervenience (the second part) described in Chapter 6. This would be impressive.

Another impressive feat would be discovering a way to eliminate some of the $O(1)$ constants which appear in the fundamental definitions of algorithmic information theory. Often, the values of these constants vary according to the choice of Universal Turing Machine, and thus they cannot in the general case be computed directly. But while the relatively small $O(1)$ discrepancy between information content measures relative to different Universal Turing Machines presents no impediment for mathematical explorations of large sets of long strings, the worry is frequently expressed that its magnitude might swamp the information content or information relationships for smaller sets of shorter strings of the sort we might use to describe real physical objects. In other words, if an $O(1)$ constant is very large in comparison to the other terms in the basic definitions, we may lose sight of the very relationships we turned to algorithmic information theory for help in quantifying.

At least two approaches could ameliorate the problem. First, we may obtain numerical upper bounds on the constants for parts of algorithmic information theory by actually working out software implementations for particular Turing Machines. Chaitin (1996) has met this challenge for one rendition, establishing for instance that n bits of axioms allow one to determine at most $n + 15{,}328$ bits of the unruly number Ω. This is fussy, however, and somewhat inelegant. A second method reformulates the basic definitions of algorithmic information theory in such a way that many constants go away. For example, information content might be defined not with respect to a particular Universal Turing Machine, but with respect to the space of all possible Universal Turing Machines. Taking the information content of a string as the average of the lengths of the shortest programs for generating it on all those different machines may then yield 'absolute' information relationships free of extra constants. Certain difficulties intrude due to the infinity of different Universal Turing Machines, and as Chaitin points out (personal communication), we may wind up replacing the arbitrariness of the $O(1)$ constant with the arbitrariness of the weighting function for calculating the average. Of course, such a measure also becomes violently noncomputable. But although I have only rough hunches about the effect of such a reformulation on the overall framework of algorithmic information theory, it may at least offer an opportunity to send a few constants packing.

Other difficulties arise from attempting to apply algorithmic information theory to real objects rather than solely to bit strings. Some people seem bothered just by the fact that information content for physical entities varies according to the specified level of description, just as that for bit strings varies according to the level of precision. However, I personally find this feature matches my intuitions perfectly. More interesting is the fact that setting a level of description may *differentially* bias the information content of different physical structures. The general effect is exactly analogous to the transitions which occur in d-diameter complexity as d varies: the particular varieties of internal structure they contain may yield complexity 'spectra' with completely different characters for distinct systems. But could such variation occlude as much as it reveals? I'm not sure, but I have tried to circumvent the worry at least a little in the course of exploring functional decomposition in Chapter 5: there, we would expect the method to return module descriptions automatically at whatever level of detail is appropriate. Metaphorically, we might imagine examining different parts of a system through a lens which can be variably zoomed: as we focus in more or less closely, patterns of structure emerge and fade, and the complexity minimisation strategy 'locks in' the right zoom levels for each part of the system. The right level is that where structural features permit overall behaviour to be captured in the most concise way possible.

One feature of information theory which hasn't received the attention it might have in this book is the formal connection between the definitions of algorithmic information theory and Bayes's theorem. Sections 4.1 and 4.2 of Chapter 8 informally touched on the selective advantages of good environmental modelling, but much more can be said by making use of the fact that, generally speaking, a minimal description of data also states the hypothesis which has the greatest probability, given the data. In other words, a program which *minimises* the description length of the data by phrasing it in terms of a succinct hypothesis and instructions for generating the data from the hypothesis also *maximises* the posterior probability of the hypothesis. (Recall that Solomonoff originally developed his version of information theory as a tool for quantifying the power of scientific theories.) So, an organism with efficiently compressed representations and mechanisms for testing hypotheses using those representations is a good Bayesian organism.[6] A whole sea of theorems waits to be proved about Bayesian properties of self models.

Following the information theory thread for the moment, the entire method of functional decomposition strikes me as a very large, perhaps too large, hammer brought down on a stubborn problem to crack it by brute force. I wonder if the whole thing can be accomplished more concisely and elegantly, without giving up the qualities of absoluteness, precision, and so on. We know already that any alternative method must either give up some features or must itself be just as noncomputable. But computability issues aside, perhaps the functional logical depth measure itself could be altered in some way so a less complicated method could be built atop it. Unfortunately, I can offer no suggestions on where to begin.

Chapter 4 showed that the mystery of Jackson's 'knowledge argument' against physicalism was in the words, not in the world. But although the book's 'world first' approach makes clear that the story can be re-told in such a way that the source of confusion pops out clearly from the background, it remains puzzling how the concept of knowledge can be explicated so as to dissolve the mystery from Jackson's original wording. Certainly it is mostly tied up with accommodating the fact that, on a physicalist view, entering new states of knowledge requires entering new physical states; this makes it easier to sort out the silly unexamined assumptions about acquiring all the knowledge about states just by reading descriptions of them. Working out the details carefully is trickier than I may have made it sound, however. This waiting task is a perfect example of what the traditional philosophical tools are best prepared to address.

Discussions of the broader relevance of interactive decoherence for philosophy of physics appeared only on the fringes of a chapter aimed mainly at rebutting some of the wilder claims of the 'consciousness physicists' and other advocates of intimate links between minds and the fundamental processes of quantum physics. But basic questions linger, and I'm especially curious about just how far the line of thought expressed in section 6 of that chapter may be taken: to what extent can interactive decoherence alone tie up some of those loose ends of interpretation which have tangled our understanding of quantum mechanics since its inception? Does it really help bring into focus a satisfying picture of the world, given that using it to fill in the picture at the macroscopic level seems to require agnosticism about microphysical reality?

In developing a framework for grounding conscious experience in the temporal evolution of an underlying cognitive structure, Chapters 6 and 8 have opened a Pandora's Box of issues about the relationship between the character of underlying cognitive processes and the character of supervening conscious experience. For instance, section 5.2 in Chapter 6 observed the obvious: vast amounts of cognitive change take place without any effect whatsoever on subjective experience. One important question thus centres on the distinction between those processes which do implement consciousness and those which don't. Just how much of the modelling of an organism's environment and its self actually implements consciousness—i.e., implements the self model—and exactly why? If all and only those processes which change the self model affect (and *effect*) conscious experience, what are those underlying cognitive mechanisms (of which there are probably vastly many) which effect or retard the spread of activity within the self model, thereby widening or narrowing the scope of consciousness? And in cases of dreamless sleep or deep general anaesthesia, for instance, is it more appropriate to say that activity within the self model has slowed to a standstill or that the self model itself has actually shrunk to the point that it ceases temporarily to exist? In such an event, do *I* cease to exist, or does there just cease to be anything it is like to be me? Are these mutually exclusive alternatives, or are they just different ways of describing the same condition?

Closely related is a concern about internal modelling itself. Although much has been said about the advantages of the kind of complex cognitive mediation between stimuli and behaviour for which I have suggested the self model is well suited, something seems peculiar about the idea within a selectionist context. The problem, which connects in some ways to the dynamics vs. computation debate, is this: given a continual loop of interaction between an organism and its environment, taking place in real time, why should a comparatively slow loop mediated by complex modelling and sophisticated cognition within the organism be preferred to a much faster loop yielding simple reactive behaviour? When I was four years old and my mother's warning against it enticed me instead to go ahead and try placing the palm of my hand on a hot iron just to see what would happen, surely no complex internal modelling had to take place before my body 'decided' to yank my hand away. When survival (or even the reduction of blistering!) depends on it, it seems to be the *reflex reaction*, and not the complex deliberative plan of attack, which carries the day. *Contra* the line of thought suggested in section 4.1 of Chapter 8, some more or less *direct* link between stimulus and response often seems, on the face of it, to do more for an organism's chances of survival and reproduction than elaborate and physiologically costly mechanisms for building, maintaining, and exploiting representational models.

One specific example through which I have often found myself thinking about this issue relates to the process of learning an intricate sensory and motor skill such as a martial art. Although a few very limited studies have been published on the psychological or behavioural effects of martial arts training (Delvatauiliili 1995, Foster 1997), to my knowledge the area has received little attention in the mainstream literature.

Among experienced martial artists, it is well known that a student has reached a significant performance milestone when she begins to deliver techniques without having to devote conscious attention to their mechanical details.[7] During early stages, a student deliberately decides which movements to execute and when, and each action must be accompanied by a good deal of conscious direction. This mirrors the process of learning the technique in the first place: a new student, confronted by some complex physical movement she has never before encountered, generally manages at best to grasp, and to model, some subset of the instructor's movements only 'from the outside'—that is, without appreciating 'from the inside' exactly how her own body would move in a similar way. For the first few thousand repetitions of a new technique, the student does not really deliver a punch, say, but instead a linked linear sequence of movements patterned after her external perception of those of her instructor: the right foot skims in an arc across the floor, while the right hand rises up slightly from the hip; the left foot pivots, shifting the heel a little to the inside; the right hip begins to rotate forward and the left hand snaps back toward the left hip; and so on. When placed in a mock self defence situation, the student at this stage of training is probably in worse shape than the complete neophyte, whose biomechanically inefficient flailings at least appear spontaneously and without the distraction of conscious attention to where the feet should go. Only much later, when the student *seems* to stop altogether any modelling of her own movements, when her actions are no longer deliberately patterned after those of her instructor and when the punch is as natural and unthinking a movement as drinking, can she really perform quickly and properly.

But has the modelling really disappeared? Given that on a strict formal view, the experienced student's brain must still contain most of the same relevant informational relationships as before, it seems instead that it continues but has simply ceased being consciously directed by the self model. While the action itself continues to affect the self model (otherwise there would be no conscious experience of performing it), the modelling required to make it happen no longer does. The action *feels* completely different. A cluster of several related questions lurks in the background. Is the main advantage of the prior conscious modelling simply that it improves learning efficiency, and could this sort of rôle by itself offer sufficient selective advantage to account for its evolutionary appearance? Architecturally speaking, do the neuronal repertoires which initially subserve conscious modelling early in the learning process eventually pass over control to other repertoires which model nonconsciously, or does their own activity eventually become nonconscious? That is, does conscious modelling act as a 'supervisor' for learning over other nonconscious modelling pathways, or do pathways initially responsible for conscious modelling themselves eventually become nonconscious pathways? If the latter is the case, do we in some sense 'lose' part of our self model when we learn a complex skill with enough proficiency that performing it no longer requires conscious modelling? Returning to the original question, what about genuinely reflexive reactions, where sensory afferents only reach the spinal cord or brain stem before efferent signals are on their way back out to the muscles? What is the tradeoff between the flexibility of behaviour moderated by modelling, conscious or otherwise, and the raw speed of reflex?

Finally, Chapter 9 puzzles me. Setting aside the subtle theoretical features of computational models based on real or rational numbers, do the capabilities of analogue and digital models actually coincide over the range of empirically relevant conditions? In other words, do the special capabilities of analogue systems appear robustly in theory but not at all under empirically relevant circumstances? Suppose someone gave us a black box which they claimed could solve the halting problem: enter any natural number whatsoever, they claim, and the black box will tell you whether the Turing program with that number halts. Could any finite sequence of experiments reject the hypothesis that the black box was just a finite state automaton designed to give the right answers for a certain subset of the natural numbers? (Recall that in the general case we couldn't even verify that answers for non-halting programs were correct!) If not, could finite experiments settle a dispute comparing the functioning of an analogue neural network, with capabilities much weaker than the black box super-solver, and an ordinary digital computer? In a similar vein, is the proposition that physical parameters in the real world only take on discrete values empirically falsifiable? If not, then is it tautological? What about the proposition that the real physical world contains some continuous values? (Platonic gasp: could the distinction between rational and real be a creation of our minds, with no bearing at all on the real world?) I do not know the answers to these questions, but it seems to me that unless we have a very firm grip on the answers to the verificationist problems in particular, we don't really have much business arguing about whether or not real world phenomena are best matched by real or rational number models. In the absence of a clear answer to the question of how we would verify empirically that real physical parameters are either continuous or discrete, defenders of the traditional 'digital is good enough' line, despite their frequently cocky self assuredness, haven't a logical leg to stand on. Nor do advocates of the 'analogue is everywhere' position.

Quite apart from the logical and mathematical details of information, function, mental states, supervenience, computation, and so on, one really big question remains about the picture of minds offered in this book. How do we build one? How *exactly* might we go about engineering not only basic hypothesis testing, but a fully integrated sensory and motor system able to model and interact with its world and possessed of the full repertoire of information theoretic properties of the self model? A theoretical account of consciousness grounded in information formalism is one thing, but nothing would better demonstrate an understanding of large tracts of the mind/body problem than actually building a *self*.

When the topic is consciousness, it has become traditional to view circumspection and pessimism as reflecting philosophical depth and maturity; apparently, only those who haven't really thought about it hard enough or who just don't understand the problem could possibly be optimistic. I disagree. I believe human beings will meet the challenge to understand, creating consciousness artificially within my lifetime. I hope to be there when the future is born.

NOTES

[1] Erasing the relationship of opposition between self and body in this way renders ailments such as Alzheimer's disease or multiple sclerosis all the more terrifying: unlike, say, a broken leg or a stomach ulcer, in compromising the normal functioning of the central nervous system, such an affliction strikes at the very foundation of the self.

[2] I mean this literally in the sense of there being substantial mutual information content between them as well as in the more fundamental sense that for each such data structure, there is something it is like to be it.

[3] Provided that we sometimes are able to do what we want, this tongue-tripping combination offers one angle on the counterfactual question of whether an agent *could have done otherwise*: given that she did do what she wanted, could she have wanted (and subsequently done) something else (what she wanted to want to do but didn't)? But in any case, shifting attention to higher order wants really turns the question into one of psychology; provided the notion of being free to want what we want is psychologically coherent—which it arguably may not be—supervenience creates no serious impediments. I don't even want (or want to want) to think about the freedom to want to want what we want to want!

[4] Given an overall goal of maximising some measure of pleasure or well being, undiluted utilitarianism appears to suggest, counterintuitively, that gross violations of the individual good may sometimes be justified by their positive effect on the general good. For instance, causing horrific agony to one individual might be justified if doing so would bring some small benefit to countless others.

[5] Actually, even if we allow the weak view as an unargued assumption, it is relatively harmless: only someone for whom we might reserve the label 'mathematically mystic fundamentalist' could believe that it actually has anything to do with *explaining* consciousness. The truth or falsity of weak mathematical mysticism by itself sheds no light whatsoever on the question of *why* seemingly conscious behaviour really does require consciousness.

[6] One snag frequently overlooked about the Bayesian connection, however, is the obscene relationship between information content and maximum string length—the $a(n)$ from section 4.1 of Chapter 3. The fact that retrieving some data sets from their minimal descriptions can take

astronomical amounts of time suggests the greatest advantage may come not with being a *perfect* Bayesian, just with being a good one.

[7] Certainly I do not write as anything of an authority on this topic. With dan level ('black belt') experience only in Ryukyu kobudo and junior grades in a few other arts, I am only just beginning to grasp in rough outline the essence of martial arts training. But I believe the general features of training which I describe here will be apparent to most students after only some months or a few years of experience. Indeed, I believe they are general features of many different learning processes, such as learning to drive a car.

References

Aberth, O. (1971) 'The Failure in Computable Analysis of a Classical Existence Theorem For Differential Equations', *Proceedings of the American Mathematical Society* 30: 151-56.

Abiteboul, S. and E. Shamir, eds. (1994) *Lecture Notes in Computer Science 820: Automata, Languages, and Programming.* Berlin: Springer-Verlag.

Aertsen, A. and H. Preissl. (1991) 'Dynamics of Activity and Connectivity in Physiological Neuronal Networks', in Schuster (1991), pp. 281-302.

Albrecht, A. (1992) 'Investigating Decoherence in a Simple System', *Physical Review D* 46: 5504-20.

Alexander, S. (1920) *Space, Time, and Deity.* London: Macmillan.

Amari, S. and M. Arbib, eds. (1982) *Lecture Notes in Biomathematics 45: Competition and Cooperation in Neural Nets.* New York: Springer-Verlag.

Amundson, R. and G. Lauder. (1994) 'Function Without Purpose: The Uses of Causal Role Function in Evolutionary Biology', *Biology and Philosophy* 9: 443-69.

Baars, B.J. (1988) *A Cognitive Theory of Consciousness.* Cambridge: Cambridge University Press.

Balcázar, J.L.; J. Díaz, and J. Gabarró. (1988) *Structural Complexity I.* Berlin: Springer-Verlag.

Banks, J.; J. Brooks, G. Cairns, G. Davis, and P. Stacey. (1992) 'On Devaney's Definition of Chaos', *American Mathematical Monthly* April: 332-34.

Barahona, F. (1982) 'On the Computational Complexity of Ising Spin Glass Models', *Journal of Physics* A15: 3241-53.

Barnsley, M. (1988) *Fractals Everywhere.* San Diego: Academic Press.

Bartlett, F.C. (1964) *Remembering: A Study in Experimental and Social Psychology.* Cambridge: Cambridge University Press.

Barwise, J., ed. (1977) *Handbook of Mathematical Logic.* New York: North Holland.

Basar, E. and T.H. Bullock, eds. (1989) *Brain Dynamics.* Heidelberg: Springer.

Beer, R.D. and J.C. Gallagher. (1992) 'Evolving Dynamic Neural Networks for Adaptive Behavior', *Adaptive Behavior* 1: 91-122.

Beckermann, A.; H. Flohr, and J. Kim, eds. (1992) *Emergence or Reduction? Essays on the Prospects of Nonreductive Physicalism.* Berlin: Walter de Gruyter.

Bekenstein, J.D. (1973) 'Black Holes and Entropy', *Physical Review* D7(8): 2333-46.

Bekenstein, J.D. (1974) 'Generalized Second Law of Thermodynamics in Black-Hole Physics', *Physical Review* D9(12): 3292-3300.

Bekenstein, J.D. (1981a) 'Universal Upper Bound on the Entropy-to-Energy Ratio for Bounded Systems', *Physical Review* D23(2): 287-98.

Bekenstein, J.D. (1981b) 'Energy Cost of Information Transfer', *Physical Review Letters* 46(10): 623-26.

Bekenstein, J.D. (1994) 'Entropy Bounds and Black-Hole Remnants', *Physical Review* D49(4): 1912-21.

Bell, J. (1990) 'Against Measurement', *Physics World* 3(8): 33-40.

Bellman, R.E., ed. (1962) *Mathematical Problems in the Biological Sciences, Proceedings of Symposia in Applied Mathematics.* Providence, Rhode Island: American Mathematical Society.

Bennett, C.H. (1973) 'Logical Reversibility of Computation', *IBM Journal of Research and Development* 17: 525-32.

Bennett, C.H. (1982) 'The Thermodynamics of Computation—A Review', *International Journal of Theoretical Physics* 21: 905-40.
Bennett, C.H. (1987a) 'Information, Dissipation, and the Definition of Organization', in Pines (1987).
Bennett, C.H. (1987b) 'Demons, Engines, and the Second Law', *Scientific American* 257(5): 88-96.
Bennett, C.H. (1990) 'How to Define Complexity in Physics, and Why', in Zurek (1990), 137-48.
Bennett, C.H. and R. Landauer. (1985) 'The Fundamental Physical Limits of Computation', *Scientific American* 253(1): 38-46.
Black, A.H. and W.F. Prokasy, eds. (1972) *Classical Conditioning II: Current Research and Theory.* New York: Appleton-Century-Crofts.
Blackmore, S. (1993) *Dying to Live.* London: Grafton.
Block, N. (1978) 'Troubles with Functionalism', in Savage (1978), 261-325.
Block, N. (1981) 'Psychologism and Behaviorism', *Philosophical Review* 90: 5-43.
Block, N. and R. Stalnaker. (1996) 'Conceptual Analysis and the Explanatory Gap', unpublished manuscript, December 1996.
Blum, L.; M. Shub, and S. Smale. (1989) 'On a Theory of Computation and Complexity Over the Real Numbers: *NP*-Completeness, Recursive Functions and Universal Machines', *Bulletin of the American Mathematical Society* 21: 1-47.
Boff, K.R.; L. Kaufman, and J.P. Thomas, eds. (1986) *Handbook of Perception and Human Performance, vol. II.* New York: Wiley.
Bogen, J. (1981) 'Agony in the Schools', *Canadian Journal of Philosophy* 11: 1-21.
Bohr, N. (1934) *Atomic Theory and the Description of Nature.* Cambridge: Cambridge University Press.
Bongard, M.M. (1970) *Pattern Recognition.* New York: Spartan.
Boolos, G.S. and R.C. Jeffrey. (1989) *Computability and Logic*, 3rd edition. Cambridge: Cambridge University Press.
Brillouin, L. (1956) *Science and Information Theory.* New York: Academic Press.
Bringsjord, S. (1992) *What Robots Can and Can't Be.* Dordrecht: Kluwer Academic Publishers.
Broad, C.D. (1925) *Mind and Its Place in Nature.* London: Routledge and Kegan Paul.
Browder, F.E., ed. (1976) *Mathematical Developments Arising from Hilbert Problems, Proceedings of Symposia in Pure Mathematics*, vol. 28. Providence, Rhode Island: American Mathematical Society.
Browne, A. and J. Pilkington. (1994) 'Variable Binding in a Neural Network Using a Distributed Representation', in Verleysen (1994), 199-204.
Carew, T.J.; R.D. Hawkins, T.W. Abrams, and E.R. Kandel. (1984) 'A Test of Hebb's Postulate at Identified Synapses Which Mediate Classical Conditioning in *Aplysia*', *Journal of Neuroscience* 4: 1217-24.
Carnap, R. (1947a) 'On the Application of Deductive Logic', *Philosophy and Phenomenological Research* 8: 143-47.
Carnap, R. (1947b) 'Reply to Nelson Goodman', *Philosophy and Phenomenological Research* 8: 461-62.
Carpenter, G.A. and S. Grossberg. (1981) 'Adaptation and Transmitter Gating in Vertebrate Photoreceptors', *Journal of Theoretical Neurobiology* 1: 1-42.
Carpenter, G.A. and S. Grossberg. (1983) 'A Neural Theory of Circadian Rhythms: The Gated Pacemaker', *Biological Cybernetics* 48: 35-59.
Cartwright, N. (1983) *How the Laws of Physics Lie.* Oxford: Oxford University Press.
Cartwright, N. (1989) *Nature's Capacities and Their Measurement.* Oxford: Oxford University Press.

Chaitin, G.J. (1966) 'On the Length of Programs for Computing Finite Binary Sequences: Statistical Considerations', *Journal of the Association for Computing Machinery* 13: 547-69.

Chaitin, G.J. (1970) 'To a Mathematical Definition of "Life"', *Association for Computing Machinery SICACT News* 4: 12-18.

Chaitin, G.J. (1974) 'Information-Theoretic Limitations of Formal Systems', *Journal of the Association for Computing Machinery* 21: 403-24.

Chaitin, G.J. (1975) 'A Theory of Program Size Formally Identical to Information Theory', *Journal of the Association for Computing Machinery* 22: 329-40.

Chaitin, G.J. (1977) 'Algorithmic Information Theory', *IBM Journal of Research and Development* 21: 350-59+.

Chaitin, G.J. (1979) 'Toward a Mathematical Definition of "Life"', in Levine and Tribus (1979), 477-98.

Chaitin, G.J. (1982) 'Gödel's Theorem and Information', *International Journal of Theoretical Physics* 22: 941-54.

Chaitin, G.J. (1987a) *Information Randomness & Incompleteness: Papers on Algorithmic Information Theory*. Singapore: World Scientific. Also see Chaitin's web page at http://www.cs.auckland.ac.nz/CDMTCS/chaitin.

Chaitin, G.J. (1987b) *Algorithmic Information Theory*. Cambridge: Cambridge University Press.

Chaitin, G.J. (1996) 'An Invitation to Algorithmic Information Theory', Presented April 1996 at a Computer Science Colloquium at the University of New Mexico. Edited transcript available from http://www.cs.auckland.ac.nz/CDMTCS/chaitin.

Chalmers, D.J. (1996a) 'Does a Rock Implement Every Finite State Automaton?', *Synthese* 108: 309-33.

Chalmers, D.J. (1996b) *The Conscious Mind: In Search of a Fundamental Theory*. Oxford: Oxford University Press.

Chalmers, D.J. (1996c) 'Reply to Mulhauser's Review of *The Conscious Mind*', *Psyche* 2(35) review-2-chalmers.

Changeux, J.-P. and A. Danchin. (1976) 'Selective Stabilization of Developing Synapses as a Mechanism for the Specification of Neuronal Networks', *Nature* 264: 705-11.

Changeux, J.-P.; T. Heidmann, and P. Patte. (1984) 'Learning by Selection', in Marler and Terrace (1984), 115-37.

Cherniak, C. (1981), 'Minimal Rationality', *Mind* 90: 161-83.

Cherniak, C. (1984) 'Computational Complexity and the Universal Acceptance of Logic', *Journal of Philosophy* 81: 739-55.

Chisholm, R.M. (1946) 'The Contrary-to-fact Conditional', *Mind* 55: 289-307.

Churchland, P.M. (1981) 'Eliminative Materialism and the Propositional Attitudes', *Journal of Philosophy* 78: 67-90.

Churchland, P.S. (1986) *Neurophilosophy: Toward a Unified Science of the Mind-Brain.* Cambridge, Massachusetts: MIT Press.

Churchland, P.S. and T.J. Sejnowski. (1992) *The Computational Brain.* Cambridge, Massachusetts: MIT Press.

Clark, A. and R. Lutz, eds. (1992) *Connectionism in Context.* New York: Springer-Verlag.

Cohen, R.S. and M. Wartofsky, eds. (1969) *Boston Studies in the Philosophy of Science*, vol. 5. Dordrecht: Reidel.

Colodny, R.G., ed. (1970) *Nature and Function of Scientific Theories.* Pittsburgh: University of Pittsburgh Press.

Cottrell, Allin. (1996) 'On the Conceivability of Zombies', unpublished manuscript, Wake Forest University, October 1996. Available from http://www.wfu.edu/~cottrell.

Coulter, J. (1995) 'The Informed Neuron: Issues in the Use of Information Theory in the Behavioral Sciences', *Minds and Machines* 15(4): 583-96.

Craik, F.I.M. and R.S. Lockhart. (1972) 'Levels of Processing: A Framework for Memory Research', *Journal of Verbal Learning and Verbal Behavior* 11: 671-84.

Craik, F.I.M. and E. Tulving. (1975) 'Depth of Processing and the Retention of Words in Episodic Memory', *Journal of Experimental Psychology: General* 104: 268-94.

Crick, F. (1994) *The Astonishing Hypothesis: The Scientific Search For the Soul.* New York: Simon & Schuster.

Cummins, R. (1975) 'Functional Analysis', *Journal of Philosophy* 72: 741-64.

Cytowic, R.E. (1989) *Synesthesia: A Union of the Senses.* New York: Springer-Verlag.

Cytowic, R.E. (1993) *The Man Who Tasted Shapes: A Bizarre Medical Mystery Offers Revolutionary Insights Into Emotions, Reasoning & Consciousness.* New York: G.P. Putnam's.

Darwin, C. (1871) *The Descent of Man, and Selection in Relation to Sex.* London: Murray.

Davidson, D. (1967) 'Causal Relations', *Journal of Philosophy* 64: 691-703.

Davidson, D. (1973) 'Radical Interpretation', *Dialectica* XXVII: 313-28.

Davidson, D. (1980) *Essays on Actions & Events.* Oxford: Clarendon Press.

Davies, M.K. and I.L. Humberstone (1980) 'Two Notions of Necessity', *Philosophical Studies* 38: 1-30.

Davis, M., ed. (1965) *The Undecidable: Basic Papers on Undecidable Propositions, Unsolvable Problems and Computable Functions.* Hewlett, New York: Raven Press.

Davis, M.; Y.V. Matijasevic, and J. Robinson. (1976) 'Hilbert's Tenth Problem. Diophantine Equations: Positive Aspects of a Negative Solution', in Browder (1976), pp. 323-78.

Davis, M., H. Putnam and J. Robinson (1961) 'The Decision Problem for Exponential Diophantine Equations', *Annals of Mathematics* 74: 425-36.

Delvatauiliili, J. (1995) 'Does Brief Aikido Training Reduce Aggression of Youth?', *Perceptual and Motor Skills* 80(1): 297-98.

Dennett, D.C. (1975) 'Why the Law of Effect Will Not Go Away', *Journal of the Theory of Social Behaviour* 5: 179-87.

Dennett, D.C. (1978) *Brainstorms.* Cambridge, Massachusetts: MIT Press.

Dennett, D.C. (1981) 'Making Sense of Ourselves', *Philosophical Topics* 12: 63-81.

Dennett, D.C. (1984) *Elbow Room: The Varieties of Free Will Worth Wanting.* Oxford: Clarendon Press.

Dennett, D.C. (1991) *Consciousness Explained.* Boston: Little, Brown.

Dennett, D.C. (1994) 'Artificial Life as Philosophy', *Artificial Life* 1: 291-92.

Dennett, D.C. (1995a) 'Get Real', *Philosophical Topics* 22: 505-68.

Dennett, D.C. (1995b) 'The Unimagined Preposterousness of Zombies', *Journal of Consciousness Studies* 2(4): 322-26.

Dennett, D.C. (1995c) *Darwin's Dangerous Idea: Evolution and the Meanings of Life.* London: Simon & Schuster.

d'Espagnat, B. (1989) *Conceptual Foundations of Quantum Mechanics*, Second Edition. Reading, Massachusetts: Addison-Wesley.

Deutsch, D. (1985) 'Quantum Theory, the Church-Turing Principle and the Universal Quantum Computer', *Proceedings of the Royal Society of London* A400: 97-117.

Devaney, R.L. (1988) *An Introduction to Chaotic Dynamical Systems.* New York: Addison-Wesley.

Dretske, F. (1981) *Knowledge and the Flow of Information.* Cambridge, Massachusetts: MIT Press.

Dretske, F. (1983) 'Précis of *Knowledge and the Flow of Information*', *Behavioral and Brain Sciences* 6: 55-63.

Dummett, M. (1976) 'Is Logic Empirical?', in Lewis (1976), pp. 45-68. Reprinted in Dummett (1978), pp. 269-89. (Reprint pagination used for references in this book.)
Dummett, M. (1978) *Truth and Other Enigmas*. London: Duckworth.
Durkheim, E. (1938) *The Rules of Sociological Method*. Chicago: Free Press.
Earman, J. (1986) *A Primer on Determinism*. Dordrecht: Reidel.
Eccles, J.C. (1964) *The Physiology of Synapses*. Berlin: Springer-Verlag.
Eccles, J.C. (1986) 'Do Mental Events Cause Neural Events Analogously to the Probability Fields of Quantum Mechanics?', *Proceedings of the Royal Society of London* B227: 411-28.
Eccles, J.C. (1990) 'A Unitary Hypothesis of Mind-Brain Interaction in the Cerebral Cortex', *Proceedings of the Royal Society of London* B240: 433-51.
Edelman, G.M. (1978) 'Group Selection and Phasic Reentrant Signalling: A Theory of Higher Brain Function', in Edelman and Mountcastle (1978), 51-100.
Edelman, G.M. (1981) 'Group Selection as the Basis for Higher Brain Function', in Schmitt *et al.* (1981), 535-63.
Edelman, G.M. (1989) *Neural Darwinism: The Theory of Neuronal Group Selection*. Oxford: Oxford University Press.
Edelman, G.M. and L.H. Finkel. (1984) 'Neuronal Group Selection in the Cerebral Cortex', in Edelman *et al.* (1984), 653-95.
Edelman, G.M.; W.E. Gall, and W.M. Cowan, eds. (1984) *Dynamic Aspects of Neocortical Function*. New York: Wiley.
Edelman, G.M. and V.B. Mountcastle, eds. (1978) *The Mindful Brain: Cortical Organization and the Group-Selective Theory of Higher Brain Function*. Cambridge, Massachusetts: MIT Press.
Edelman, G.M. and G.N. Reeke, Jr. (1982) 'Selective Networks Capable of Representative Transformation, Limited Generalizations, and Associative Memory', *Proceedings of the National Academy of Sciences* 79: 2091-95.
Edelman, S. (forthcoming) 'Representation is Representation of Similarities', to appear in *Behavioral and Brain Sciences*. Available from http://www.ai.mit.edu/~edelman/mirror.
Eeckman, F.H. and W.J. Freeman. (1991) 'Asymmetric Sigmoid Nonlinearity in the Rat Olfactory System', *Brain Research* 557: 13-21.
Eigen, M. (1992) *Steps Towards Life: A Perspective on Evolution*. Oxford: Oxford University Press.
Einstein, A.; B. Podolsky, and N. Rosen. (1935) 'Can Quantum-Mechanical Description of Reality be Considered Complete?', *Physical Review* 47: 777-80.
Ettlinger, G. (1981) 'The Relationship Between Metaphorical and Cross-Modal Abilities: Failure to Demonstrate Metaphorical Recognitin in Chimpanzees Capable of Cross-Modal Recognition', *Neuropsychologia* 19: 583-586.
Ettlinger, G. and W.A. Wilson. (1990) 'Cross-Modal Performance: Behavioral Processes, Phylogenetic Considerations and Neural Mechanisms', *Behavioral Brain Research* 40: 169-92.
Evans, G. (1979) 'Reference and Contingency', *The Monist* 62: 161-89.
Feller, W. (1950) *An Introduction to Probability Theory and its Applications*, vol. 1. New York: John Wiley & Sons.
Feynman, R.P. (1967) *The Character of Physical Law*. Cambridge, Massachusetts: MIT Press.
Finkel, L.H. and G.M. Edelman. (1985) 'Interaction of Synaptic Modification Rules Within Populations of Neurons', *Proceedings of the National Academy of Sciences* 82: 1291-95.
Finkel, L.H.; G.N. Reeke, Jr., and G.M. Edelman. (1989) 'A Population Approach to the Neural Basis of Perceptual Organization', in Nadel *et al.* (1989), 146-79.
Fodor, J.A. (1975) *The Language of Thought*. New York: Thomas Crowell.

Fodor, J.A. (1980) 'Methodological Solipsism Considered as a Research Strategy in Cognitive Science', *Behavioral and Brain Sciences* 3: 63-109.
Fodor, J.A. (1992) 'The Big Idea: Can There Be a Science of Mind?', *Times Literary Supplement*, 3 July: 5.
Fodor, J.A. and B.P. McLaughlin. (1990) 'Connectionism and the Problem of Systematicity: Why Smolensky's Solution Did Not Work', *Cognition* 35: 183-204.
Fodor, J.A. and Z. Pylyshyn. (1988) 'Connectionism and Cognitive Architecture: A Critical Analysis', *Cognition* 28: 3-71.
Ford, K.M. and P.J. Hayes. (1991) *Reasoning Agents in a Dynamic World: The Frame Problem.* Greenwich: JAI Press.
Foster, Y.A. (1997) 'Brief Aikido Training Versus Karate and Golf Training and University Students' Scores on Self-Esteem, Anxiety, and Expression of Anger', *Perceptual and Motor Skills* 84(2): 609-10.
Freeman, W.J. (1964) 'A Linear Distributed Feedback Model for Prepyriform Cortex', *Experimental Neurology* 10: 525-47.
Freeman, W.J. (1979) 'Nonlinear Gain Mediation of Cortical Stimulus Response Relations', *Biological Cybernetics* 33: 237-47.
Freeman, W.J. (1987) 'Simulation of Chaotic EEG Patterns With Dynamic Model of the Olfactory System', *Biological Cybernetics* 56: 139-50.
Freeman, W.J. (1988) 'Strange Attractors That Govern Mammalian Brain Dynamics Shown by Trajectories of Electroencephalographic (EEG) Potential', *IEEE Transactions on Circuits and Systems* 35: 781-83.
Freeman, W.J. (1989) 'Analysis of Strange Attractors in EEGs With Kinesthetic Experience and 4-D Computer Graphics', in Basar and Bullock (1989).
Freeman, W.J. and C.A. Skarda. (1985) 'Spatial EEG Patterns, Nonlinear Dynamics and Perception: The Neo-Sherringtonian View', *Brain Research Reviews* 10: 147-75.
Fukushima, K. (1980) 'Neocognitron: A Self-Organizing Neural Network Model for a Mechanism of Pattern Recognition Unaffected by Shift in Position', *Biological Cybernetics* 36: 193-202.
Fujii, H.; H. Ito, K. Aihara, N. Ichinose, M. Tsukuda. (1996) 'Dynamical Cell Assembly Hypothesis—Theoretical Possibility of Spatiotemporal Coding in the Cortex', *Neural Networks* 9(8): 1303-50.
Gács, P. and J. Körner. (1973) 'Common Information Is Far Less Than Mutual Information', *Problems in Control and Information Theory* 2(2): 149-62.
Gallistel, C.R. (1990) *The Organization of Learning.* Cambridge, Massachusetts: MIT Press.
Gardner, M. (1979) 'The Random Number Ω Bids Fair to Hold the Mysteries of the Universe', *Scientific American* 241(5): 20-34.
Gaudiano, P. and S. Grossberg. (1991) 'Vector Associative Maps: Unsupervised Real-Time Error-Based Learning and Control of Movement Trajectories', *Neural Networks* 4: 147-83.
Gazzaniga, M.S., ed. (1984) *Handbook of Cognitive Neuroscience.* New York: Plenum.
Gell-Mann, M. and J.B. Hartle. (1990) 'Quantum Mechanics in the Light of Quantum Cosmology', in Zurek (1990), 425-459.
Gerdes, V.G.J. and R. Happee. (1994) 'The Use of an Internal Representation in Fast Goal-Directed Movements', *Biological Cybernetics* 70(6): 513-24.
Geschwind, N. (1965) 'Disconnexion Syndromes in Animals and Man', *Brain* 88: 237-94+.
Gluck, M.A. and D.E. Rumelhart, eds. (1990) *Neuroscience and Connectionist Theory.* New Haven: Lawrence Erlbaum.
Gödel, K. (1931) [1965] 'Uber Formal Unentscheidbare Sätze der *Principia Mathematica* und Verwandter Systeme I', *Monatshefte für Mathematic und Physik* 38: 173-98. Translated by E. Mendelson and reprinted in English as 'On Formally Undecidable Propositions of *Principia Mathematica* and Related Systems I', in Davis (1965), pp. 5-38.

Gödel, K. (1936) [1965] in *Ergebnisse Eines Mathematischen Kolloquiums*, 7: 23-24. Translated by M. Davis and reprinted in English as 'On the Length of Proofs', in Davis (1965), pp. 82-83.

Godfrey-Smith, P. (1996) *Complexity and the Function of Mind in Nature.* Cambridge: Cambridge University Press.

Goldman, S. (1953) *Information Theory.* New York: Prentice-Hall.

Good, I.J., ed. (1961) *The Scientist Speculates: An Anthology of Partly-Baked Ideas.* London: Heinemann.

Goodman, N. (1947a) 'The Infirmities of Confirmation Theory', *Philosophy and Phenomenological Research* 8: 149-51.

Goodman, N. (1947b) 'The Problem of Counterfactual Conditionals', *Journal of Philosophy* 44: 113-28.

Goodman, N. (1955) *Fact, Fiction, and Forecast.* Cambridge, Massachusetts: Harvard University Press.

Gould, S.J. and E. Vrba. (1981) 'Exaptation: A Misssing Term in the Science of Form', *Paleobiology* 8:4-15.

Greenfield, P.M. (1992) 'Language, Tools, and Brain: The Ontogeny and Phylogeny of Hierarchically Organized Sequential Behavior', *Behavioral and Brain Sciences* 14: 531-95.

Griffiths, R.B. (1984) 'Consistent Histories and the Interpretation of Quantum Mechanics', *Journal of Statistical Physics* 36: 219-272.

Griffiths, R.B. (1995) Review of Omnès (1994), *Foundations of Physics* 25(8): 1231-36.

Grossberg, S. (1973) 'Contour Enhancement, Short-Term Memory, and Constancies in Reverberating Neural Networks', *Studies in Applied Mathematics* 52: 217-57.

Grossberg, S. (1976a) 'Adaptive Pattern Classification and Universal Recoding, I: Parallel Development and Coding of Neural Feature Detectors', *Biological Cybernetics* 23: 121-34.

Grossberg, S. (1976b) 'Adaptive Pattern Classification and Universal Recoding, II: Feedback, Expectation, Olfaction, and Illusions', *Biological Cybernetics* 23: 187-202.

Grossberg, S. (1978) 'A Theory of Human Memory: Self-Organization and Performance of Sensory-Motor Codes, Maps, and Plans', in Rosen and Snell (1978), 233-374.

Grossberg, S. (1980) 'How Does a Brain Build a Cognitive Code?', *Psychological Review* 87: 1-51.

Grossberg, S., ed. (1981a) *Mathematical Psychology and Psychophysiology.* Providence, Rhode Island: American Mathematical Society.

Grossberg, S. (1981b) 'Adaptive Resonance in Development, Perception, and Cognition', in Grossberg (1981a), 107-56.

Grossberg, S. (1982a) *Studies of Mind and Brain: Neural Principles of Learning, Perception, Development, Cognition, and Motor Control.* Boston: Reidel Press.

Grossberg, S. (1982b) 'A Psychophysiological Theory of Reinforcement, Drive, Motivation, and Attention', *Journal of Theoretical Neurobiology* 1: 286-369.

Grossberg, S. (1982c) 'Processing of Expected and Unexpected Events During Conditioning and Attention: A Psychophysiological Theory', *Psychological Review* 89: 529-72.

Grossberg, S. (1982d) 'Associative and Competitive Principles of Learning and Development: The Temporal Unfolding and Stability of STM and LTM Patterns', in Amari and Arbib (1982), 295-341.

Grossberg, S. (1984) 'Some Psychophysiological and Pharmacological Correlates of a Developmental, Cognitive, and Motivational Theory', in Karrer *et al* (1984), 58-151.

Grossberg, S., ed. (1987a) *The Adaptive Brain I.* Amsterdam: North-Holland.

Grossberg, S. (1987b) 'Competitive Learning: From Interactive Activation to Adaptive Resonance', *Cognitive Science* 11: 23-63.

Grossberg, S. (1988) 'Nonlinear Neural Networks: Principles, Mechanisms, and Architectures', *Neural Networks* 1: 17-61.
Grossberg, S. (1995) 'Are There Universal Principles of Brain Computation?', in Mira and Sandoval (1995), 1-6.
Grossberg, S. and M. Kuperstein. (1989) *Neural Dynamics of Adaptive Sensory Motor Control: Expanded Edition.* Elmsford, New York: Pergamon Press.
Grush, R. and P.S. Churchland. (1995) 'Gaps in Penrose's Toilings', *Journal of Consciousness Studies* 2(1): 10-29.
Grzegorczyk, A. (1955) 'Computable Functionals', *Fundamentals of Mathematics* 42: 168-202.
Grzegorczyk, A. (1957) 'On the Definitions of Computable Real Continuous Functions', *Fundamentals of Mathematics* 44: 61-71.
Gunderson, K., ed. (1975) *Language, Mind, and Knowledge.* Minnesota Studies in the Philosophy of Science, vol. 7. Minneapolis: University of Minnesota Press.
Gustafsson, M.; L. Asplund, O. Gällmo, and E. Nordstöm. (1992) 'Pulse Coded Neural Networks for Hardware Implementation', Presented September 1992 at the *First Swedish National Conference on Connectionism* in Skövde, Sweden.
Hameroff, S.R. (1994) 'Quantum Coherence in Microtubules: A Neural Basis for Emergent Consciousness?', *Journal of Consciousness Studies* 1: 98-118.
Hameroff, S.R. and R. Penrose. (1995) 'Orchestrated Reduction of Quantum Coherence in Brain Microtubules: A Model for Consciousness', *Neural Network World* 5(5): 793-804.
Hameroff, S.R. and R. Penrose. (1996) 'Conscious Events as Orchestrated Space-Time Selections', *Journal of Consciousness Studies* 3(1): 36-53.
Hameroff, S.R. and R.C. Watt. (1982) 'Information Processing in Microtubules', *Journal of Theoretical Biology* 98: 549-61.
Hare, R.M. (1952) *The Language of Morals.* Oxford: Oxford University Press.
Harnad, S., ed. (1987a) *Categorical Perception: The Groundwork of Cognition.* Cambridge: Cambridge University Press.
Harnad, S. (1987b) 'The Induction and Representation of Categories', Chapter 18 in Harnad (1987a).
Harnad, S. (1990) 'The Symbol Grounding Problem', *Physica* D42: 335-46.
Harnad, S. (1992) 'Connecting Object to Symbol in Modeling Cognition', in Clark and Lutz (1992), 75-90.
Harnad, S. (1993a) 'Grounding Symbols in the Analog World with Neural Nets', *Think* 2(1): 12-78.
Harnad, S. (1993b) 'Problems, Problems: The Frame Problem as a Symptom of the Symbol Grounding Problem', *Psycoloquy* 4(34) frame-problem.11.harnad.
Harnad, S. (1995) 'Grounding Symbolic Capacity in Robotic Capacity', in Steels and Brooks (1995), 277-286.
Harnad, S.; S.J. Hanson, and J. Lubin. (1991) 'Categorical Perception and the Evolution of Supervised Learning in Neural Nets', in Powers and Reeker (1991), 65-74.
Harvey, I. (forthcoming) 'Evolving Robot Consciousness: The Easy Problems and the Rest', in Mulhauser (forthcoming).
Haugeland, J. (1985) *Artificial Intelligence: The Very Idea.* Cambridge, Massachusetts: MIT Press.
Hayek, F.A. (1952) *The Sensory Order: An Inquiry Into the Foundations of Theoretical Psychology.* Chicago: University of Chicago Press.
Hebb, D.O. (1949) *The Organization of Behavior.* New York: Wiley.
Heil, J. and A. Mele, eds. (1993) *Mental Causation.* Oxford: Clarendon Press.
Hellman, G. and F. Thompson. (1975) 'Physicalism: Ontology, Determination and Reduction', *Journal of Philosophy* 72: 551-64.

Hempel, C.G. (1965) *Aspects of Scientific Explanation.* New York: Free Press.
Hempel, C.G. (1966) *Philosophy of Natural Science.* Englewood Cliffs, New Jersey: Prentice Hall.
Hempel, C.G. and P. Oppenheim (1948) 'Studies in the Logic of Explanation', *Philosophy of Science* 15: 135-75. Reprinted in Hempel (1965) as Chapter 9.
Hille, B. (1984) *Ionic Channels of Excitable Membranes.* Sunderland, Massachusetts: Sinauer Associates.
Hobson, E.W. (1923) *The Domain of Natural Science.* Aberdeen University Studies no. 89. Aberdeen.
Hofstadter, D.R. (1979) *Gödel, Escher, Bach: An Eternal Golden Braid.* New York: Basic Books.
Holland, J.H. (1992) 'Complex Adaptive Systems', *Daedalus*, Winter: 25.
Holland, J.H.; K.J. Holyoak, R.E. Nisbett, and P.R. Thagard. (1986) *Induction: Processes of Inference, Learning, and Discovery.* Cambridge, Massachusetts: MIT Press.
Hook, S., ed. (1960) *Dimensions of Mind.* New York: New York University Press.
Hookway, C. and D. Peterson, eds. (1993) *Philosophy and Cognitive Science.* Cambridge: Cambridge University Press.
Horgan, T. (1987) 'Supervenient Qualia', *Philosophical Review* 96: 491-520.
Horgan, T. (1993) 'From Supervenience to Superdupervenience: Meeting the Demands of a Material World', *Mind* 102: 555-86.
Horgan, T. and J. Tienson, eds. (1991) *Connectionism and the Philosophy of Mind.* Dordrecht: Kluwer Academic Press.
Horgan, T. and J. Tienson. (1993) 'Levels of Description in Nonclassical Cognitive Science', in Hookway and Peterson (1993), 159-88.
Horwich, P. (1990) *Truth.* Oxford: Basil Blackwell.
Horwich, P. (1995) 'Meaning, Use and Truth', *Mind* 104(414): 355-68.
Hubel, D.H. (1988) *Eye, Brain and Vision.* New York: Scientific American Library.
Hubel, D.H. and T.N. Wiesel. (1962) 'Receptive Fields, Binocular Interaction, and Functional Architecture in the Cat's Visual Cortex', *Journal of Physiology* 160: 106-54.
Humphrey, K.; R.C. Tees, and J. Werker. (1979) 'Auditory-Visual Integration of Temporal Relations in Infants', *Canadian Journal of Psychology* 33: 347-52.
Humphreys, P.W. (1989) 'Scientific Explanation: The Causes, Some of the Causes, and Nothing But the Causes', in Kitcher and Salmon (1989), 283-306.
Imamizu, H.; Y. Uno, and M. Kawato. (1995) 'Internal Representations of the Motor Apparatus—Implications from Generalization in Visuomotor Learning', *Journal of Experimental Psychology: Human Perception and Performance* 21(5): 1174-98.
Ingvar, D.H. (1985) '"Memory of the Future": An Essay on the Temporal Organization of Conscious Awareness', *Human Neurobiology* 4: 127-36.
Jackendoff, R. (1983) *Semantics and Cognition.* Cambridge, Massachusetts: MIT Press.
Jackendoff, R. (1987) 'The Status of Thematic Relations in Linguistic Theory', *Linguistic Inquiry* 18: 369-411.
Jackendoff, R. (1990) *Semantic Structures.* Cambridge, Massachusetts: MIT Press.
Jackson, F. (1982) 'Epiphenomenal Qualia', *Philosophical Quarterly* 32: 127-36.
Johnson, D.S. (1983) 'The *NP*-Completeness Column: An Ongoing Guide', *Journal of Algorithms* 4: 87-100.
Jones, E.G. and A. Peters, eds. (1985) *Association and Auditory Cortices.* New York: Plenum Press.
Jones, J.P. (1978) 'Three Universal Representations of Recursively Enumerable Sets', *Journal of Symbolic Logic* 43: 335-51.
Jones, J.P. (1982) 'Universal Diophantine Equation', *Journal of Symbolic Logic* 47: 549-71.

Jones, J.P. and Y.V. Matijasevic. (1984) 'Register Machine Proof of the Theorem on Exponential Diophantine Representation of Enumerable Sets', *Journal of Symbolic Logic* 49: 818-29.

Joos, E. and H.D. Zeh. (1985) 'The Emergence of Classical Properties Through Interaction with the Environment', *Zeitschrift für Physik* B 59: 223+.

Judd, K.T. and K. Aihara. (1993) 'Pulse-Propagation Networks: A Neural Network Model that Uses Temporal Coding by Action Potentials', *Neural Networks* 6(2): 203-15.

Kalckar, J. (1967) 'Niels Bohr and His Youngest Disciples', in Rozental (1967), 227-239.

Karp, R.M. and R.J. Lipton. (1980) 'Some Connections Between Nonuniform and Uniform Complexity Classes', in *Proceedings of the 12th ACM Symposium on Theory of Computing*, pp. 302-9.

Karrer, R.; J. Cohen, and P. Tueting, eds. (1984) *Brain and Information: Event Related Potentials.* New York: New York Academy of Sciences.

Kelso, S.R. and T.H. Brown. (1986) 'Differential Conditioning of Associative Synaptic Enhancement in Hippocampal Brain Slices', *Science* 232: 85-87.

Kiefer, C. (1991) 'Interpretation of the Decoherence Functional in Quantum Cosmology', *Classical and Quantum Gravity* 8: 379-91.

Kim, J. (1984) 'Concepts of Supervenience', *Philosophy and Phenomenological Research* 45: 153-76.

Kim, J. (1987) '"Strong" and "Global" Supervenience Revisited', *Philosophy and Phenomenological Research* 48: 315-26.

Kim, J. (1989) 'Mechanism, Purpose, and Explanatory Exclusion', *Philosophical Perspectives* 3: 77-108.

Kim, S.W. and S.M. Kim. (1995) 'Turing Computability and Artificial Intelligence: Gödel's Incompleteness Results', *Kybernetes* 24(6): 57.

Kirk, R. (1979) 'From Physical Explicability to Full-Blooded Materialism', *Philosophical Quarterly* 29: 229-37.

Kitcher, P. (1989) 'Explanatory Unification and the Causal Structure of the World', in Kitcher and Salmon (1989), 410-505.

Kitcher, P. (1992) 'The Naturalist's Return', *Philosophical Review* 101: 53-113.

Kitcher, P. and W.C. Salmon (1987) 'Van Fraassen on Explanation', *Journal of Philosophy* 84: 315-30.

Kitcher, P. and W.C. Salmon, eds. (1989) *Scientific Explanation.* Minnesota Studies in the Philosophy of Science, vol. 13. Minneapolis: University of Minnesota Press.

Kleene, S.C. (1952) *Introduction to Metamathematics*. Amsterdam: North Holland.

Kleinschmidt, J. and J.E. Dowling. (1975) 'Intracellular Recordings from Gekko Photoreceptors During Light and Dark Adaptation', *Journal of General Physiology* 66: 617-48.

Kneale, W. (1950) 'Natural Laws and Contrary-to-Fact Conditionals', *Analysis* 10: 123.

Kohonen, T. (1984) Self-Organization and Associative Memory. New York: Springer-Verlag.

Kolmogorov, A.N. (1965) 'Three Approaches to the Quantitative Definition of Information', *Problems of Information Transmission* 1(1): 1-7.

Kornblith, H., ed. (1994) *Naturalizing Epistemology,* Second Edition. Cambridge, Massachusetts: MIT Press.

Kosko, B. (1992) *Neural Networks and Fuzzy Systems: A Dynamical Approach to Machine Intelligence.* Englewood Cliffs, New Jersey: Prentice-Hall International.

Kostrikin, A.I. and Yu. I. Manin. (1989) *Linear Algebra and Geometry*. London: Gordon and Breach.

Kripke, S. (1980) *Naming and Necessity*. Cambridge, Massachusetts: Harvard University Press.

Kripke, S. (1982) *Wittgenstein on Rules and Private Language*. Cambridge, Massachusetts: Harvard University Press.

Kuffler, S.W. (1953) 'Discharge Patterns and Functional Organization of Mammalian Retinas', *Journal of Neurophysiology* 16: 37-68.
Kuffler, S.W.; J.G. Nicolls, and A.R. Martin. (1984) *From Neuron to Brain.* Sunderland, Massachusetts: Sinauer Associates.
Lacombe, D. (1955a) 'Extension de la Notion de Fonction Récursive aux Fonctions d'Une ou Plusieurs Variables Réelles I', *C.R. Académie de Science, Paris* 240: 2478-80.
Lacombe, D. (1955b) 'Extension de la Notion de Fonction Récursive aux Fonctions d'Une ou Plusieurs Variables Réelles II, III', *C.R. Académie de Science, Paris* 241: 13-14+.
Landauer, R. (1991) 'Information is Physical', *Physics Today* 44(5): 23-9.
LeDoux, J.E.; L. Barclay, and A. Premack. (1980) 'The Brain and Cognitive Sciences: Conference Report', *Annals of Neurology* 4: 391-8.
LeDoux, J.E. and W. Hirst, eds. (1986) *Mind and Brain: Dialogues in Cognitive Neuroscience.* Cambridge: Cambridge University Press.
Leff, H.S. and A.F. Rex. (1990) *Maxwell's Demon: Entropy, Information, Computing.* Princeton: Princeton University Press.
Levine, J. (1983) 'Materialism and Qualia: The Explanatory Gap', *Pacific Philosophical Quarterly* 64: 354-61.
Levine, R.D. and M. Tribus, eds. (1979) *The Maximum Entropy Formalism.* Cambridge, Massachusetts: MIT Press.
Lewis, H.D., ed. (1976) *Contemporary British Philosophy*, 4th series. London.
Locke, J. (1690) [1975] *An Essay Concerning Human Understanding.* Edited by P.H. Nidditch. Oxford: Clarendon Press.
London, F. and E. Bauer. (1939) 'La Théorie de l'Observation en Mécanique Quantique', No. 775 of *Actualités Scientifiques et Industrielles: Exposés de Physique Générale.* Published under the direction of Paul Langevin, Paris: Hermann. English translation 'The Theory of Observation in Quantum Mechanics' from reconciled versions of A. Shimony, of J.A. Wheeler and W.H. Zurek, and of J. McGrath and S. McLean McGrath, in Wheeler and Zurek (1983), 217-59. (Wheeler and Zurek pagination used for references in this book.)
Lucas, J.R. (1961) 'Minds, Machines and Gödel', *Philosophy* 36: 112-27.
Lund, J.S.; R.D. Lund, A.E. Hendrickson, A.H. Bunt, and A.F. Fuchs. (1975) 'The Origin of Efferent Pathways From the Primary Visual Cortex, Area 17, of the Macaque Monkey as Shown by Retrograde Transport of Horseradish Peroxidase', *Journal of Comparative Neurology* 164: 287-304.
Lycan, W.G. (1981) 'Form, Function, and Feel', *Journal of Philosophy* 78: 24-49.
Lycan, W.G. (1987) *Consciousness.* Cambridge, Massachusetts: MIT Press.
Lynch, G. (1986) *Synapses, Circuits, and the Beginnings of Memory.* Cambridge, Massachusetts: MIT Press.
Lynch, G. and M. Baudry. (1984) 'The Biochemistry of Memory: A New and Specific Hypothesis', *Science* 224: 1057-63.
Lynch, G.; R. Granger, J. Larson, and M. Baudry. (1989) 'Cortical Encoding of Memory: Hypotheses Derived from Analysis and Simulation of Physiological Learning Rules in Anatomical Structures', in Nadel *et al.* (1989), 180-224.
Lynch, J.C. (1980) 'The Functional Organization of Posterior Parietal Association Cortex', *Behavioral and Brain Sciences* 3: 485-534.
Marks, L.E.; R.J. Hammeal, M.H. Bornstein, and L.B. Smith. (1987) *Perceiving Similarity and Comprehending Metaphor.* Chicago: University of Chicago Press.
Marler, P. and H.S. Terrace, eds. (1984) *The Biology of Learning.* New York: Springer-Verlag.
Marrocco, R.T. (1986) 'The Neurobiology of Perception', in LeDoux and Hirst (1986), 33-88.

Maass, W. and P. Orponen. (1996) 'On the Effect of Noise in Discrete-Time Analog Computations', unpublished manuscript, University of Jyväskylä, Finland, 26 September 1996. Available from http://www.math.jyu.fi/~orponen/papers.

Maudlin, T. (1989) 'Computation and Consciousness', *Journal of Philosophy* 86: 407-32.

McCulloch, G. (1995) *The Mind and Its World.* London: Routledge.

McCulloch, W.S. and E. Pitts. (1943) 'A Logical Calculus of the Ideas Immanent in Nervous Activity', *Bulletin of Mathematical Biophysics* 5: 115-33.

McLaughlin, B. (1992) 'The Rise and Fall of British Emergentism', in Beckermann *et al.* (1992), pp. 49-93.

McNaughton, B.L. (1989) 'Neuronal Mechanisms for Spatial Computation and Information Storage', in Nadel *et al.* (1989), 286-350.

McNaughton, B.L.; B. Leonard, and L. Chen. (1989) 'Cortical-Hippocampal Interactions and Cognitive Mapping: A Hypothesis Based on Reintegration of the Parietal and Inferotemporal Pathways for Visual Processing', *Psychobiology* 17: 236-76.

McNaughton, B.L. and R.G.M. Morris. (1987) 'Hippocampal Synaptic Enhancement Within a Distributed Memory System', *Trends in Neurosciences* 10: 408-15.

McNaughton, B.L. and L. Nadel. (1990) 'Hebb-Marr Networks and the Neurobiological Representation of Action in Space', in Gluck and Rumelhart (1990), 1-63.

Merzenich, M.M.; J.H. Kaas, J.T. Wall, R.J. Nelson, M. Sur, and D.J. Felleman. (1983a) 'Topographic Reorganization of Somatosensory Cortical Areas 3b and 1 in Adult Monkeys Following Restricted Deafferentation', *Neuroscience* 8: 33-55.

Merzenich, M.M.; J.H. Kaas, J.T. Wall, R.J. Nelson, M. Sur, and D.J. Felleman. (1983b) 'Progression of Change Following Median Nerve Section in the Cortical Representation of the Hand in Areas 3b and 1 in Adult Owl and Squirrel Monkeys', *Neuroscience* 10: 639-65.

Merzenich, M.M.; R.J. Nelson, M.P. Stryker, M. Cynader, A. Schoppman, and J.M. Zook. (1984) 'Somatosensory Cortical Map Changes Following Digit Amputation in Adult Monkeys', *Journal of Comparative Neurology* 224: 591-605.

Metzinger, T. (1993) 'Subjectivity and Mental Representation', presented 4 July at the *Second Annual Conference of the European Society for Philosophy and Psychology* in Sheffield, England.

Miall, R.C. and D.M. Wolpert. (1996) 'Forward Models for Physiological Motor Control', *Neural Networks* 9(8): 1265-79.

Miller, G.A. (1953) 'What is Information Measurement?', *American Psychologist* 8: 3-11.

Millikan, R.G. (1984) *Language, Thought and Other Biological Categories.* Cambridge, Massachusetts: MIT Press.

Minsky, M. (1962) 'Problems of Formulation for Artificial Intelligence', in Bellman (1962), 35.

Minsky, M. (1967) *Computation: Finite and Infinite Machines.* Englewood Cliffs, New Jersey: Prentice-Hall.

Minsky, M. and S. Papert. (1969) *Perceptrons: An Introduction to Computational Geometry.* Cambridge, Massachusetts: MIT Press.

Mira, J. and F. Sandoval, eds. (1995) *Lecture Notes in Computer Science 930: From Natural to Artificial Neural Computation.* Berlin: Springer-Verlag.

Moore, C. (1990) 'Unpredictability and Undecidability in Dynamic Systems', *Physical Review Letters* 64(20): 2354-57.

Moore, C. (1991) 'Generalized Shifts: Unpredictability and Undecidability in Dynamic Systems', *Nonlinearity* 4(2): 199-230.

Morán, F.; A. Moreno, J.J. Merelo, and P. Chacón, eds. (1995) *Lecture Notes in Artificial Intelligence 929: Advances in Artificial Life.* Berlin: Springer-Verlag.

Moreno-Díaz, R. and J. Mira-Mira, eds. (1995) *Brain Processes, Theories and Models.* Cambridge, Massachusetts: MIT Press.

Morton, A. (1993) 'Mathematical Models: Questions of Trustworthiness', *British Journal for the Philosophy of Science* 44(4): 659-74.

Mulhauser, G.R. (1992) 'Computability in Neural Networks', Presented September 1992 at the meeting of the *British Society for the Philosophy of Science* in Durham, England.

Mulhauser, G.R. (1993) 'What is it Like to be Nagel?', *The Philosopher: Journal of the Philosophical Society of England* April: 19-24.

Mulhauser, G.R. (1993b) 'Chaotic Dynamics and Introspectively Transparent Brain Processes', Presented July 1993 at the *Second Annual Conference of the European Society for Philosophy and Psychology* in Sheffield, England.

Mulhauser, G.R. (1993c) 'Computability in Chaotic Analogue Systems', Presented July 1993 at the *International Congress on Computer Systems and Applied Mathematics* in St. Petersburg, Russia.

Mulhauser, G.R. (1993d) 'Cognitive Transitions and the Strange Attractor: A Reply to Peter Smith', Presented September 1993 at the *Fifth Joint Council Initiative Summer School in Cognitive Science and Human Computer Interaction* in Edinburgh, Scotland.

Mulhauser, G.R. (1995a) 'Materialism and the "Problem" of Quantum Measurement', *Minds and Machines* 5(2): 207-17.

Mulhauser, G.R. (1995b) 'To Simulate or Not to Simulate: A Problem of Minimising Functional Logical Depth', in Morán *et al.* (1995), 530-43.

Mulhauser, G.R. (1995c) 'Suppose a Wavefunction Collapsed in the Forest', in Pylkkänen and Pylkkö (1995), 134-41.

Mulhauser, G.R. (1995d) 'What Philosophical Rigour Can and Can't Be', review of Bringsjord (1992), *Psycoloquy* 6(28) robot-consciousness.15.mulhauser.

Mulhauser, G.R. (1995e) 'On the End of a Quantum Mechanical Romance', *Psyche* 2(19) decoherence-1-mulhauser.

Mulhauser, G.R. (1995f) 'Seating Conscious Sensation in a Materially Instantiated Data Structure', in Moreno-Díaz and Mira-Mira (1995), 98-106.

Mulhauser, G.R. (1996) '"Bridge Out" on the Road to a Theory of Consciousness', review of Chalmers (1996b), *Psyche* 2(34) review-1-mulhauser.

Mulhauser, G.R. (1997a) 'In the Beginning, There Was Darwin', *Philosophical Books* 38(2): 81-89. Discussion/review of Dennett (1995c), with Dennett's reply pp. 89-92.

Mulhauser, G.R. (1997b) 'Quantum Theories of Mind and Decoherence', Presented June 1997 at the IEE colloquium on *Quantum Computing: Theory, Applications and Implications* in London, England. Contained in IEE Digest No. 97/145, pp. 6/1-3.

Mulhauser, G.R., ed. (forthcoming) *Evolving Consciousness.* Amsterdam: John Benjamins.

Mundale, J. and W. Bechtel. (1996) 'Integrating Neuroscience, Psychology, and Evolutionary Biology Through a Teleological Conception of Function', *Minds and Machines* 6(4): 481-505.

Nadel, L.; L.A. Cooper, P. Culicover, and R.M. Harnish, eds. (1989) *Neural Connections, Mental Computations.* London: MIT Press.

Nadel, L. and L. MacDonald. (1979) 'Hippocampus: Cognitive Map or Working Memory?', *Behavioral and Neural Biology* 29: 405-9.

Nagel, E. (1961) *The Structure of Science: Problems in the Logic of Scientific Explanation.* London: Routledge & Kegan Paul.

Nagel, T. (1974) 'What is it Like to Be a Bat?', *Philosophical Review* 83: 435-50. Reprinted in Nagel (1979), 165-80.

Nagel, T. (1979) *Mortal Questions.* Cambridge: Cambridge University Press.

Neander, K. (1991) 'Functions as Selected Effects: The Conceptual Analyst's Defense', *Philosophy of Science* 58: 168-84.

Neisser, U. (1967) *Cognitive Psychology.* New York: Appleton-Century-Crofts.
Newell, A. (1980) 'Physical Symbol Systems', *Cognitive Science* 4: 135-83.
Nolfi, S. and O. Miglino. (forthcoming) 'Studying the Emergence of Grounded Representations: Exploring the Power and the Limits of Sensory-Motor Coordination', in Mulhauser (forthcoming).
O'Keefe, J. (1989) 'Computations the Hippocampus Might Perform', in Nadel *et al.* (1989), 225-84.
O'Keefe, J. and L. Nadel. (1978) *The Hippocampus as a Cognitive Map.* Oxford: Oxford Univrersity Press.
O'Malley, G. (1964) *Shelley and Synesthesia.* Evanston, Illinois: Northwestern University Press.
Oldfield, R.C. and O.L. Zangwill. (1942a) 'Head's Concept of the Schema and its Application in Contemporary British Psychology, II, Critical Analsyis of Head's Theory', *British Journal of Psychology* 33: 58-64.
Oldfield, R.C. and O.L. Zangwill. (1942b) 'Head's Concept of the Schema and its Application in Contemporary British Psychology, III, Bartlett's Theory of Memory', *British Journal of Psychology* 33: 111-29.
Omnès, R. (1990) 'From Hilbert Space to Common Sense: A Synthesis of Recent Progress in the Interpretation of Quantum Mechanics', *Annals of Physics* 20: 354-447.
Omnès, R. (1994) *The Interpretation of Quantum Mechanics.* Princeton: Princeton University Press.
Orponen, P. (1996) 'A Survey of Continuous-Time Computation Theory', unpublished manuscript, University of Jyväskylä, Finland. Available from http://www.math.jyu.fi/~orponen/papers.
Palmer, S.E. (1978) 'Fundamental Aspects of Cognitive Representation', in Rosch and Lloyd (1978), pp. 259-303.
Pandya, D. and E.H. Yeterian. (1985) 'Architecture and Connections of Cortical Association Areas', in Jones and Peters (1985).
Papineau, D. (1987) *Reality and Representation.* Oxford: Blackwell.
Paris, J. and L. Harrington. (1977) 'A Mathematical Incompleteness in Peano Arithmetic', in Barwise (1977).
Paz, J.P. and S. Sinha. (1992) 'Decoherence and Back Reaction in Quantum Cosmology—Multidimensional Minisuperspace Examples', *Physical Review D* 45: 2823-42.
Paz, J.P.; S. Habib, and W.H. Zurek. (1993) 'Reduction of the Wave Packet: Preferred Observable and Decoherence Time Scale', *Physical Review D* 47: 488-501.
Peacocke, C. (1992) *A Study of Concepts.* Cambridge, Massachusetts: MIT Press.
Penrose, R. (1989) *The Emperor's New Mind: Concerning Computers, Minds and the Laws of Physics.* Oxford: Oxford University Press.
Penrose, R. (1994) *Shadows of the Mind.* Oxford: Oxford University Press.
Penrose, R. and S.R. Hameroff. (1995) 'What Gaps? Reply to Grush and Churchland', *Journal of Consciousness Studies* 2(2): 99-112.
Pepper, S.C. (1926) 'Emergence', *Journal of Philosophy* 23: 241-45.
Pérez, R.; L. Glass and R. Shlaer. (1975) 'Development of Specificity in the Cat's Visual Cortex', *Journal of Mathematical Biology* 1: 275-88.
Pines, D., ed. (1987) *Emerging Syntheses in Science.* Reading, Massachusetts: Addison-Wesley.
Pinker, S. and A. Prince. (1988) 'On Language and Connectionism: Analysis of a Parallel Distributed Model of Language Acquisition', *Cognition* 28: 73-193.
Pittendrigh, C.S. (1958) 'Adaptation, Natural Selection, and Behavior', in Roe and Simpson (1958).

Place, U.T. (1956) 'Is Consciousness a Brain Process?', *British Journal of Psychology* 47:44-50.

Pomerantz, J.R. and M. Kubovy. (1986) 'Theoretical Approaches to Perceptual Organization: Simplicity and Likelihood Principles', in Boff *et al.* (1986).

Popper, K.R. and J.C. Eccles. (1977) *The Self and Its Brain: An Argument for Interactionism.* Berlin: Springer-Verlag.

Post, E. (1944) 'Recursively Enumerable Sets of Positive Integers and Their Decision Problems', *Bulletin of the American Mathematical Society* 50: 284-316. Reprinted in Davis (1965), pp. 305-37.

Poundstone, W. (1985) *The Recursive Universe: Cosmic Complexity and the Limits of Scientific Knowledge.* New York: Morrow.

Pour-El, M.B. (1974) 'Abstract Computability and Its Relation to the General Purpose Analog Computer (Some Connections Between Logic, Differential Equations and Analog Computers)', *Transactions of the American Mathematical Society* 199: 1-28.

Pour-El, M.B. and J.I. Richards. (1979) 'A Computable Ordinary Differential Equation Which Possesses No Computable Solution', *Annals of Mathematical Logic* 17: 61-90.

Pour-El, M.B. and J.I. Richards. (1981) 'The Wave Equation With Computable Initial Data Such That Its Unique Solution is Not Computable', *Advances in Mathematics* 39: 215-39.

Pour-El, M.B. and J.I. Richards. (1989) *Computability in Analysis and Physics.* Berlin: Springer-Verlag.

Powers, D.W. and L. Reeker, eds. (1991) *Working Papers of the AAAI Spring Symposium on Machine Learning of Natural Language and Ontology.*

Putnam, H. (1960) 'Minds and Machines', in Hook (1960), 148-79.

Putnam, H. (1969) 'Is Logic Empirical?', in Cohen and Wartofsky (1969), pp. 216-41.

Putnam, H. (1975a) *Mind, Language and Reality.* Cambridge: Cambridge University Press.

Putnam, H. (1975b) 'The Meaning of "Meaning"', in Gunderson (1975), 131-93.

Putnam, H. (1988) *Representation and Reality.* Cambridge, Massachusetts: MIT Press.

Pylkkänen, P. and P. Pylkkö, eds. (1995) *New Directions in Cognitive Science.* Helsinki: Finnish Artificial Intelligence Society.

Pylyshyn, Z.W. (1980a) 'Computation and Cognition: Issues in the Foundations of Cognitive Science', *Behavioral and Brain Sciences* 3: 111-169.

Pylyshyn, Z.W. (1980b) 'The "Causal Power" of Machines', *Behavioral and Brain Sciences* 3: 442-44.

Pylyshyn, Z.W. (1984) *Computation and Cognition.* Cambridge, Massachusetts: MIT Press.

Quine, W.V.O. (1951) 'Two Dogmas of Empiricism', *Philosophical Review* 60: 20-43.

Quine, W.V.O. (1962) 'Paradox', *Scientific American* 206(4): 84-96.

Quine, W.V.O. (1970) *Philosophy of Logic.* Englewood Cliffs, New Jersey: Prentice-Hall.

Rabin, M. (1960) 'Computable Algebra, General Theory and Theory of Computable Fields', *Transactions of the American Mathematical Society* 95: 341-60.

Radcliffe-Brown, A.R. (1952) *Structure and Function in Primitive Society.* Chicago: Free Press.

Railton, P. (1981) 'Probability, Explanation, and Information', *Synthese* 48: 233-56.

Ralston, Z.T. (1976) *Synesthesia in Gide's 'La Symphonie Pastorale'.* Charleston, South Carolina: Citadel.

Rescher, N. (1970) *Scientific Explanation.* New York: Free Press.

Rescorla, R.A. and A.R. Wagner. (1972) 'A Theory of Pavlovian Conditioning: Variations in the Effectiveness of Reinforcement and Nonreinforcement', in Black and Prokasy (1972).

Roe, A. and C.G. Simpson, eds. (1958) *Behavior and Evolution.* New Haven: Yale University Press.

Rogers Jr., H. (1967) *Theory of Recursive Functions and Effective Computability.* New York: McGraw-Hill.
Rosch, E. and B.B. Lloyd, eds. (1978) *Cognition and Categorization.* Hillsdale, New Jersey: Lawrence Erlbaum Associates.
Rosen, R. and F. Snell, eds. (1978) *Progress in Theoretical Biology*, vol. 5. New York: Academic Press.
Rosenblatt, F. (1962) *Principles of Neurodynamics.* New York: Spartan.
Ross, K.A. and C.R.B. Wright. (1988) *Discrete Mathematics*, Second Edition. Englewood Cliffs, New Jersey: Prentice-Hall International.
Rozental, S., ed. (1967) *Niels Bohr: His Life and Work as Seen By His Friends and Colleagues.* Amsterdam: North-Holland.
Rueger, A. and W.D. Sharp (1996) 'Simple Theories of a Messy World: Truth and Explanatory Power in Nonlinear Dynamics', *British Journal for the Philosophy of Science* 47: 93-112.
Rumelhart, D.E.; J.L. McClelland, and the PDP Research Group. (1986) *Parallel Distributed Processing: Explorations in the Microstructure of Cognition*, vol. 1. Cambridge, Massachusetts: MIT Press.
Russell, B. (1908) 'Mathematical Logic as Based on the Theory of Types', *American Journal of Mathematics* 30: 222-54. Reprinted in van Heijenoort (1967), 150-82.
Sakurai, J.J. (1994) *Modern Quantum Mechanics*, Revised Edition. Reading Massachusetts: Addison-Wesley.
Salmon, W.C. (1970) 'Statistical Explanation', in Colodny (1970), 173-231.
Salmon, W.C. (1984) *Scientific Explanation and the Causal Structure of the World.* Princeton: Princeton University Press.
Sanides, F. (1975) 'Comparative Neurology of the Temporal Lobe in Primates Including Man with Reference to Speech', *Brain and Language* 2: 396-419.
Saunders, S. (1993) 'Decoherence, Relative States, and Evolutionary Adaptation', *Foundations of Physics* 23: 1553-85.
Savellos, E.E. and U.D. Yalçin, eds. (1995) *Supervenience: New Essays.* Cambridge: Cambridge University Press.
Savage, C.W., ed. (1978) *Perception and Cognition: Issues in the Foundations of Psychology.* Minnesota Studies in the Philosophy of Science, vol. 9. Minneapolis: University of Minnesota Press.
Savitt, S. (1982) 'Searle's Demon and the Brain Simulator Reply', *Behavioral and Brain Sciences* 5: 342-43.
Scheffler, I. (1963) *The Anatomy of Inquiry: Philosophical Studies in the Theory of Science.* New York: Alfred A. Knopf.
Schmajuk, N.A. and C.V. Buhusi. (1997) 'Spatial and Temporal Cognitive Mapping: A Neural Network Approach', *Trends in Cognitive Sciences* 1: 109-14.
Schmajuk, N.A. and A.D. Thieme. (1992) 'Puposive Behavior and Cognitive Mapping: An Adaptive Neural Network', *Biological Cybernetics* 67: 165-74.
Schmajuk, N.A.; A.D. Thieme, and H.T. Blair. (1993) 'Maps, Routes, and the Hippocampus: A Neural Network Approach', *Hippocampus* 3: 387-400.
Schmitt, F.O.; F.G. Worden, G. Adelman, and S.G. Dennis, eds. (1981) *Organization of the Cerebral Cortex.* Cambridge, Massachusetts: MIT Press.
Schrödinger, E. (1935) [1980] 'Die gegenwärtige Situation in der Quantenmechannic', *Naturwissenschaften* 23: 807-12+. Translated (1980) as 'The Present Situation in Quantum Mechanics', *Proceedings of the American Philosophical Society* 124: 323-38.
Schuster, H.G., ed. (1991) *Nonlinear Dynamics and Neuronal Networks: Proceedings of the 63rd W.E. Heraeus Seminar.* New York: VCH Publishers.
Seager, W.E. (1991) *Metaphysics of Consciousness.* London: Routledge.

Searle, J.R. (1980) 'Minds, Brains and Programs', *Behavioral and Brain Sciences* 3: 417-24.
Searle, J.R. (1990) 'Is the Brain a Digital Computer?', *Proceedings and Addresses of the American Philosophical Association* 64: 21-37.
Selman, A. (1982a) 'Reductions on *NP* and *P*-Selective Sets', *Theoretical Computer Science* 19: 287-304.
Selman, A. (1982b) 'Analogues of Semi-Recursive Sets and Effective Reducibilities to the Study of *NP* Complexity', *Information and Control* 52: 36-51.
Shannon, C.E. and W. Weaver. (1949) *The Mathematical Theory of Communication.* Urbana, Illinois: University of Illinois Press.
Shepard, R.N. (1968) Review of Neisser (1967), *American Journal of Psychology* 81: 285-89.
Shepard, R.N. and S. Chipman. (1970) 'Second-Order Isomorphism of Internal Representations: Shapes of States', *Cognitive Psychology* 1: 1-17.
Shepherd, G.M., ed. (1990) *The Synaptic Organization of the Brain*, 3 ed. Oxford: Oxford University Press.
Siegelmann, H.T. (1994) 'On the Computational Power of Probabilistic and Faulty Neural Networks', in Abiteboul and Shamir (1994), 23-34.
Siegelmann, H.T. (1995) 'Computation Beyond the Turing Limit', *Science* 268: 545548.
Siegelmann, H.T. and E.D. Sontag. (1992) 'On the Computational Power of Neural Nets', in *Proceedings of the 5th ACM Workshop on Comutational Learning Theory*, 440-49.
Siegelmann, H.T. and E.D. Sontag. (1994) 'Analog Computation Via Neural Networks', *Theoretical Computer Science* 131: 331-60.
Siegelmann, H.T. and E.D. Sontag. (1995) 'Computational Power of Neural Networks', *Journal of Computer System Sciences* 50(1): 132-50.
Skarda, C.A. and W.J. Freeman. (1987) 'How Brains Make Chaos In Order to Make Sense of the World', *Behavioral and Brain Sciences* 10: 161-95.
Skryms, B. (1980) *Causal Necessity.* New Haven: Yale University Press.
Smart, J.J.C. (1959) 'Sensations and Brain Processes', *Philosophical Review* 68:141-56.
Smith, E.E. and D.L. Medin. (1981) *Categories and Concepts.* Cambridge, Massachusetts: Harvard University Press.
Smith, P. (1993) Commentary on Mulhauser (1993b). Presented July 1993 at the *Second Annual Conference of the European Society for Philosophy and Psychology* in Sheffield, England.
Smith, P. (forthcoming) *Explaining Chaos.* To appear 1997 from Cambridge University Press.
Smolensky, P. (1988a) 'On the Proper Treatment of Connectionism', *Behavioral and Brain Sciences* 11: 1-74.
Smolensky, P. (1988b) 'The Constituent Structure of Connectionist Mental States: A Reply to Fodor and Pylyshyn', *Southern Journal of Philosophy* 26(supplement): 137-62.
Smolensky, P. (1990) 'Tensor Product Variable Binding and the Representation of Symbolic Structures in Connectionist Systems', *Artificial Intelligence* 46: 159-216.
Solomonoff, R.J. (1964) 'A Formal Theory of Inductive Inference', *Information and Control* 7(1): 1-22+.
Sommerer, J.C. and E. Ott. (1993) 'A Physical System With Qualitatively Uncertain Dynamics', *Nature* 365: 138-40.
Spencer, H. (1855) *Principles of Psychology.* London: Longman, Brown, and Green.
Stabler, E.P. (1987) 'Kripke on Functionalism and Automata', *Synthese* 70: 1-22.
Steels, L. and R. Brooks, eds. (1995) *The Artificial Life Route to Artificial Intelligence: Building Situated Embodied Agents.* New Haven: Lawrence Erlbaum.
Sterelny, K. (1990) *The Representational Theory of Mind.* Oxford: Basil Blackwell.
Stich, S.P. (1994) 'Could Man Be an Irrational Animal?', in Kornblith (1994), 337-57.
Stich, S.P. and T.A. Warfield, eds. (1994) *Mental Representation.* Oxford: Blackwell.

Strawson, P. (1985) 'Causation and Explanation', in Vermazen and Hintikka (1985), 115-36.
Stuart, G.J. and B. Sakmann. (1994) 'Active Propagation of Somatic Action Potentials Into Neocortical Pyramidal Cell Dendrites', *Nature* 367: 69-72.
Swanson, L.W.; T.J. Teyler, and R.F. Thompson. (1982) *Hippocampal Long-Term Potentiation: Mechanisms and Implications for Memory.* Neurosciences Research Program Bulletin 20. Cambridge, Massachusetts: MIT Press.
Szilard, L. (1929) [1983] 'Uber die Entropieverminderung in einem thermodynamischen System bei Eingriffen intelligenter Wesen', *Zeitschrift für Physik* 53: 840-56. Translated by A. Rapoport and M. Knoller (1969) as 'On the Decrease of Entropy in a Thermodynamic System By the Intervention of Intelligent Beings', *Behavioral Science* 9: 301-10. Reprinted in Wheeler and Zurek (1983), 539-48.
Tallal, P. and J. Schwartz. (1980) 'Temporal Processing, Speech Perception and Hemispheric Asymmetry', *Trends in Neurosciences* 3: 309-11.
Tani, J. (1996) 'Model-Based Learning for Mobile Robot Navigation from the Dynamical System Perspective', *IEEE Transations on Systems, Man, and Cybernetics* 3: 421-36.
Taylor, R. (1963) *Metaphysics.* Englewood Cliffs, New Jersey: Prentice-Hall.
Thompson, E. (1992) 'Novel Colours', *Philosophical Studies* 68: 321-49.
Tolkien, J.R.R. (1991) *The Lord of the Rings.* London: Harper Collins. First published in a single volume, 1968.
Tsuda, I. (1994) 'From Micro-Chaos to Macro-Chaos: Chaos Can Survive Even in Macroscopic States of Neural Activities', *Psycoloquy* 5(12) eeg-chaos.3.tsuda.
Turing, A.M. (1936) 'On Computable Numbers, with an Application to the Entscheidungsproblem', *Proceedings of the London Mathematical Society* 42: 230-65. (Corrections in Turing 1937.)
Turing, A.M. (1937) 'On Computable Numbers, with an Application to the Entscheidungsproblem. A Correction', *Proceedings of the London Mathematical Society* 43: 544-46. (Corrections to Turing 1936.)
Turing, A.M. (1939) 'Systems of Logic Based on Ordinals', *Proceedings of the London Mathematical Society*, series 2, 45: 161-228.
Turing, A.M. (1950) 'Computing Machinery and Intelligence', *Mind* 59: 433-60.
van Brakel, J. (1994) 'Supervenience and Anomalous Monism', Presented November 1994 at the International Seminar on Davidson's Philosophy at the University of Utrecht.
van Fraassen, B. (1980) *The Scientific Image.* Oxford: Oxford University Press.
van Gelder, T. (1995a) 'What Might Cognition Be if Not Computation?', *Journal of Philosophy* 91: 345-81.
van Gelder, T. (1995b) 'Connectionism, Dynamics, and the Philosophy of Mind', in proceedings of *Philosophy and The Sciences of Mind, The Third Pittsburgh-Konstanz Colloquium in the Philosophy of Science.*
van Gelder, T. (forthcoming) 'The Dynamical Hypothesis in Cognitive Science', to appear in *Behavioral and Brain Sciences.*
van Heijenoort, J., ed. (1967) *From Frege to Gödel: A Source Book in Mathematical Logic, 1879-1931.* Cambridge, Massachusetts: Harvard University Press.
van Inwagen, P. (1983) *An Essay on Free Will.* Oxford: Clarendon Press.
Vergis, A.; K. Steiglitz, and B. Dickinson. (1986) 'The Complexity of Analog Computation', *Mathematics and Computers in Simulation* 28: 91-113.
Verleysen, M., ed. (1994) *European Symposium on Artificial Neural Networks 1994.* Brussels: D facto.
Vermazen, B. and M. Hintikka, eds. (1985) *Essays on Davidson: Actions and Events.* Oxford: Clarendon Press.
von der Malsburg, C. (1973) 'Self-Organization of Orientation Sensitive Cells in the Striate Cortex', *Kybernetik* 14: 85-100.

von Neumann, J. (1932) [1955] *Mathematische Grundlagen der Quantenmechanik.* Berlin: Springer. Translated by R.T. Beyer (1955) as *Mathematical Foundations of Quantum Mechanics.* Princeton: Princeton University Press.
von Neumann, J. (1966) *Theory of Self-Reproducing Automata.* Urbana, Illinois: University of Illinois Press. Edited and completed by A.W. Burks and published posthumously.
Wagner, S.; E. Winner, D. Cichetti, and H. Gardner. (1981) '"Metaphorical" Mapping in Human Infants', *Child Development* 52: 728-31.
Wheeler, J.A. and W.H. Zurek, eds. (1983) *Quantum Theory and Measurement.* Princeton: Princeton University Press.
Wheeler, R.H. (1920) *The Synaesthesia of a Blind Subject.* Eugene, Oregon: The University Press.
Wheeler, R.H. and T.D. Cutsforth. (1922) 'The Synaesthesia of a Blind Subject with Comparative Data from an Asynaesthetic Blind Subject', *University of Oregon Publications* 1(10), June.
Widrow, G. and M.E. Hoff. (1960) 'Adaptive Switching Circuits', *Institute of Radio Engineers, Western Electronic Convention Record*, Part 4: 96-104.
Wigner, E.P. (1961) 'Remarks on the Mind-Body Question', in Good (1961), 284-302.
Wigner, E.P. (1963) 'The Problem of Measurement', *American Journal of Physics* 31: 6-15.
Wigner, E.P. (1967) *Symmetries and Reflections.* Bloomington, Indiana: Indiana University Press.
Wigstrom, H.; B.L. McNaughton, and C.A. Barnes (1982) 'Long-term Synaptic Enhancement in Hippocampus is Not Regulated by Post-Synaptic Membrane Potential', *Brain Research* 233: 195-99.
Wilkins, W.K. and J. Wakefield. (1995) 'Brain Evolution and Neurolinguistic Preconditions', *Behavioral and Brain Sciences* 18: 161-82.
Williams, G.C. (1966) *Adapation and Natural Selection.* Princeton: Princeton University Press.
Wittgenstein, L. (1953) *Philosophical Investigations.* Oxford: Oxford University Press.
Wolfram, S. (1985) 'Origins of Randomness in Physical Systems', *Physical Review Letters* 55: 449-52.
Wolpert, D.M.; Z. Ghahramani, and M.I. Jordan. (1995) 'An Internal Model for Sensorimotor Integration', *Science* 269: 1880-82.
Wright, J.J.; R.R. Kydd, and D.T.J. Liley. (1993) 'EEG Models: Chaotic and Linear', *Psycoloquy* 4(60) eeg-chaos.1.wright.
Wright, J.J.; R.R. Kydd, and D.T.J. Liley. (1994) 'Noise is Crucial to EEG Dynamics', *Psycoloquy* 5 (19) eeg-chaos.5.wright.
Wright, L. (1973) 'Functions', *Philosophical Review* 82: 139-68.
Yao, Y. and W.J. Freeman. (1990) 'Model of Biological Pattern Recognition With Spatially Chaotic Dynamics', *Neural Networks* 3: 153-70.
Yasuhara, A. (1971) *Recursive Function Theory and Logic.* New York: Academic Press.
Zeh, H.D. (1993) 'There Are No Quantum Jumps, Nor Are There Particles', *Physics Letters* A172: 189-92.
Zurek, W.H. (1991) 'Decoherence and the Transition from Quantum to Classical', *Physics Today* 44(10): 36-44.
Zurek, W.H. (1993) 'Negotiating the Tricky Border Between Quantum and Classical', *Physics Today* 46(4): 13-15+.
Zurek, W.H., ed. (1990) *Complexity, Entropy, and the Physics of Information.* Reading, Massachusetts: Addison-Wesley.

Index

—#—

—A—

—B—

—D—

—E—

—F—

—G—

—H—

—I—

—J—

—K—

—L—

—S—

—T—

—U—

—V—

—W—

—Y—

—Z—

STUDIES IN COGNITIVE SYSTEMS

1. J.H. Fetzer (ed.): *Aspects of Artificial Intelligence.* 1988
   ISBN 1-55608-037-9; Pb 1-55608-038-7
2. J. Kulas, J.H. Fetzer and T.L. Rankin (eds.): *Philosophy, Language, and Artificial Intelligence.* Resources for Processing Natural Language. 1988 ISBN 1-55608-073-5
3. D.J. Cole, J.H. Fetzer and T.L. Rankin (eds.): *Philosophy, Mind and Cognitive Inquiry.* Resources for Understanding Mental Processes. 1990 ISBN 0-7923-0427-6
4. J.H. Fetzer: *Artificial Intelligence: Its Scope and Limits.* 1990
   ISBN 0-7923-0505-1; Pb 0-7923-0548-5
5. H.E. Kyburg, Jr., R.P. Loui and G.N. Carlson (eds.): *Knowledge Representation and Defeasible Reasoning.* 1990 ISBN 0-7923-0677-5
6. J.H. Fetzer (ed.): *Epistemology and Cognition.* 1991 ISBN 0-7923-0892-1
7. E.C. Way: *Knowledge Representation and Metaphor.* 1991 ISBN 0-7923-1005-5
8. J. Dinsmore: *Partitioned Representations.* A Study in Mental Representation, Language Understanding and Linguistic Structure. 1991 ISBN 0-7923-1348-8
9. T. Horgan and J. Tienson (eds.): *Connectionism and the Philosophy of Mind.* 1991
   ISBN 0-7923-1482-4
10. J.A. Michon and A. Akyürek (eds.): *Soar: A Cognitive Architecture in Perspective.* 1992
    ISBN 0-7923-1660-6
11. S.C. Coval and P.G. Campbell: *Agency in Action.* The Practical Rational Agency Machine. 1992 ISBN 0-7923-1661-4
12. S. Bringsjord: *What Robots Can and Can't Be.* 1992 ISBN 0-7923-1662-2
13. B. Indurkhya: *Metaphor and Cognition.* An Interactionist Approach. 1992
    ISBN 0-7923-1687-8
14. T.R. Colburn, J.H. Fetzer and T.L. Rankin (eds.): *Program Verification.* Fundamental Issues in Computer Science. 1993 ISBN 0-7923-1965-6
15. M. Kamppinen (ed.): *Consciousness, Cognitive Schemata, and Relativism.* Multidisciplinary Explorations in Cognitive Science. 1993 ISBN 0-7923-2275-4
16. T.L. Smith: *Behavior and its Causes.* Philosophical Foundations of Operant Psychology. 1994
    ISBN 0-7923-2815-9
17. T. Dartnall (ed.): *Artificial Intelligence and Creativity.* An Interdisciplinary Approach. 1994
    ISBN 0-7923-3061-7
18. P. Naur: *Knowing and the Mystique of Logic and Rules.* 1995 ISBN 0-7923-3680-1
19. P. Novak: *Mental Symbols.* A Defence of the Classical Theory of Mind. 1997
    ISBN 0-7923-4370-0
20. G.R. Mulhauser: *Mind Out of Matter.* Topics in the Physical Foundations of Consciousness and Cognition. 1998 ISBN 0-7923-5103-7

KLUWER ACADEMIC PUBLISHERS – DORDRECHT / BOSTON / LONDON